CAR DESIGN
STRUCTURE & ARCHITECTURE

Other TAB books by the author:

No. 2037 *Streamlining and Car Aerodynamics*
No. 2046 *Modern Diesel Cars*
No. 2069 *The Complete Handbook of Automotive Power Trains*
No. 2090 *The MICHELIN Magic*

CAR DESIGN
STRUCTURE & ARCHITECTURE

BY JAN P. NORBYE

TAB BOOKS Inc.
BLUE RIDGE SUMMIT, PA. 17214

FIRST EDITION

FIRST PRINTING

Printed in the United States of America

Library of Congress Cataloging in Publication Data

Norbye, Jan P.
Car design.

Includes index.
1. Automobiles—Design and construction. I. Title.
II. Title: Car design, structure and architecture.
TL240.N66 1984 629.2'31 83-4922
ISBN 0-8306-2104-0 (pbk.)

Contents

Introduction

A SUBJECT SUCH AS CAR DESIGN HOLDS DEEP fascination for generations of drivers. Schoolboys have been known to let classroom instruction go in one ear and out the other while enthusiastically filling their copy books with drawings of cars. For a small minority, car design has become a profession.

By its narrowest definition, car design is simply styling. The styling aspect is what awakens design talent in youngsters. But wait! Look at their sketches. From the most primitive two-dimensional profiles to the most elaborate rendering in three-dimensional perspective, they bear witness to a remarkable awareness of functionality. The kids generally give some thought to practical matters such as the disposition of the mechanical elements in the chassis and the seating space and arrangement of the doors. Perhaps, subconsciously, they are doing more than giving exterior form to an existing engineering product; they are engaging in vehicle architecture.

The artist in some of them gets carried away and begins to exaggerate certain dimensions—the length of the hood, the size of the wheels, and so on. That's not to be interpreted as failure to see the true proportions of a car. Those techniques are the same that have always been used by advertising illustrators trying to glorify the client's product. And it's not without effect.

Most boys stop drawing cars at some point in their adolescence or early manhood. That does not mean they lose interest in the subject. Many will remain car enthusiasts all their lives, avid readers of car magazines and books, members of car clubs, and some even become car collectors. It is to them that this book is dedicated.

The scope of this book is wide. It takes in all aspects of car design, including body engineering, choice of construction materials and methods, and other matters of intimate concern to the professional car designer. This book examines the reasons behind the positioning of seats, engines, and other necessary elements, and it explains how these choices of space allocation affect the styling freedom of the body designer.

Few things can be properly understood or evaluated without a knowledge of the historical background. With car design, prior art has much to contribute in the form of lessons on what not to do as well as in terms of inspiration for new creations. Merely looking at car exteriors, however, provides no more than half the picture. It is vitally important to look also at the technical evolution of vehicle structures, from the horse-drawn carriage with its suspended body and the bicycle with its lack of springs to the modern unit-construction body shell. The first cars took their influence from the carriage or bicycle. Gradually, the influences began to mix and interesting hybrid constructions appeared. A third source of technical influence came into play after 1920—the aeroplane.

What of the future? Car design is a greater challenge than ever because of the pressure of conflicting requirements such as reconciling the most effective space utilization with the envelope having the best aerodynamic form. The task of making it look so good that people will want to buy it remains uppermost in the stylist's consciousness, increasingly troubled by the need to consider ease of manufacturing and production cost.

In this book, I have tried to present all relevant facts on car design never before available between two covers. The result is a book I would have loved to have had in my boyhood and would find useful at any age. I hope my readers will find that it not only answers their questions but elucidates the interdependence of technical and esthetic choices throughout the anatomy of the car.

Part 1
Car Design

Chapter 1

Economy and Function

NEVER BEFORE HAS THE CAR OF THE FUTURE come in for closer examination within the auto industry, in government circles, by the fuel-and-energy producers, by the financial world, and at the consumer level. The shape of future cars will depend on fuel prices and availability, the cost of raw materials and labor, road and traffic conditions, and regulations concerning auto safety, the effects of automobile use on the environment, technological advances, and competition for the markets.

ELEMENTS OF FUNCTION

Design is dictated by function. If that rule is not observed, gross inefficiencies will result. The function of a car as a transport medium depends, above all, on the intended payload. That includes the number of passengers, the amount of luggage/cargo (in terms of weight, size, and shape), and other items forming part of the payload (Fig. 1-1).

A second part of a car's function is the speed with which it is intended to carry its payload. Speed costs money. Sending a telegram costs more than sending a first-class letter. An airplane ticket is more expensive than bus fare. Inevitably, a fast car costs more than a slow one.

Speed is inseparable from acceleration. It's not just the cruising speed that matters, but also the time it takes to reach the intended travel speed from standstill or to resume normal speed after driving at reduced speed.

Higher performance demands more power. That means a more expensive engine—bigger or more highly stressed. In either case it will cost more. It will also consume more energy in operation. If it's bigger, it will be heavier and demand more installation space. Those are serious considerations for the car designer and engineer alike. Body design must also be practical for the operator's needs, whether it's a taxicab, commuter car, family wagon, or sports car.

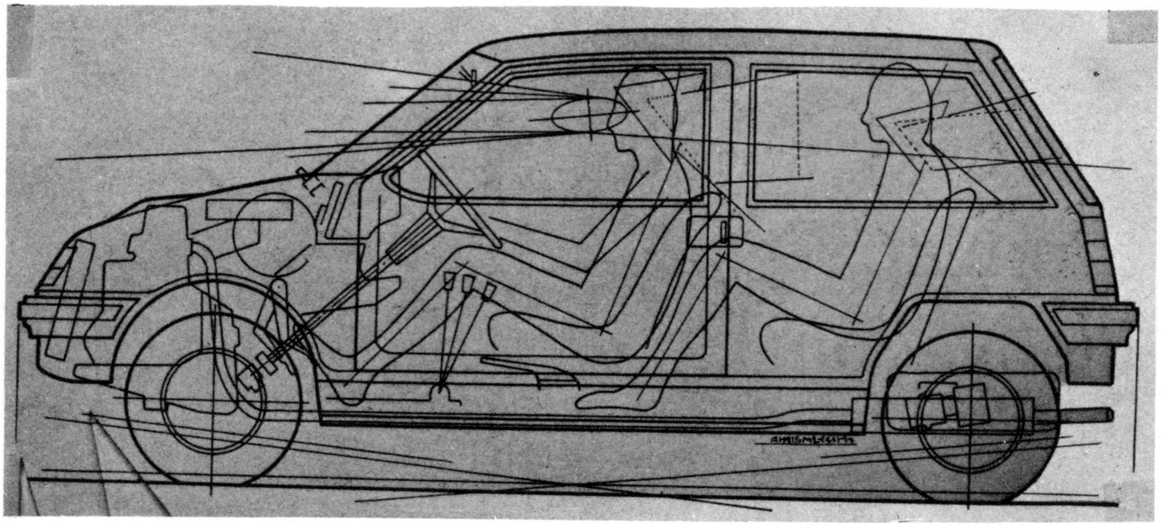

Fig. 1-1. Example of packaging a small four-seater sedan (Austin Metro).

The body must have convenient openings for passengers to get in and out and for loading and unloading luggage. Step height must not be too high and door sills must be sufficiently wide. Doors must be positioned so that they correspond to where the seats are, and nothing must obstruct the doorsill area. Doors must be wide enough and high enough and swing far enough.

Trunks should be shaped to resemble a cube or two cubes joined (as closely as possible) so as to accept the bulkiest pieces of luggage. The liftover height should be low, and the lid must correspond in size to the trunk capacity so that any object that will fit in the trunk can actually be put there.

Examples abound—modern and not so modern—of cars failing to meet one or more of these basic requirements. The designer's difficulties do not end with the practical needs of driver, passengers, and cargo. Body design must also allow for logical and space-efficient accommodation of all the mechanical elements needed to make the car go (such as engine and transmission, suspension systems, steering and brakes).

Due attention must be paid to the dynamic safety of the car, unladen or with maximum payload. Just as cars start and stop, they also change direction of travel. In every turn, the car and its payload are subject to centrifugal force.

Consequently, the center of gravity should be as close to the ground as permitted by the design concept. Not just the heavy components, such as the engine and the transmission must be considered, but also the payload. Equal or nearly equal front/rear weight distribution is also regarded as a desirable feature (Fig. 1-2).

WAYS OF SAVING ENERGY

Making the car economical is extremely important. Since 1975, the fuel economy of cars sold in the United States has been subject to sales-weighted limits set by the Energy Policy and Conservation Act. Congress has set AFES (Average Fuel Economy Standards) rising to 27.5 miles per gallon CAFE (Corporate Average Fuel Economy) for 1985. Cars that fail to reach that figure may still be legal, but only if made by a company that builds a

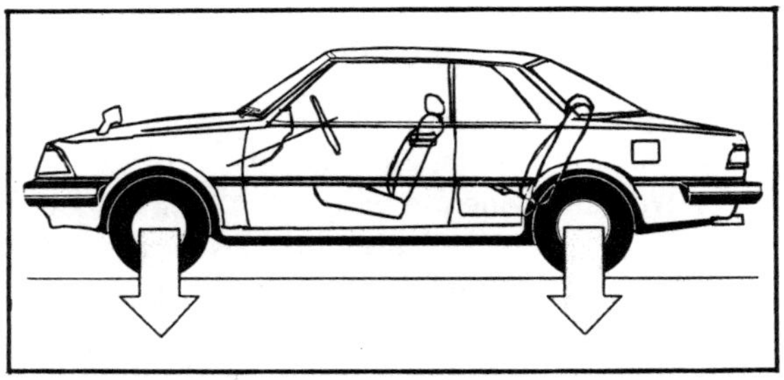

Fig. 1-2. Weight distribution is greatly affected by layout and design.

sufficient proportion of cars giving even higher fuel mileage.

As a result of AFES, the auto industry has undertaken successive phases of a general downsizing weight-shedding and streamlining process that will enable the companies to meet the 1985 figure ahead of schedule (Fig. 1-3). Even if no further AFES are set, the energy picture will force the industry towards further improvement, while also refining its products to give better satisfaction in other respects. During the rush to downsize, conventional considerations such as comfort and practicality were given secondary importance. And some cars suffer for embodying less intelligent solution than others.

Dr. Ulrich Seiffert, Volkswagen's research director, tried to classify the automaker's options in this statement: "The fuel economy of a passenger car with a given powerplant and performance is determined significantly by the vehicle's tractive resistance which essentially is composed of inertial resistance, aerodynamic drag, and rolling resistance. While rolling resistance problems can be handled best by the tire industry, the car maker can take appropriate measures that affect the other two resistances. Inertia resistance largely is a function of vehicle weight, while aerodynamic drag can be affected by the configuration and ducting of the air flow. Further vehicle weight reduction in the present European cars using the standard materials by design modifications is limited. Therefore, alternative materials would have to be used, which, in all probability, would entail higher cost. The aerodynamic drag may be reduced substantially by proper styling, which might give the vehicle a somewhat new appearance but is not likely to result in significant additional cost."

Other factors than fuel must also be taken into account at the design stage. Oil consumption and tire wear are also part of the total consumption picture. Lubricating oil is more expensive than gasoline or diesel fuel, and it's important to make the best use of it (which means smaller oil pans and longer oil-change intervals).

Tire life is another important part of the overall running costs. Car tires are made mainly from synthetic rubber (based on petrochemicals). Longer tire life helps conserve oil resources in addition to easing the cost per mile to the owner.

Since the enactment of AFES, the emission-control effort has been relegated to a status-quo role. The debate continues about the relationship between the hastily adopted Federal Emission-Control Standards (Act of 1967 and 1970) and their true benefit in terms of overall air quality. A rollback remains a possibility.

A thoughtful Frenchman and former Renault executive, Marc Ouin, now head of France's Chambre Syndicale des Constructeurs d'Autombiles (equivalent to Detroit's Motor Vehicle Man-

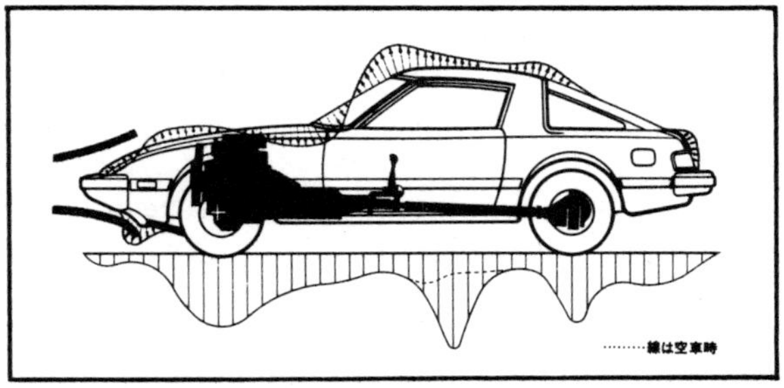

Fig. 1-3. Aerodynamic pressure map shows the importance of design for smooth air flow.

ufacturers Association) points out that improvements in air quality do follow the industry's drive to improve fuel economy. Why? Because emissions are by-products of a fuel-burning process, burning less fuel will produce lower levels of pollutants. From this point of view, what the motor manufacturers have been doing all along is the right path both environmentally and economically. This is perhaps difficult for the general public to understand.

According to Marc Quin, "For the man in the street, the automobile is the immediate symbol of waste of oil, maybe just because the automobile is on the street, and not so the trains, the planes, the domestic and industrial boilers, which he does not see constantly circulate in front of him. It is not surprising, under these conditions, that public opinion is sometimes shocked when it hears that the oil must be reserved in priority for transport (and for the petrochemical industry), other energy sources being assigned to fixed installations. Note that the renewal of the existing car park by the removal of older cars which consume more, replacing them with new cars consuming less, is a fundamental factor in saving money. The decision-makers must take account of this fact."

As long as the energy crunch is with us, there will be strong pressure for improvement on all fronts.

WEIGHT-SAVING MATERIALS

Energy consumption means more than just fuel consumption. It starts at the manufacturing stage. How much energy is spent in building the car? Energy is consumed in making the steel, aluminum, and other materials used. And of course it takes energy to run the assembly line. You can add up all the energy inputs for the total car and draw certain conclusions. A change of materials for one or more applications might be indicated or there might be a need for a different method of assembling the car.

The use of lightweight materials to reduce car weight is to some extent counteracted by the greater energy input needed to produce workable auto parts from such materials.

"Weight-saving can make a big contribution to improving a car's fuel economy," stated Gordon Bashford, former chief of research for BL Technology, speaking to a meeting of the Automobile Division of the Institute of Mechanical Engineers in 1980. He went on, "First, there is the primary saving gained by substituting a part of lightweight material for one of heavier construction. For every such part there is a secondary saving in the form of lighter structural loads, which is known as a 'knock-on' or 'ripple' effect. For instance, a lighter engine opens the way for also lightening the drive train, suspension system, frame and/or body shell, and so on.

"Because of the higher energy input required to produce aluminum components for replacing steel ones, the light car of the future will consume more energy during its manufacture," said Bashford. "It takes 21,000 kilowatt-hours to produce a one-liter car now, but the car of the future will demand 29,000 kilowatt-hours. Because the future car demands less running energy, however, it will get a net energy credit early in its life. Over the life of the car, say 10,000 miles, the higher energy input will be paid back several times over."

AFTER THE CAR HATERS

After a period (1966-72, approximately) of frantic and misguided search to replace the car by public transit, from "people-movers" to high-speed magnetic levitation trains, most planners recognize that there is no substitute for the automobile.

The car makers were accused of lacking imagination in finding solutions to all the problems that faced the automobile. Critics spouted off, uninvitedly, about the need for a revolution in the automobile and found a ready audience. They called for fundamental and brutal changes in the product (without more than a superficial understanding of the problems.

The anticar lobby blamed the automobile for all possible and impossible ills of a modern industrial society, from killing our wildlife to bankrupting the railroads. Myron S. Huckle, a retired colonel, did not mean to be funny when he stated that "Young people use the automobile to escape all supervision and control and develop new refinements in delinquency and crime."

To such a mentality, the car also became the target of accusations of guilt in traffic accidents involving an intoxicated driver. When traffic engineering fails and congestion results, the same sort of warped thinking put the blame on the car. With the same perverted logic, the ugliness of junkyards also was made out to be the fault of the car, and not the proprietors.

CARS ARE INDISPENSABLE

Those ill-advised, misguided voices have by and large been silenced. Saner spirits have come to prevail, and genuine solutions for alleviating the friction on the interface of the car with society are being adopted. Better-designed cars are part of the picture.

As Sergio Pininfarina said in 1977, "There is no doubt that public transport will have to be improved and vehicles made more comfortable and attractive, but I do not think they will ever manage to supersede the car which has remained the favorite means of transport, in spite of all the criticism. Most people have lost nothing of their feeling for the personal car; it is also clear that styling continues to play a decisive role in the choice of a car, exactly as it did in the past. It can also be maintained that the appearance of a car is important not only on its own account but also insofar as it helps to create the manufacturer's image."

Dr. Ulrich Seiffert of Volkswagen wrote in 1980: "The automobile has become a product that is indispensable in today's highly technological world. It is a product that is best suited to meet the increased requirements for independent personal mobility. Therefore, it has become necessary for the automobile industry to take into account in their planning not only the actual market situation but also increasingly the various sociopolitical aspects. Volkswagenwerk's Research Division has been involved in a multitude of development projects and has found solutions to the problems, further reducing the motor vehicle's detrimental effects on the environment, decreasing energy consumption and increasing the safety level."

A year later, Prof. Werner Breitschwerdt, technical director of Mercedes-Benz, told the press at a meeting in Frankfurt: "In my view, we should not attempt to effect improvements by discriminating against *one* form of transport and belittling what it has to offer in order to enhance the attractions of another. For if we decide to restrict the natural development of a single element and don't try to change our whole transport system on the basis of technical progress and the most intelligent solution, then we may not only prevent the very progress we wish to achieve but may even be encouraging a step backwards."

CAR TYPES AND SIZES

Some auto industry spokesmen have voiced concern that AFES should cause all cars to become small and similar. There is no factual basis for such fears. One of the first to see the situation clearly was Sergio Pininfarina, who said in 1977: "As I see it, the cars of the future will not be standardized but highly diversified. Let me be more specific:

1. The small, silent urban car with low pollution level.

2. The car for long journeys, comfortable and roomy.

3. The universal car, a compromise between the first two.

4. The special car that can be used for work and touring, highly suitable for the developing countries.

5. The sports car, for the joy of driving, for those who love speed."

Pierre Tiberghien, who is in charge of car design and research for Renault, sees future cars as continuing to be made for general purposes, with different levels of price, performance, equipment and carrying capacity. Today's concept of the mass-production family car will prevail. What about specialized commuter cars and other special-purpose vehicles? Tiberghien does not give them much chance in Europe. There consumers cannot easily be forced into multicar ownership because even those who can afford cars run into parking problems. From the manufacturer's point of view, the market for special-purpose vehicles is limited by the investment needed to produce them (which means higher unit costs).

Will the small car get smaller still? "For us," said Tiberghien, "the R-5 is the smallest size we are now considering for the future. That doesn't mean we won't make other, lower-cost cars of the same size, perhaps powered by something other than a water-cooled four-cylinder engine. But less roomy, no."

The challenge to designers, engineers, inventors, and innovators in any automotive area of science has never been greater. They have to formulate the right concepts, plan the right product, find the right ways to build it, and get it to the market at the right time. The picture of how these tasks are divided and new cars created are probably somewhat blurred in the minds of many readers. I admit I have myself been confused by the term "product planning."

PRODUCT PLANNING

"Product planning" does not mean the planning of a specific new car. Product planning stops before the new car project is defined in terms of type, size, layout, componentry, and price. The task of what is called product planning comprises information-gathering and data analysis with a view to predicting market trends and opportunities. The product planner, therefore, as a specialist in identification of trends and interpretation of marketing data, with the aim of providing an economic map of long-range developments affecting the automobile.

Into the product planner's map go data on the evolution of consumer habits, the national economy (interest rates, rate of inflation, savings as a percentage of personal income, and the growth rate of the gross industrial product). This map is then used by the car corporation's higher management to draw certain conclusions about the new products needed for the corporation to meet its long-range objectives.

Product planning does not lead to clear plans for a new product or a family of new products, but instead its ultimate result is a marketing strategy. Out of this strategy, a product-definition committee will formulate specific objectives for a series of new models (each corresponding to a perceived future demand and each scheduled for introduction at the most propitious time).

Once a project has been fully defined and scheduled for production, it is handed over to engineers and designers, working in close coordination. They have to work to a cost target. That means coordination with production specialists at an early stage. Decisions must be made quickly as to the extent to which existing components can be used (for instance, an engine from a current model, the floor pan of another, and the front suspension from a third).

DESIGN AND ENGINEERING

Whenever new mechanical components are needed, the decision must be made whether to manufacture them inhouse or buy them from outside or associated suppliers. If a new engine is needed, that's a matter of such great scope that it goes beyond the individual car project. An engine needs a long time for testing and development, and a long lead time for tooling after the design has been frozen. The investment in a new engine line is so heavy that car companies try to keep them in use in 20 years or more.

The car type, approximate size, and certain other specifications are contained in the concept for the program. The engineers start by packaging. That means accommodating the desired number of passengers and the mechanical components into an envelope that corresponds to the size, weight, and performance targets.

This envelope is the basis for the car designers' work (Fig. 1-4). The designer is responsible for the artistic development of the final product (which has an enormous bearing on its acceptability in the market place. Of course, the designers must coordinate their work with structures specialists, body engineers, and manufacturing engineers who can rule on the feasibility of innovations proposed by the designers. Exterior shapes and details are submitted to aerodynamicists for wind tunnel testing.

Most people who buy cars think that the work of the designers and design engineers ends when production begins. That's not true for any car company. But in most of them, it's a different group of engineers that is in charge of continuing the development once the car has been released on the market. The engineers responsible for creating the car usually go on to new projects. For one company, Peugeot, the same engineers follow the car they develop until the day it is taken out of production.

LESSONS FROM PEUGEOT

A small point, a little detail of internal organization, reveals the core of Peugeot's product philosophy. Right from the concept stage, account is taken of a model's evolution up to the time of its replacement. That includes body style variations, new powertrain options, and hundreds of smaller modifications. The reasons for changes stem from a multitude of sources: feedback from the service and warranty people, changes in the supply situation or purchasing policy, new technical developments from inside Peugeot or outside sources, requests from the sales and marketing staff, or demands from the manufacturing experts.

Fig. 1-4. Elaborate color drawings are made early in the design stage.

It does not mean that the production run is limited in time from the startup date; Peugeot allows wide flexibility in the phasing out of old models. New models are rarely introduced as direct successors to preceding ones, but they are created to fill gaps in the continuously evolving range. This encourages a period of overlap during which old and new cars are built side by side (e.g., 203 and 403 in the 1950s, 403 and 404 in the 1960s, 404 and 504 in the 1970s, and the 504 and 505 today). This enables

Peugeot to get the fullest use out of its tooling. Production of old models can be stretched out as far as the demand makes it profitable. Still, Peugeot engineers think in terms of only 10 to 12 years as the planned production life for a new car.

The initial phase is called Pre-Project Study. Pre-Project Studies are handled by an office known as DSV (Departement Synthese Voiture = Car Definition Office) where duties include all aspects of product planning design and engineering. The DSV must make feasibility studies and arrive at a synthesis of the various design objectives and the limits they impose on each other. Further, the DSV must assure the integration of the work. All elements must be prepared with due regard to the whole car and be ready at the same time. The DSV sets a deadline for the lead time.

The styling center is part of the DSV. This section's work ends with prototype construction at about 30 months prior to the production date. The project is then handed over to another office, DEI (Departement d'Etudes Industrialisation = Production Design Office).

The DEI takes the DSV prototype and turns it into a production model. This section is responsible for development work, testing, and quality (but not concerned with production). The DEI prepares all likely versions of the body structure from the start even if not all are to be produced until later. The 504, for example, was built for 12 years as sedans and station wagons before Peugeot added the pick-up version.

The 304 sedan went into production in September, 1969, followed by a sports coupe and convertible (replacing the 204 sports models) in April, 1970. The 304 wagon was added in October, 1970. Peugeot introduced the 305 sedan in November, 1977, adding a wagon in March, 1980. Sports versions of the 305 have not yet appeared, and the 505 so far exists only in its basic form (a four-door sedan).

A second office is engaged in each project as consultants to the DSV and suppliers to the DEI. It is called DGO (Departement Groupes d'Organes = Mechanical Components Office) and is responsible for the engineering of power units, transmission systems, suspension elements, steering and brakes, plus associated systems (cooling, exhaust, wheels, and tires).

It is the DEI, however, that is responsible for meeting the cost targets, in production, which will permit the car to be sold at a profit in the intended market segment. This responsibility extends throughout the car's production life, even when a model range is expanded with special versions spanning several market segments. The 104, new in 1972, is a case in point where an upgrading process brings new versions each year as well as providing a base for new models from the other car divisions (Citroen Visa and Talbot Samba).

The DEI has direct inputs from the sales and marketing staffs once a model goes into production, and subsequent development is to a great extent market oriented. But the DEI also has the never-ending task of finding ways to improve quality while simultaneously searching for means to reduce production cost.

Peugeot has a long record of offering a wide range of models. Its cars have not undergone any general down sizing. Nevertheless, the dimensions of the smallest model have been reduced, step by step, ever since 1965.

In 1974, Dr. Friedrich Goes, a Volkswagen engineer who worked on the original Golf and was stationed in America for some years while the Westmoreland plant was tooling up for Rabbit production, had some salient observations to make with regard to car size. "The change in the average size of vehicles and thus of small car size is affected by the standard of living of the population which needs the means of transportation," said Goes. But clearly the correlation between standard of living and vehicle size only holds good up to a certain density of traffic. Then the inconvenience of large vehicles in dense traffic tends to make smaller cars more popular, despite a further rise in living standards. The greater affluence also brings a demand for second and third cars which, because they need not satisfy the total family transportation requirements, are in the main, small cars."

CHALLENGES TO THE DESIGNER

The car designer is by definition a creative person. He works with concepts about 5 to 10 years advanced, depending on whether he is engaged on a current project or an advanced study. But his creativity is continually placed under the combined restraints of cost, feasibility, and regulations (Fig. 1-5).

Ford's recently retired vice president of design, Gene Bordinat, stated in 1965 that "The men and women who create the shape, color and texture of these objects (cars) are involved in an art form of a very high order, and they combine this art with the practical considerations of function, manufacturing feasibility, consumer acceptance, and cost." See Fig. 1-6.

Fig. 1-5. A scale model in the making.

In 1979, GM engineer Richard E. Fancy said, "Market trends, investment limits, manufacturing capabilities, future regulation, and other forces all serve to focus the designer towards reality, yet the designer continually challenges both planners and engineers to see the limitations in a new light and push back the limits of reality."

Who is subordinate to whom? That varies from company to company. Generally, in European companies it is the designers that have the least authority, but the rank of the designer has been brought up several steps in recent times. In America, the designers used to have free play (in those days they were called stylists). But the demands of the public, the lawmakers, and the corporation have eroded the designers' power of decision-making.

Europe and the United States are about to converge on a policy that has been practiced by Volkswagen since the early 1970s. In cases of conflict between the designers and the engineers, technicians, aerodynamicists, and production men, it is top management that acts as jury and judge, ruling on the overall balance of demands from the sales side with the reservations of the quality-control men and cost accountants.

CARS OF THE FUTURE

Five major influences are shaping the car of the future, according to an analysis made by Fredrich W. Lohr, director of engineering for Opel. These influences are superimposed on the present concept of the passenger car. Lohr does not list them in order of importance because conditions are continually changing and thereby altering the balance among the factors. One influence can be defined as what the customer is used to, modified by certain consumer expectations.

Fig. 1-6. The finished product invariably ends up looking less radical than the clay mockup.

A second influence comes under the heading of research and development; and a third is lead-time requirement for development. Legislation constitutes a fourth influence, and the fifth can be summed up as availability of energy and raw materials.

Consumer customs and expectations do not simply pull the future car concept in one direction; the picture is full of crosscurrents. It has a sociological aspect, says Lohr, in terms of ownership and image of a car. Next, car buyers have individual mobility needs and use their cars for different purposes. Their demands for comfort and ease of operation also differ a great deal.

What the consumer is accustomed to and what he expects relates to three principal criteria: payload, maximum rate of acceleration, and top speed. The car manufacturer is somewhat bound by the type, character, and performance of his past products and cannot unilaterally make basic alternations without the risk of losing customer loyalty.

Probes of consumer awareness of the need for wide-sweeping changes, in response to changes in the cost and availability of energy (both for producing the car and for operating it), and in legislation show that consumer expectations have been deflected in the direction of more economical cars.

"Our task," says Fredrich Lohr, "is to minimize our consumption of raw materials while maximizing its recovery rate by recycling, and simultaneously to find substitution materials of lower energy content."

According to John Oldfield, Ford vice president in charge of car product planning in Europe, the average European car of the 1990s could be powered by a turbocharged three-cylinder diesel engine and give 80 mpg at a steady 55 mph. He was addressing an audience of 250 officials from Common-Market governments, scientists and engineers at a 1981 energy symposium organized by the British Open University at Milton Keynes.

The average car, John Oldfield continued, could also have an electronically controlled, continuously variable transmission (stepless drive). In addition, he stated, "by extensive use of lightweight materials and advanced airflow management techniques, the typical family car could provide the same amount of passenger accommodation and luggage space as in present-day models. It will have narrow-section tires giving low rolling resistance, and feature computer-controlled, driver-information systems."

It has been estimated that, between 1973 and 1979, the average fuel consumption of cars offered in European markets diminished by 10 percent. Moreover, the European manufacturers have undertaken to further reduce this consumption by at least 10 percent between now and 1985.

"The effort will, naturally, be maintained beyond 1985, " Marc Quin points out, "and it is probably not exaggerated to imagine that in the year 2000, automobiles will consume half the amount of energy they use today, while affording the consumer better service with regard to comfort, safety, and reliability."

Chapter 2

Styling: More Than an Art Form

THERE IS GENERAL AGREEMENT AMONG CAR manufacturers that tomorrow's design trends will be dominated by a combination of economics and technology. Shortages of energy and materials, changes in life styles, and transportation requirements will increasingly impinge on the stylist's task. Therein lie opportunities as well as problems. One factor that will continue to play a central part in auto design evolution for the rest of the 20th century is aerodynamics. The concepts of air flow and drag have been known since the early days of auto racing. But it took OPEC and a tenfold increase in crude oil prices to get the industry to put its knowledge to full use.

Today it's inconceivable that a new model could go into production without the benefit of wind-tunnel testing and development (Fig. 2-1). "Aerodynamic development of a production car is an exercise in optimizing a compromise," says Uwe Bahnsen, vice president of design for Ford of Europe, "The most efficient aerodynamic shape hardly makes a practical motorcar, while the package that's most efficient as a container for people, their luggage, and the required mechanical elements, does not necessarily produce good aerodynamics."

Of course, the compromise involves several other factors: legal requirements and safety standards, passenger comfort and convenience, and a variety of product-engineering and production-engineering considerations.

Uwe Bahnsen is one of a new breed of designers whose work is guided by a sense of practical values rather than a desire to visually astound the viewer. Born in Hamburg in 1930, Bahnsen studied at the Academy of Fine Arts in Hamburg. His first job was in advertising and graphic design. He became involved with motor racing and even built his own racing car. This led to his joining the Ford-Werke in Cologne, Germany early in 1958 as a

Fig. 2-1. Air flow is studied in the wind tunnel.

member of the design staff (then headed by Wesley Dahlberg). Bahnsen worked in Dearborn for a year (1962-63) and then returned to Germany. During the 1967-70 period he was at Basildon, in England, as chief designer of Ford's advanced and interior design. Back in Cologne, he created the 1975 Escort body, and was appointed vice president of design for Ford of Europe in the spring of 1976, with his office at Basildon.

MARKET DEMANDS

The public asks a lot more from the looks of a car than aerodynamic efficiency. An automobile must also be attractive, well balanced, and coherent. It should have an element of innovation (something that makes it look new) and at the same time it must preserve some sort of continuity with the models that went before—as a symbol of marque identity (Fig. 2-2).

"The styling of a car is one of the most complex design projects there is," says Bruno Sacco, head of the Mercedes-Benz styling department. "The total of all the interrelated problems has to be tackled and solved while bearing in mind the entire concept

Fig. 2-2. A scale model and the finished product in the Mercedes-Benz design studio.

at which we are aiming." Sacco has been a body designer with Daimler-Benz since the age of 25, after completing his studies in Torino in 1957. He has held his current post since January, 1975. Sacco is one stylist who recognizes the importance of a technical approach to car design. "Technical concepts which are well-balanced, fully developed, and made to last, almost inevitably imply that styling will have to create optically congenial outward appearances for these concepts," says Sacco. "The results of a team effort are clearly recognizable styling families, which will allow the man in the street to identify a Mercedes at first glance, not only as a Mercedes, but at the same time as the perfect combination of safety, quality, durability, and economy."

STYLING FAMILIES AND CONTINUITY

Aside from continuity of overall appearance in a succession of cars, there is the family resemblance within the range at any one time. Some of this family resemblance is inevitable because the pressure of manufacturing cost control dictates a sharing of basic bodies among the greatest number of models. Perhaps it is sometimes desirable to establish the identity of a new model. Therefore, two-door and four-door sedans of the same make and model are essentially the same car. Converti-

BODY	CHEVROLET	PONTIAC	OLDSMOBILE	BUICK	CADILLAC
A	Celebrity	A-6000	Ciera		
B	Impala		Delta 88		
C			"98"	Electra	Sedan de Ville
E			Toronado	Riviera	Eldorado
F	Camaro	Firebird			
G	Malibu	Bonneville	Cutlass	Regal	
J	Cavalier	J-2000	Firenza	Skyhawk	Cimarron
K					Seville
T	Chevette	T-1000			
X	Citation	Phoenix	Omega	Skylark	

Fig. 2-3. Sharing of body types between the domestic car divisions of General Motors.

bles and sport coupes are often derived from the basic sedan, but several makers prefer to offer specific body shells for these types. Station wagons, on the other hand, are almost invariably sedan-based models.

The sharing of bodies also cuts across marque distinctions. Within General Motors, for instance, all car divisions share a number of basic bodies (Fig. 2-3). Styling studios of the individual divisions are restricted to devising variations on front and rear ends, and to some extent, the "skin" (exterior sheet metal) for fenders, doors, and hoods. Parallel situations exist within Ford Motor Company and Chrysler Corporation.

SHARING FOR ECONOMY

Peugeot (with its Citroen and Talbot subsidiaries), Fiat (with Lancia and Autobianchi), Volkswagen, and Audi enjoy a certain sharing of basic body shells. As in other European organizations, it is mainly in the area of mechanical components that intragroup as well as intergroup parts rationalization has been pursued.

Intergroup sharing exists between Renault and Peugeot, for instance, but not at the complete bodyshell level. Within British Leyland there was once so much sharing of bodies, engines, and chassis that the common source (Austin or Morris) of a car with a more pretentious name (Riley, MG, or Wolseley) could not be concealed from the most myopic viewer. "Badge-engineering," said the British public, who began to develop a strong taste for imported cars.

Mercedes-Benz started out in the prosperity of the 1950s and 1960s by building sports models with completely different bodies from the sedans that made up the bulk of the company's car production. First came the 300 SL, in 1954, and then the 190 SL two years later. Both were replaced in 1963 with the arrival of the 230 SL; it also had its own distinctive body. Since 1961, the company has offered a sports coupe derived from the sedan body (220 SE). This solution has proved to be the preferred one, replacing the remaining special sports-car bodies in 1981. See Fig. 2-4.

The 220 series, launched in October, 1958, had its first offspring the 220 SE coupe—introduced in April, 1961. The W-108 next-generation family car (250-280 series) was in turn based on the 220 SE, as were the 230 SL, 250 SL, and 280 SL sports cars with distinctive roadster bodies. The 1971 SLC (W-107) was a main ingredient in the new S-class (W-116) for 1972, which led to the latest S-class (W-126) in 1978. The new 380 SEC and 500 SEC replace the SLC (and belong in the W-126 family). Specific sports-car bodies were discontinued in 1981.

MORE PERSPIRATION THAN INSPIRATION?

In the eyes of the nonprofessional car enthusiast, stylists and designers have acquired a reputation for ingenious intuition that lets itself be guided solely by the inspiration of the moment. Sindelfingen (home of Mercedes-Benz body development, and the firm's main assembly plant) rests on a very firm foundation. The creative work is based on a design philosophy that can be expressed in a few words: car design is only possible with team work. At Mercedes-Benz, this team has always taken care to keep its work as near to reality as possible by

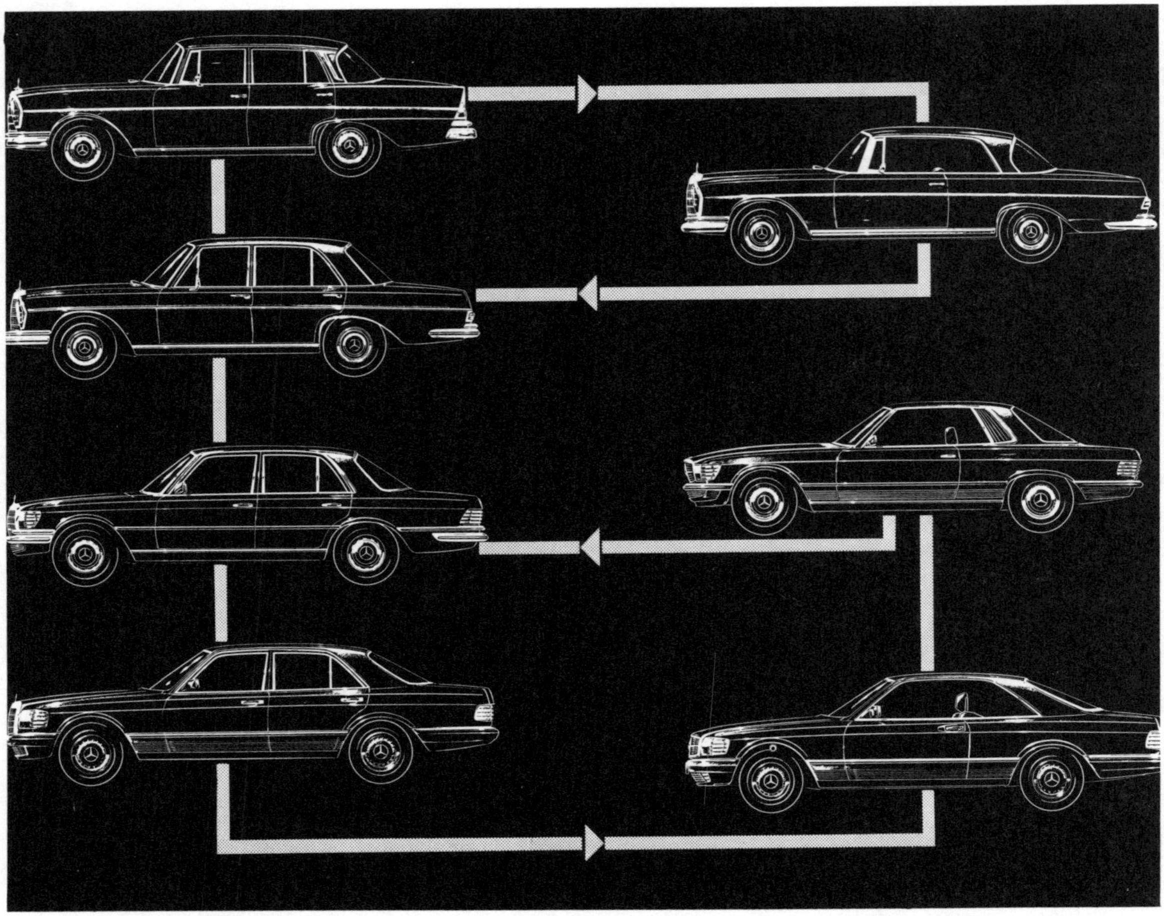

Fig. 2-4. The relationship between the sports coupes and four-door sedans in the Mercedes-Benz range 1959-1982.

closely cooperating with the firm's technical departments and by constantly updating—bearing in mind the classic appearance of earlier models.

Bruno Sacco and his assistants do not start with a clean sheet of paper or a little clay model. They start with a set of specifications handed over from the engineering department. That imposes restrictions on the imagination and creativity of the stylist, but at the same time constitutes a challenge. It puts the designer into a frame of mind that obliges him not to stray from answering the demands made on the vehicle and what is required of it. The purpose for which it will be used, the field of vision, the amount of room it should offer, economic, ecological, and legislative factors limit the stylists' freer flights of fancy even before the first line is drawn.

At this early stage, ideas are developed that will be of importance for the final shape of the car. It is now that the graphic sketches on a 1/5 scale are drawn as based for discussion for determining which direction the styling will take.

FROM SKETCH TO FINISHED MODEL

After the rough sketch phase, efforts are directed to what is for the most part a wearisome struggle to find the best possible shape. Promising suggestions are modeled in clay on a 1/5 scale. When their aerodynamic properties are examined in the wind tunnel, it can then be improved. At the same time, a dimensional concept is worked on in which the dimensions of the passenger compartment, luggage, and engine compartments are already determined on a 1:1 scale.

Ideas for the new color and material samples are thought out in other studios. While development work is going on, they make proposals for the design of ergonomically suitable and aesthetically attractive instrument panels, seats, operating controls, paneling, and padding for the interior.

Next, a selection is made of the 1/5-scale models that offer the best chance of being successful from the visual point of view. They are then measured with the aid of an electronically controlled scanner and data stored.

These outer skin measuring data can now be used automatically by the same piece of equipment for outlining the silhouette for the actual-size transfer on to the true-to-scale model, and also for cutting a prepared block of clay or a blank which finally becomes a realistic, actual-size replica of the model.

If the enlarged shape is not quite so pleasing as the attractive, smaller original, it is mainly because of the alteration in the angle from which it is viewed. This can sometimes cause the stylists to rethink and rearrange the design concept. The next step—extensive tests and revision work done in the wind tunnel—help to resolve this sort of problem. At the end of this phase of development, the design concepts, which were being worked out alongside, are submitted to the critical gaze of the management board. The board will make the decision as to which vehicle design will be taken a stage further.

Once again the scanner collects countless dimensional data for the selected model. The data will all be included in the prepared elevation drawing (that is to say in the actual size outline).

The model that concludes the design process (and which a person can actually get into) looks deceptively like a real car. In it, all the component parts of the exterior shape and the interior fittings—which up till this point had been put together separately—are amalgamated to form a likeness to the final product that will later on be an addition to or a continuation of the Mercedes-Benz range.

This finished model then has to be examined by the management board for a final decision. At this point, alterations to some of the details can still be made, but the overall design is frozen. The selected model thus forms the basis for all the subsequent development work. The final elevation drawing resulting from the scanning of the approved model represents the only valid combination of data for building initial and ultimate prototypes and standard production cars.

At this point, the designers usually have new tasks awaiting their attention—another program, be it an all-new model or a facelift for an older one. As for the job they just finished, they merely "go along with the model". That means participating in constant discussion and co-operation with other departments such as engineering, production, sales, and service to ensure that the results are visually and functionally satisfactory.

CREATIVE THINKING AT CITROEN

Something of a contrast—in approach, mentality, purpose, and style—is in existence at Citroen. Perhaps it results from the lightheartedness of life in Paris, its frivolity and glitter, as opposed to the methodical stolidness and pedantic following of instructions one's prejudices might lead one to suspect are behind the creations from Stuttgart. Actually the Germans are highly imaginative as well as eminently practical. The French can get hung up on insignificant details and lose sight of the key factors. Be that as it may.

According to Jacques Wolgensinger, "The body of a Citroen car is initially a creation of the mind—before being transposed into reality." See Fig. 2-5. "It is the exacting task of the creative people in the styling section to invent the pure, functional lines of the vehicle-to-be. Their creative thinking, must, however, rest on concrete data: the overall dimensions of the mechanical parts and the passenger-space obligations with which they have to comply." See Fig. 2-6.

"While not forbidden to dream," says Wolgensinger, "they must always retain at the back of their minds these hampering imperatives, together with a strict set of specifications, itself defined in accordance with the findings of market surveys, national and international regulations, not to mention economic considerations. These are the basic data from which spring the stylists' research and the various stages in the design of a car body."

The stylist does rough sketches showing the overall lines of the car as he sees it. From the initial sketches stem more refined drawings in which the

Fig. 2-5. Citroen styling concepts are defined more and more clearly in successive sketches.

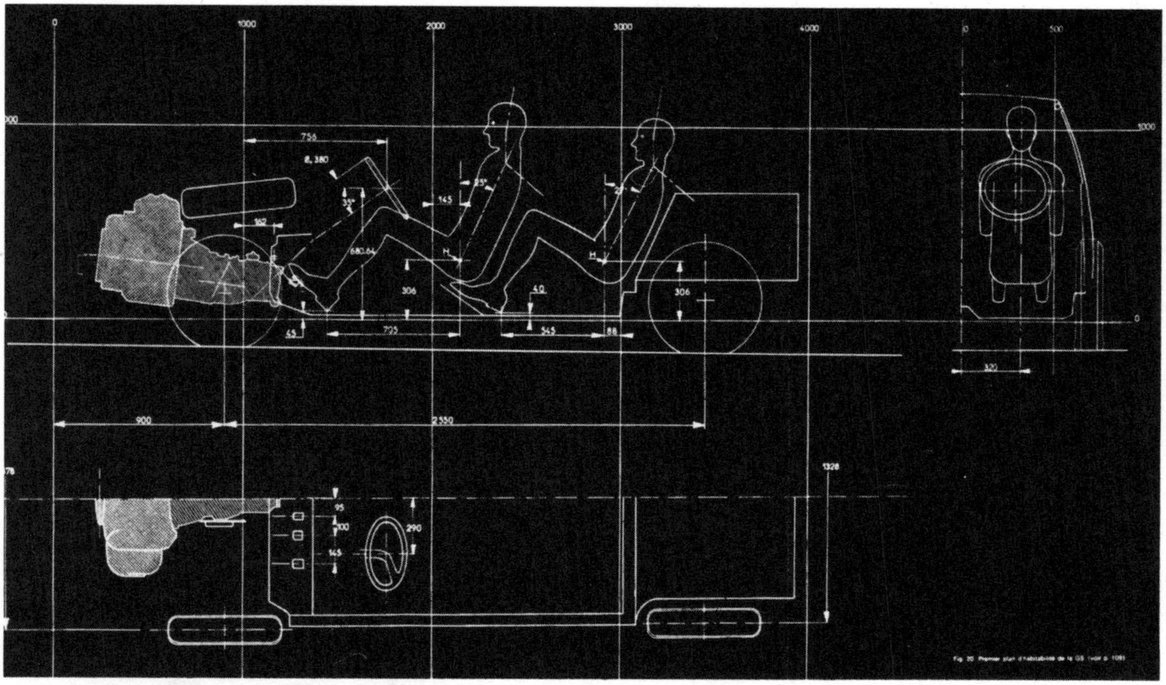

Fig. 2-6. The "floor plan" must be rigorously respected by the body designers.

imagined vehicle can be represented full scale. After the roughs comes the active studio phase. During this phase, the stylist and his team work on a full-size model made up of a wooden framework supporting a wire netting armature on which plaster is cast.

In the Citroen modeling studio, several stylists simultaneously do research, in parallel, on several plaster models for one and the same body. Each of these models usually deals with two proposals (one for each half of the vehicle). See Fig. 2-9. From this exploratory work often emerges a result far removed from the shapes suggested in the initial drawing.

THE CAR IS A WHOLE

When the model is back in the styling section, after electronic measurements for the computer, a hollow mold or matrix is made by applying a layer of plastic reinforced with a wooden armature. In order to make withdrawal of the parts possible, the car is broken down into large subareas (individual impressions of which are taken). As a general rule these areas are right and left side panels, rear portion, roof, and forward part.

From these matrices, the outside parts of the body also will be taken by applying with a brush a 0.2-inch layer of araldite plastic. These parts are cut out and assembled for a "negative" or "female" araldite model of the vehicle to be fitted with the interior upholstery of the prototype for use especially in testing for spaciousness and accessibility.

"A good stylist," says Trevor Fiore, Citroen's chief of styling, "should first and foremost have a deep understanding of the object he is working on. A car is something complex. It is at the same time a

Fig. 2-7. Modular system of clay mockup construction allows alterations to a single module without affecting the others.

Fig. 2-8. What looks like slapdash playing with modeling clay is actually a high-precision task.

symbol of prestige, a working instrument, and an extension of your personality. It is a whole."

Trevor Fiore began as a part-time styling consultant to Citroen, in 1979, and in mid-1980 was appointed head of the styling section. He was born, in England, in 1937, son of a French mother and a British father whose name betrayed his Italian origins. As a graduate student, Trevor Fiore worked with Raymond Loewy Associates in Paris. He developed contacts with the Italian coachbuilders and began a long-lasting collaboration with Fissore.

It was Trevor Fiore who designed the TVR Trident and the Elva-BMW. He also created the shape of the Monteverdi Hai, and acted as a styling consultant for DAF, Aston Martin, Bristol Cars, and De Tomaso Automobili. After coming to Citroen, he designed the pyramidical Karin coupe and the far more interesting Xenia sedan/wagon see Figs. 2-10 and 2-11.

LONG LEAD TIME

"On the other hand," continues Trevor Fiore, "the stylist must take into account the technical imperatives of overall dimensions and space. Very often it is from contradictory elements that we have to conceive and design a beautiful car—without, of course, neglecting the taste and expectations of the public, with the additional difficulty that we must foresee their preferences and expectations as accurately as possible five years in advance.

"I think that the future for car design lies in innovation," proclaims Trevor Fiore. "These last 20 years, every stylist has lived under constraints dictated by questions of cost, manufacturing methods, and increasingly strict regulations. But such a state of affairs has had few consequences in the sense of standardization. If anything, it is marketing research that tends to make cars all alike. In the auto industry, there are such huge investments at stake that the people in charge are afraid to launch a new car that is not like the others. In Europe, our cars are now more and more similar, and the public is starting to realize it. Herein lies an opportunity that Citroen ought to seize, for we have always had the reputation of making avant-garde cars, going beyond established norms, and that's exactly what the public is looking for these days."

TO 1999 AND BEYOND

Over the past 25 years, much has been written about the car of the year 2000. Some of it has been science fiction and some of it plain nonsense. Now that we are in the last quarter-century before that magic number, it's time to stop fantasizing and look at the facts. According to former Renault President Bernard Vernier—Palliez, speaking in 1977, the car of the year 2000 isn't going to be all that different from the cars we are driving today. He was addressing a meeting of the French Society of Automobile Engineers in Paris, on the occasion of the

Fig. 2-9. By splitting each car in half, four different designs can be compared on two clay mockups.

Fig. 2-10. The hood and windshield are in the same plane on the Citroen Xenia.

Fig. 2-11. The Citroen Xenia is half sedan and half station wagon.

occasion of the Society's 50th anniversary, and he explained his reasoning with a clarity that pulled the rug from under the daydreamers and illusion-makers who might have wanted to argue.

"Look here," said V-P (as he is known inside the Renault organization), "a new car has a long gestation period, say 4 or 5 years. We design cars for a production life of 10 to 15 or more years. Cars run for 10 years and longer. And that means the cars we are preparing right now will still be around in the year 2000."

What will the new Renaults look like? Design Director Robert Opron received me in his office along with his two assistants: Gaston Juchet for short-term projects and Jacques Nocher for long-term projects.

Born in 1932, Robert Opron was trained as an architect but found that he wanted to design cars more than houses. He was with Simca for three years and came to Citroen in 1962 where he put a new nose on the DS before designing the SM, GS, and CX. He has been at Renault since February, 1975, and now has overall styling responsibility for all products of the Renault group (including trucks, buses, and farm tractors).

Listening to Opron talk about his work is like standing at the foot of a veritable Niagara of fresh ideas. "Today we have new preoccupations," he began. "Due to the pressure of cost considerations, we must seek a maximum of modularity, right down to exterior body parts. We must work on the basics of market research, and seek a balance between safety, aerodynamics, maneuverability, and style."

EACH CAR FOR ITS OWN MOMENT IN TIME

Opron went on: "The product can no longer be a *trompe-l'oeil* (still-life deception), but it must *talk*. It must explain itself. Dynamic sense is the key to car styling. Fashion is important. We must break old habits and create a new fashion. But most of all a notion of time is needed. The car must be right for its time. We have to make each car for its present, not for the past or for the future."

What does that mean in terms of actual design? "Strong specific styling, for each model, combined with maximum use of common components. Each new project has to fit into a general scheme, with the primary goal of being competitive in modernity with existing models while also meeting a secondary goal of creating a new fashion."

Does Opron and his staff originate new-car projects, or does he receive "orders" from other departments? "Styling makes proposals after looking at market research," he explains, "taking into account the lead time, how society evolves, and how we expect the future to differ from the present. Each new-car project is designed with a notion of time—a time at which it will be modern, creating a new mode without being futuristic."

MODERNITY AND FUNCTIONALISM

Cost control does place restraints on a stylist. "We have new preoccupations now," explains Robert Opron, "We strive for a maximum of modularity, within a model range, and across model lines, right down to shared body panels and other body parts." As for aerodynamics, Opron points out that "streamlining must be tempered by maneuverability and aggressiveness considerations."

I asked Opron how he begins a new design. Does it always begin on paper or sometimes with a scale model? "Models come later," answered Opron, "the big question is whether a design should start with a silhouette or a plan view. Traditionally, what comes first is the profile. But for some cars, I want to start with the plan view, from the top.

"When the time has come for decision-making, from among the variety of materialized, three-dimensional studies, we select that particular concept which promises to yield the most harmonious compromise between functionalism and aesthetics, the economic factor being of course taken into consideration as well."

Functionalism is one of the watchwords of Robert Opron; another is modernity. Opron says modernity is the art of being contemporary. "Modernity is the present, not the future or the past. And the present, in essence, cannot be seized," he grinned, "in contrast with the past or the future, which can be apprehended by their duration.

"Since not long ago it seems that modernity can take other aspects," he says, "and that new forms are emerging in the creation/consumption system. New forms of organic character seem to find justification at a creative level as well as at a consumer level. We are rediscovering, through organic forms, the esthetic interest of the components of an object, of materials, and of organs. Objects created this way are not conceived as a whole, but as parts of functioning systems. Instead of designing the whole, we design each basic component according to its mechanical, ergonomic or architectural principles, and assemble them afterwards."

With regard to functionality, Opron maintains that the elements of a design should indicate the functions. He believes that earlier trends in industrial design resulted from an overreaction to the ugliness of industrial products. This lead designers to make every effort to disguise the true functions of the objects under study.

On specific features Opron had this to say: "The grille? Kill it. We don't know what more to do with it." What do you do instead? "Make an authentic air intake, that indicates its function." Hub caps? "Don't put hub caps on wheels that are both functional and pretty." In answer to my question whether he agreed that trying to disguise the function of certain design elements, stylists were sometimes guilty of harming their ability to function, he gave a strong yes.

What can we conclude from this? Future Renaults will look practical, smart, original but not weird, modern but not futuristic. The relationship between the car of the year 2000 and today's cars will be evolutionary, not revolutionary.

THE BRITISH EXPERIENCE

"I often hear it said, in a slightly derogatory way, that automobile manufacturers are in the fashion business," said David Bache, former director of styling, Leyland Cars, in the summer of 1977. He was the man behind two generations of Rover bodies before Rover was swallowed up by British Leyland, in 1968, who later was responsible for the Rover 3500, the Princess, and the Mini-Metro (but not the TR7), before leaving the Leyland organization in 1981.

David Bache continued: "The answer to that, quite simply, is "you're damn right we are! and that comment is not good only for stylists—for its necessary to recognize that fashion applies not only to appearance of a vehicle but also to its technical specification. *Furthermore we cannot afford not to be.*

"This is underlined," Bache went on, "by examining the past decade over which there has been a considerable change in the basic criteria by which car design is judged. We are now moving away from the rather simplistic view of the early '60s that complexity equals excellence. It is no longer true that a simple car must, of necessity, be a mediocre one. We now take a much more objective view of the overall performance of a car, in areas such as primary and secondary safety, durability, reliability and versatility. From a manufacturer's point of view, it may be more difficult to achieve high standard with a simpler design, because more intensive development is needed, but the results of this effort are well worthwhile because of the easier manufacturing and servicing which follows.

"It is critically important, however, when planning a new model to interpret and enlarge on future cosmetic and technical fashion trends correctly—to capture public imagination one must produce a design which is seen not merely to be acceptable but to have positive qualities as well, hopefully, as a degree of panache. This will mean going beyond any legislative requirements since with few exceptions, legal stipulations usually lag behind market and economic pressures."

THE BURDENS OF INDUSTRIALIZATION

Sacco, Opron, and Bache have one thing in common. They all work (or worked) for car companies selling in high volume. Car design is a different proposition for specialist builders who are not committed to mass markets and have a minimal investment in industrial plant. They have much more freedom of choice every step of the way through the design process, body engineering, development, and manufacturing. Mass production acts as a brake throughout the project's progress from the idea stage to the shop floor.

Specialist coachbuilders who turn out small series of custom bodies beat the metal into shape on a wooden work form (Fig. 2-12). For mass-produced steel bodies, all panels are stamped in dies mounted in hydraulic presses. Plaster casts made in the styling shops are also used by the specialists who plan, design, and order the mass-production tools.

The styling studios in all big organizations possess complete shop services to turn their ideas into hardware. From the wood and metal shops comes the precision armature or framework on which stylists shape the full-size clay model of the future cars. See Fig. 2-13.

Many specialist craftsmen work with the stylists to make the mockup look as real as possible before it is presented to management. Paint-shop personnel apply colors to the smooth clay surfaces. Other workers form clear plastic windows,

Fig. 2-12. A work model for Bertone's Alfa Romeo Giulietta Sprint.

Fig. 2-13. A styling proposal for Ford takes shape in the Ghia studio in Torino.

windshields, and backlights. Precision machinists furnish bright metal parts and engravers create jewel-like ornaments.

A clay model has short life. It deteriorates quickly and loses its accuracy. Plastic molds are taken from the clay as soon as a mockup has been approved, and these are the molds that serve to cast plaster duplicates for use in developing a whole family of body styles and variations on the approved clay mockup.

THE MASTER DRAFT

A General Motors executive, K. E. Ruff, a department head in Manufacturing Development at the corporate Technical Center, more than hints at the inertia of designing cars for mass production: "Body design organizations are ancient and venerable societies. They have existed for several decades, and during this time have gradually evolved complex codes of conduct, intricate communication patterns, groups and subgroups of specialists, recognized measures of performance, and functional differentiation.

"Very few, if any, individuals exist who understand how or why the whole organization functions. The basic unit of working information within such a group is the engineering drawing. The most prevalent type of recognized work is drafting. Problems are worked out graphically on the board. Solutions are communicated by means of drawings. A single key line may be copied from one drawing to another more than 30 times. Progress is marked by the release of drawings. Informal communications consist of conferences about a drawing at a drafting table.

"Typically the master draft of the product is the culminating product of the design effort. It portrays the results of design problem solutions and provides a complete definition of the part.

"In fact, however, the draft is usually not a complete definition of the part *surface*. This final detail of the product design is accomplished by the skilled hands of the model makers, whose three-dimensional rendering of the surfaces is the only complete master document. In many organizations it is the final expression of the design intent. The production of the master model is in itself a complex process of information transfer. Templates made to match the lines of the master draft are used to transfer the information to the wood model. Since the lines cannot absolutely match, the model maker interprets the mismatching lines scribed in wood, blending or splining a good surface through and among them."

DIE DESIGN: A BLACK ART

"Just as the language of the body engineer uses a vocabulary of fine lines on mylar", says K.E. Ruff, "the language of the model shop is a tactical language through which good surface is *felt* as well as seen. Model makers take pride in the skill of their hands and their feeling for surface rather than line. They build the object which has been suggested by the master draft. They expose all the errors and inconsistencies of the draft. They solve some of the transition problems neglected by the designer. Typically it is through an examination of their product that the validity of the design is assessed by those responsible for final release. The models of the several components of the car are cubed up (stacked up as a car) to provide a proof of panel-to-panel coordination.

"It has been suggested that good styling is a gift from heaven, body design is a favor of technology, and die design is a black art. In any case, the design of die addenda is sometimes more a matter of a consensus of opinion than an application of consistent principles. Science has not told us enough about plastic deformation to provide a simple approach to die design, and too little communication has taken place between die makers and industrial engineers for the present mysterious process to be documented except for the simpler cases."

COLLABORATION WITH INDEPENDENTS

It helps the stylist to be fully aware of the industrial implications of what he is doing in the studio—on paper, with clay models, and with rear hardware. Increasingly, the industry promotes stylists who have some engineering training and experience.

Does this mean the end of the road for the outside consultant, the industrial designer, and the independent coachbuilder? Far from it. Both Volkswagen and Fiat, with big styling centers of their own, continue to rely on outside studios for specific jobs.

"It is logical for Fiat to go to other famous Italians for exploring styling possibilities," in the opinion of Gian Paolo Boano, director of Fiat's styling center. "Sometimes we give the same assignment internally and to Pininfarina, for instance, at the same time, and then compare the results. We always keep in close contact with the specialist coachbuilders to watch the trends."

From Sergio Pininfarina's viewpoint, this sort of relationship cannot be precisely defined: "We all walk the razor's edge between electricism and ingenuity, improvisation and science, adventure and intuition. We are in that world, still mysterious, that Vittorio Gregotti once defined as . . . 'the relationship between art and industry.' To the designer, this translates into a creative moment impossible to describe without using contradictions. The results must speak for themselves as they are and as they endure."

When Fiat contracts with Bertone for the X-1/9 and with Pininfarina for the 124 Spider, it is not because the internal studios are incapable of creating a good-looking sports car. The 124 sport coupe was an inside job.

"When we assign the styling for a car to an outside organization," explains Gian Paolo Boano, "it is because we want the car to be 'autographed' by that organization. They follow their own style completely. We merely follow the project, and may suggest changes. But we prefer to to let the designer's touch prevail."

The case of the Panda was a little different. It was mainly because of a lack of capacity of Fiat's styling center at the given time for the project that it was given to Ital Design. "We had to farm it out to get results quickly, and we were certain that Giugiaro could do it."

The Fiat 126, 127, 131 (Brava) and Ritmo (Strada) were designed under Boano's leadership. The Ritmo was launched in April, 1978, and it represents a clean break with the family style that permeates the earlier models to a greater or lesser extent. But Boano disclaims sole credit for its shape.

"The lines of the car are not the creation of one person, but the fruit of a slow maturing process resulting from teamwork. After designing and evaluating a variety of different shapes of similar dimensions, it was not until 1975 that the car's definitive shape had evolved."

Are we to understand that the Ritmo style will be followed in future Fiats? "No, it is not representative of our future styling theme for family cars," says Sergio Sartorelli, who is in charge of advanced design at Fiat's styling center, "But some of its features, such as the use of protective rubber or plastic on the lower sheet metal all around the car, will be used again."

Fiat Auto owns Lancia, but all Lancia models are designed by outside firms. "I will not touch Lancia," says Boano, "because it must not be perceived by the public as a Fiat design." The Lancia must have its own look. The latest addition to the range, the Delta, was designed by Giugiaro (Ital Design).

What do the men of Fiat's styling center see as their worst problems? "Inflation," comes the immediate answer from Boano. "Higher costs. We must set a cost limit, and style for inflation. That's true even for our high-line models. Today, European cars have far higher cost than American ones. Styling is not a major part in that, but we want to eliminate styling as a factor in price increases," states Boano firmly.

Chapter 3

The Design Process

MOTORISTS WITH NO PARTICULAR ENGINEERING interests tend to feel that an automobile body—having been modeled by some artist's hand—is then somehow welded together out of separate pieces, with the result that it finally emerges handsome and good, safe and low priced.

But there is more to car design than the glamorous aspect of playing with sketch pads and scale models. There is also the tedious recording of exact dimensions and angles. This information is essential for the body engineers and the tool and die makers, who will make possible the reproduction of the designer's creation on an industrial scale.

Under the old order, body designers integrated their solutions in a master layout. Now their work is brought together in the form of a mathematical model in the memory of a computer. Computer tape (Fig. 3-1) has replaced tons and tons of paperwork; the measurements and calculations that used to occupy highly skilled personnel for countless hours can now be performed in minutes. Computers have made a major contribution towards shortening the lead time in the industry needed for a new model to progress from the drawing board to the assembly line.

THE COMPUTER REVOLUTION

While attending the SAE convention in Detroit in February, 1973, I was impressed with the demonstrations by exhibitors and lecturers on the computer invasion of the auto industry. Electronic data processing had already penetrated into all areas—from design and testing to production control and diagnostic equipment—including on-board computers to control vehicle systems independently of or in conjunction with driver commands. Computers also deal with a wide range of spheres extending from finance through market analysis, the processing of orders, after-sales service, spare parts, and the research and development sectors.

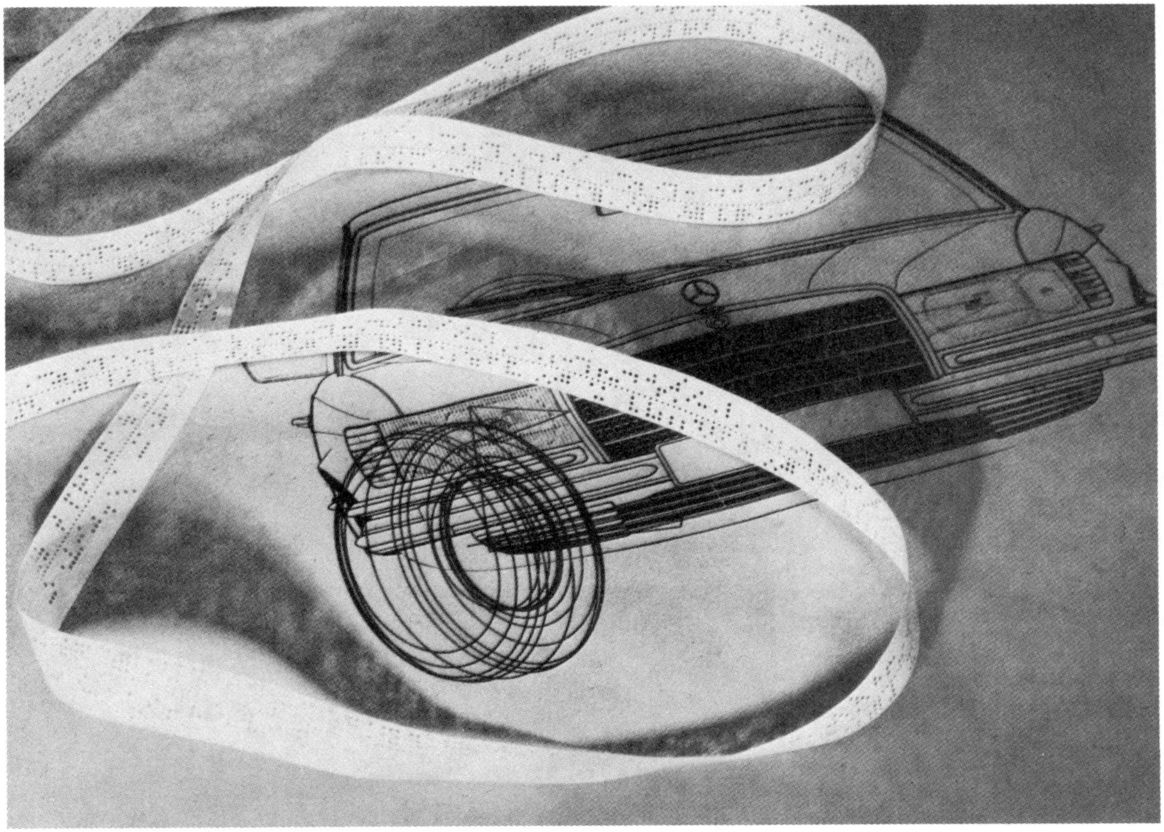

Fig. 3-1. Computer tape describes every dimension, every angle, and every curve, in minute detail.

I looked in awe at the equipment exhibited by Textron's MB Electronics and Electronic Associates, Inc. and listened attentively as A. J. Foland and R. Razi from Bendix delivered a paper on their automated design systems (also on display).

As I talked with auto industry engineers about computer applications, I began to understand the basic attraction of computers. *They work fast.* If properly programmed and operated, they can perform man-years of work in just seconds. Under the pressure of competition to produce new and improved cars and with increasing international competition, design staff work load and engineering capacity are strained.

The early attempts to apply computer technology to the body design and tooling activities were viewed as a technical challenge rather than a management problem. For several years, the research and development efforts dominated the scene. Impressive work was done by mathematicians in surface representation, finally demonstrating that all the complex sculptured surfaces of an automobile could be defined by a mathematical model.

Now is the moment when I should point out another. The computer does not replace designers or engineers. See Figs. 3-2 and 3-3. It is simply an aid. When you feed a computer relevant information, it can spit out answers far quicker than any

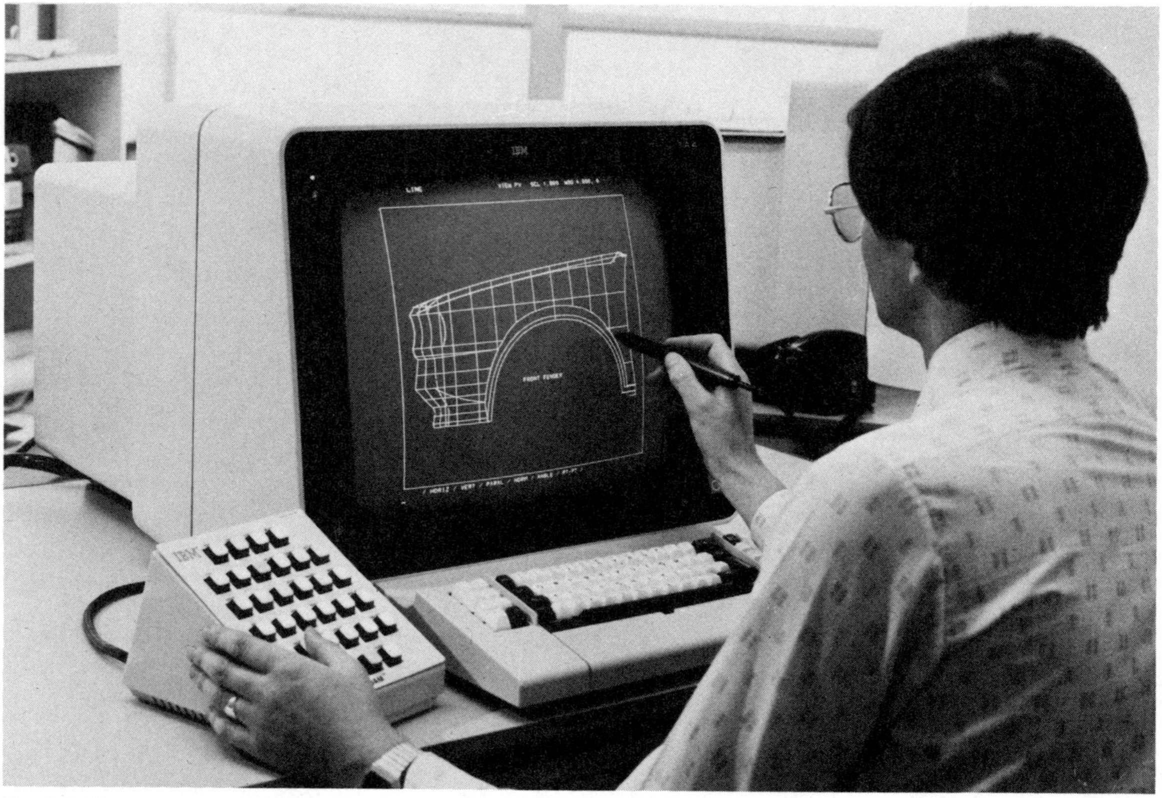

Fig. 3-2. The computer model from the design office can serve also for tool and die design.

group of engineers can calculate their way through the data. What's more, computers performing skilled tasks can be operated by relatively unskilled personnel.

Computer installation and programming requires specialists, and this opens new career opportunities inside the auto industry. Computers actually open the way to promotions at all automotive industry levels. From a company president's or general manager's point of view, the purchase of a computer or renting computer time can be seen as buying information or skill with a delivery date far in advance of any date possible by conventional means (expanding the staff, instituting training programs).

The cost? The computer might well be more expensive. It doesn't matter. What matters is that the executive officer might not have a choice, because results must be obtained by a certain date. And only computers make it possible. "The computer has revolutionized engineering," said William A. McConnel, director of systems research for the Ford engineering staff. He added, "Eventually it will be designing the complete vehicles and minimizing the need for exploratory tests."

COMPUTER-AIDED DRAFTING

The draftsman's pencil is replaced by a luminous spot that engineers can use to "draw" on a screen;

afterwards, a plotter reproduces the drawing on paper fully automatically.

There is an extensive range of previously drawn lines, curves, scales and components ready for use. Drafting is now more like a process of "composition," known as "packaging," than like traditional drafting. The highest degree of sophistication is reached in the drafting of three-dimensional mechanical parts that can be arranged in space as the actual physical parts, without using the traditional cross-sectional view procedures.

Because the computer's memory stores a great number of standard parts that have been previously drafted and perfected (bolts, screws, brackets, sundry components), drafting operations are much faster and easier.

The advantages afforded by computer-aided drafting are not limited to improved productivity and drafting quality. The availability of an extensive range of technical solutions ready for use makes it possible to arrive promptly at the optimal decision. Variations can be introduced without difficulty. In many cases, this leads to a radical cut in the time it takes to complete the entire engineering process.

Fig. 3-3. A three-dimensional sketch can be turned for viewing from any angle at any height.

The instruments of computer-aided drafting—including hardware and software components—are designed so as to be compatible with design and model building equipment. This leads to a fuller integration of drafting and design operations.

Inevitably, the arrival of the computer has caused fundamental changes in the ways new models are created (Fig. 3-4). Stylists, designers and engineers constitute a team that realizes the project, but under completely different conditions from those formerly prevailing. In the technical sphere, testing used to be the principal method of detecting constructional weaknesses and of eliminating them (Fig. 3-5). Today, with the help of computers, calculations can be made well in advance and the first steps towards the creation of an optimal prototype can be made as early as at the planning stage.

Not only can the computer carry out the most complicated calculations within an incredibly short space of time, but it also can store the results so that they can be extracted at any time in printed form or on a screen. Still there are limits to what the computer is capable of doing due to the numerical nature of the data it handles.

The main problem arising from the use of electronic data in the construction of automobile bodies is to define in mathematical terms the body that has been developed from the functional and aesthetic aspects. The computer can do nothing with a beautiful contour. It is only when this comprises the geometric elements dot—straight line—circle that an outline is produced that can be scanned and stored in numerical form on a punched or magnetic tape. In the first geometric data processing phase, therefore, a full-size or 1:5 scale

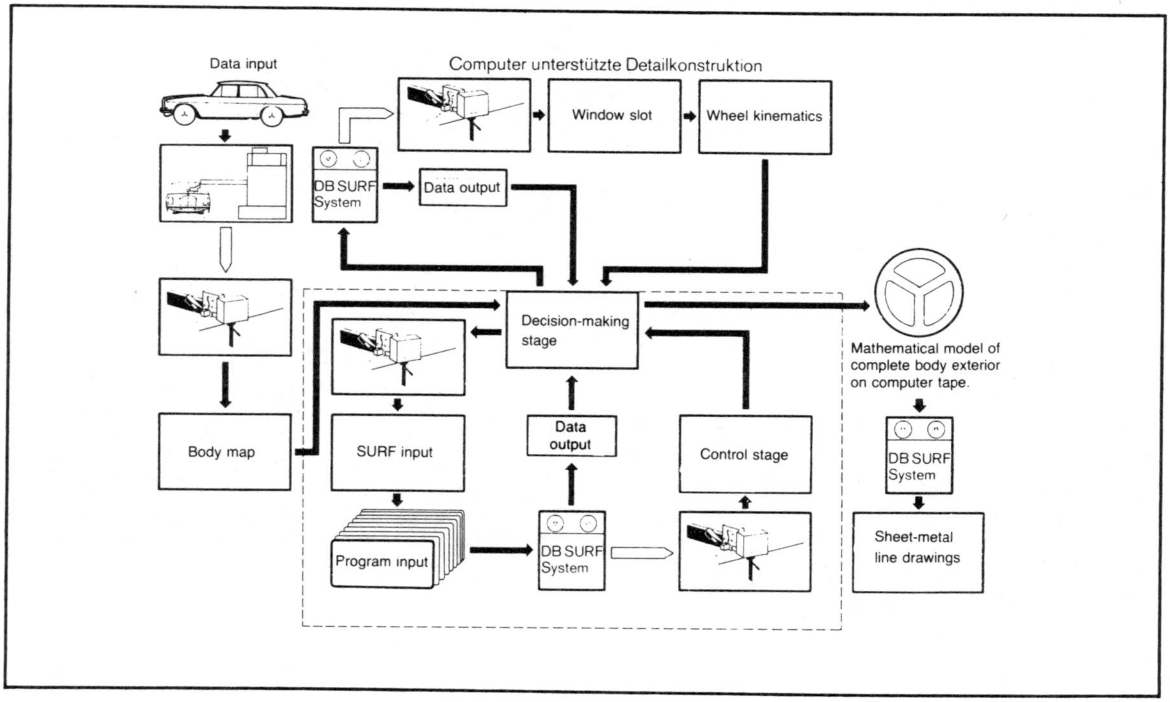

Fig. 3-4. The development of a computer-designed body.

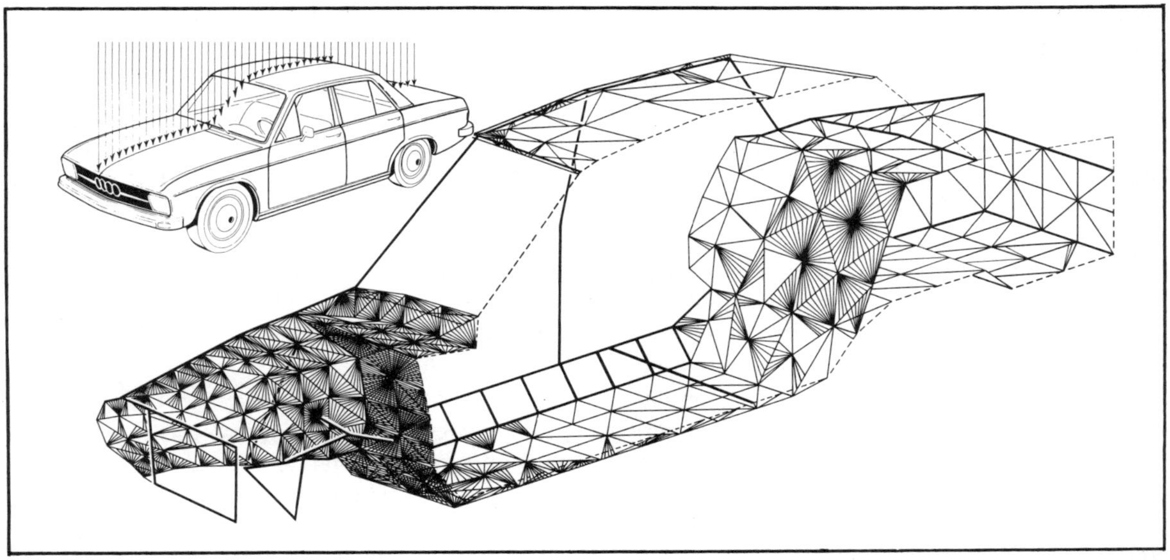

Fig. 3-5. A survey of stress zones along the center line in a head-on collision as produced by a computer with an attached drafting machine.

Fig. 3-6. Measuring instruments for the computer input data are more accurate than any previous method.

model of the future car is produced by the stylist (as a form of initial information for the computer, as it were). A three-dimensional scanning device (Fig. 3-6) then runs over the surface of the model—measuring lengths, widths, and heights automatically. In this process, the location of every point within the space is fixed as a numerical series and stored on punched tape.

The information contained on this medium is then "fed" into a numerically controlled plotter that draws fully automatically, and to any desired scale, what the scanning device has measured accurately. In this way, the first full-size diagrams are produced from the 1:5 scale model. From these diagrams a car can be constructed in which one could sit in order to judge whether the draft comes up to expectations. As the small model made by the stylist and usually made of clay or plastic is not exact in every detail, the full-size model is then scanned and measured again. This seems a complicated process but it is necessary in order to obtain accurate measurements for the body surfaces.

STRUCTURAL ANALYSIS

Computers can design lighter, stronger auto bodies than the most experienced engineers. The main reason is that the body engineers can't afford to spend the necessary time on structural analysis.

While the basic mathematics for calculating the components of a body have long been familiar, what the computer does today in hours would take years by traditional methods. It was formerly quite impossible to attempt optimum evaluation according to any particular principle (e.g., rigidity, weight, price, safety). So the general practice was to go by experience and rule of thumb, and to eliminate weakness pragmatically in the ensuing road tests.

In an investigation of body strength, the purpose of electronic data processing is to obtain the most effective combination of structural elements, the optimal use of different gauges of sheet steel, and the computation of cross sections for longitudinal and cross members, posts, and frames—the total structure that supports the whole car. See Figs. 3-7 and 3-8.

By computer analysis, three-dimensional animated mode shapes can be generated very rapidly. These mode shapes help the engineer to determine areas that require additional stiffness and also areas where weight can be taken out and thereby optimize the design.

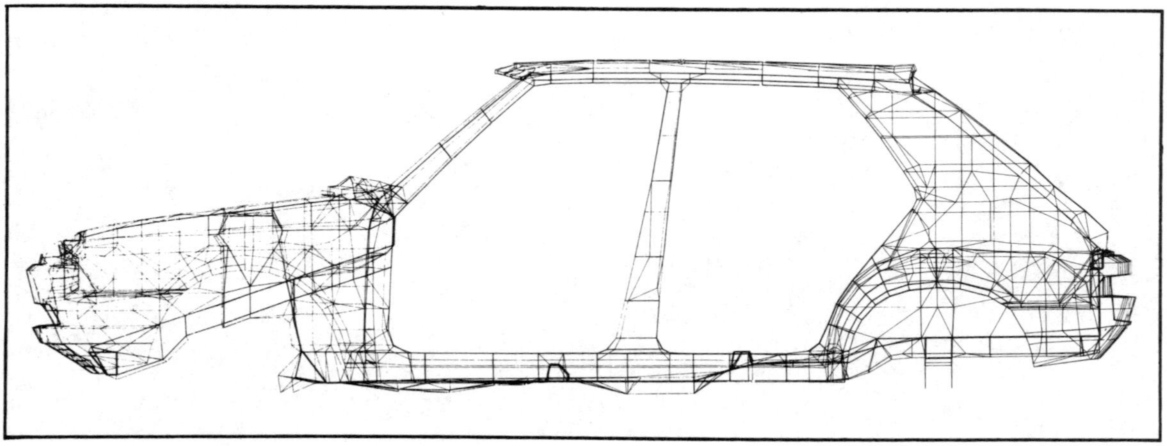

Fig. 3-7. A mathematical model of a complete car body.

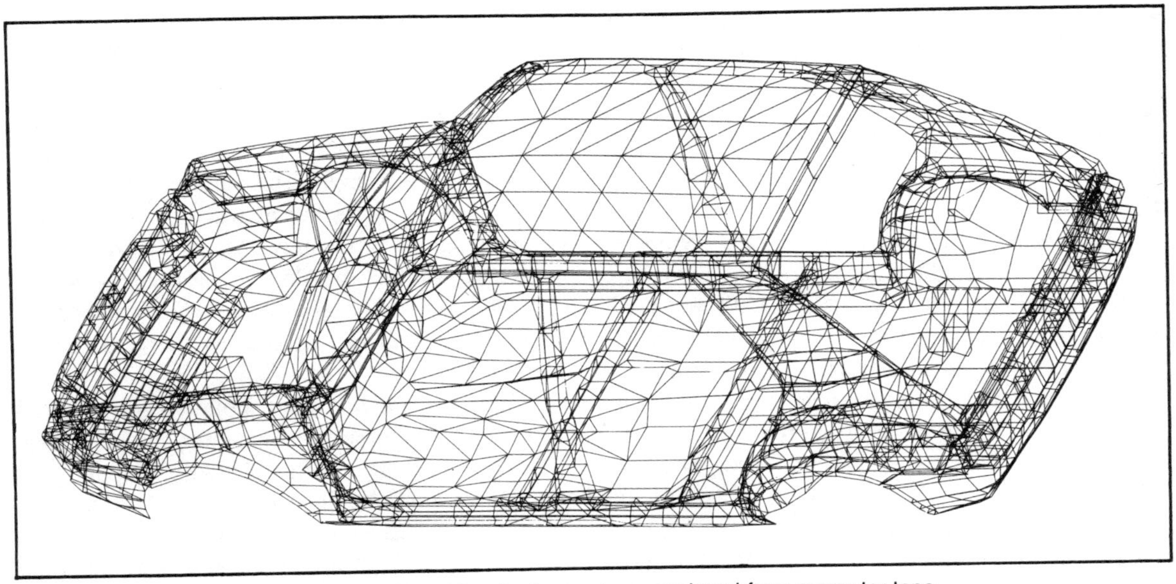

Fig. 3-8. A three-dimensional drawing of complete body structure, produced from computer tape.

What the computer analyzes is a mathematical model of the structure. Use of such models for body strength and torsional rigidity evaluation are carried out according to methods that can be grouped under the title of finite element. The finite element model means taking a complete unit and making it manageable in the computer by reducing it to very small elements.

Finite elements are basically physical building blocks in the forms of lines, areas, and spaces. A structure can consist of a thousand or more such elements. The physical characteristics of these building blocks are given by the person analyzing the problem. The finite-element method is used for the solution of static, dynamic, thermal, and acoustic problems (potential theory). It can provide accurate data on force and tension distribution under static loading, natural frequencies, and forms under dynamic loading, temperature, and sonic-pressure patterns for thermal and acoustical work.

The principle of the computation is found in a lattice work where diagonals stiffen rectangular frames. By way of comparison an attentive observer might consider the truss system of steel bridges. For this purpose, within the longitudinal members, posts, and struts of the future automobile (existing only as drawings so far), the supporting metal surfaces (such as the roof, the wheel housings, and the bulkhead between passenger and engine compartment) are broken up into triangles.

The intersections and connecting bars form a grid system. This shows the stiffening effect of the sheet-metal surfaces in the body. This division into sections permits predicting the stress at each point of the car body, whether under operating conditions (load of passengers and cargo, body torque) or in a simulated accident (crash test). The designer gets a set of dependable figures even before the first rough body is produced by time-consuming and costly hand labor in the development shops.

The computer delivers these figures point by point, triangle by triangle, in the form of lists that accumulate in an imposing but unfortunately bewildering stack. Accordingly a graph, a strategic map, is needed to show at a glance where the strategic focal points (stress zones) are located.

The computer with a drafting maching linked to it projects on paper a perspective image of the body from any desired angle. To represent the surfaces, it is programmed to show division as triangles.

This equipment translates the load values of the individual triangles, determined in the electronic calculation, into shadings, with 100% of possible load (a value fed into the computer) represented by a fully shaded triangle. Partial shading corresponds to lower percentages of surface load.

In conjunction with finite-element analysis of the body for the first Cadillac Seville (a modified X-body derived from the Chevrolet Nova), the engineers created another valuable tool utilized in this program. It was a ⅜-scale plastic model of the body (front sheet metal, frame, and bumpers built up as the parts were designed).

Plastic parts were vacuum formed over wooden models and assembled as a car. Through the use of the proper scaling factors, both static and dynamic tests could be performed on the model. One of the most useful applications of the model was for design and development engineers to place an appropriate load on a component and observe the reaction. Due to the transparency of the material, it was very easy to see the structually weak areas. Changes could be made and tests could be performed in a very short time.

COMPUTER-DESIGNED INTERIORS

At the end of 1974, Toyota announced the creation of a novel "Verification, Modification" method to computerize vehicle interior design procedures. The new method (Fig. 3-9) enhanced the quality of design work, gave engineers and technicians many more options, and substantially reduced the amount of time spent on designing.

The development was regarded as a significant advance toward the company's goal of computerizing every possible aspect of vehicle design. Toyota had earlier computerized the measurement and graphic representation of clay models and calculations related to the details of exterior body contours.

In the new computerized method for vehicle interiors, most of the mechanical work involved in the planning and designing stage could be handled by computer. This condensed or telescoped the lead time necessary and allowed Toyota to make quick alterations in response to changing public tastes.

All useful graphic-related information is stored in a computer. Information needed in the design work can be readily taken out of the computer and displayed on a graphic display unit. Previously, much time was used in evaluating all technical and design factors to get the best combination of both. Designers can verify and modify through simulation the graphic details of an interior design from any angle. Prior to this, verification and modification of interiors had to be done manually, using many types of measuring devices and actual scale models.

For example, the verification of a design for reflection-free glass that formerly took between one and two days can now be completed in less than a half hour. Similarly, the layout and design of an instrument panel or related individual devices can now be accomplished within minutes; the procedure formerly involved many time-consuming steps, including production of actual mockups.

BEGINNINGS OF NUMERICAL CONTROL

Numerical control (NC) is a technology that links computers and tape-operated machines to the development of car surfaces and the manufacture of major dies used to form sheet-metal parts. It provides the potential to produce high-quality dies and resultant high-quality automotive bodies in less time than with older design and manufacturing systems.

Numerical control is based on a blend of sci-

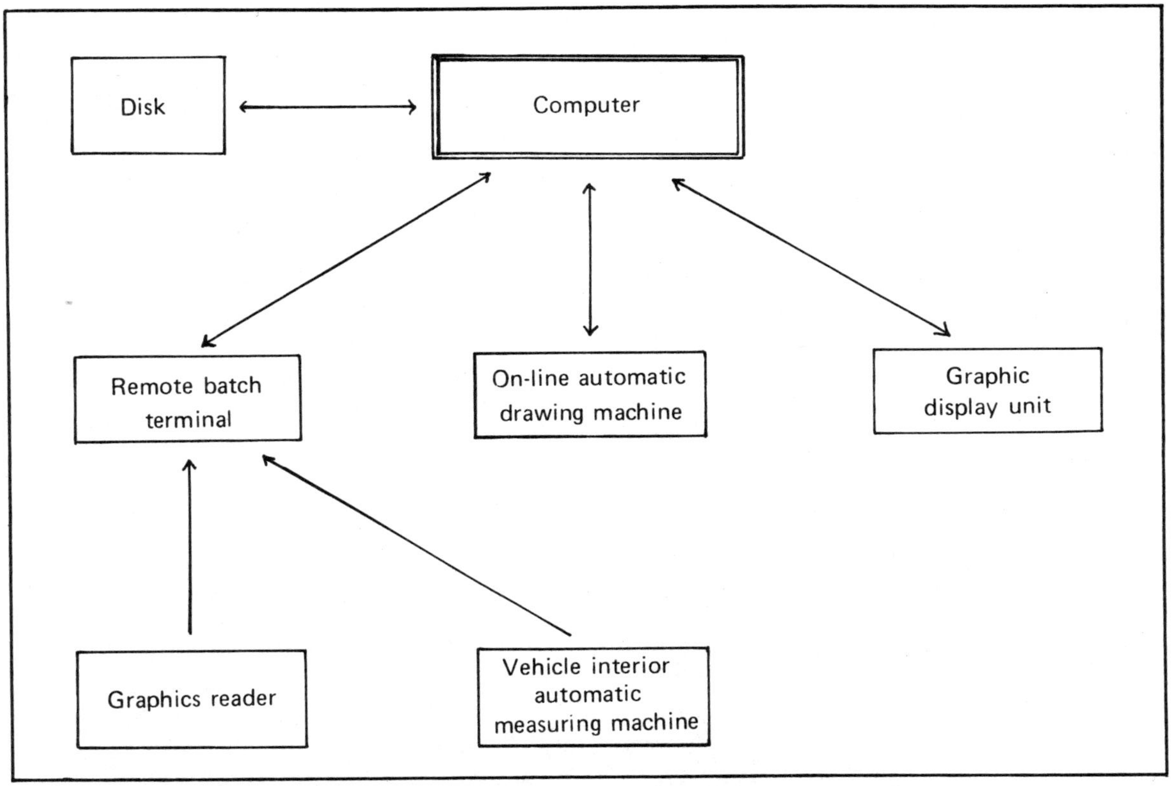

Fig. 3-9. The Toyota system for verification and modification of interior design with the aid of a computer.

ence and technology that converts surface contours—such as doors, hoods and roofs—to mathematics. In brief, NC die making works as follows. Mathematical coordinates of contour lines are fed into a computer memory unit along with recorded machining and processing requirements. Computation produces a tape that "tells" tape-controlled machines how to cut, guiding the cutting tool along the proper paths at optimum speeds and feeds.

The NC work begins with the development of a clay model that is electronically scanned. Contour information is numerically drafted on plastic sheets and then refined onto aluminum plates. The drawings are rescanned by a television camera on a coordinatorgraph. The operator views predetermined layout points on a television screen, that magnifies at a 15:1 ratio, and depresses a record button.

A numerically controlled drafting machine is used to make orthographic and perspective drawings and to check the computer cutter path before cutting dies. The computations are converted to machine-control punched paper tape, as required by particular machines. The final step is machining on multiple-axis, numerically controlled machines.

The first primitive types of numerical control were exhibited at the 1955 machine tool show in Chicago. It took only a few years until NC machines had been developed into a reliable and economical production tool for small-batch and medium-volume production of discrete parts. By 1965, scores of commerical models had been established, and thousands of machines were ordered by several industries.

Formerly a considerable part of die design was a plastic art carried out on the shop floor through modifications and additions to the wood die model of the part. The new approach works best, however, when the die design is developed by die designers working with powerful computer assistance, and represented again in the memory of the computer as a mathematical model. The combined mathematical model of part surface and die addenda now allows the numerical control elements of the system under the control of die machining experts to produce NC tapes capable of controlling machine tools in sculpturing dies of greater quality and precision than any yet produced. Formerly the tasks of machining a die from the complete die model involved a torturous, precise debilitating series of plastic casting replications. These culminated in the rough machining of a die by tracing a contorted Keller model (Fig. 3-10).

After five years of concentrated research and development work by Ford engineers, mathematicians, programmers, and technicians, the first production die resulting from NC was made for the 1963 Ford glove compartment door. The first large NC body die was for the 1964 Falcon rear deck lid, produced in late 1962. NC helped Ford bring the Mustang to market early in 1964. It was also used to

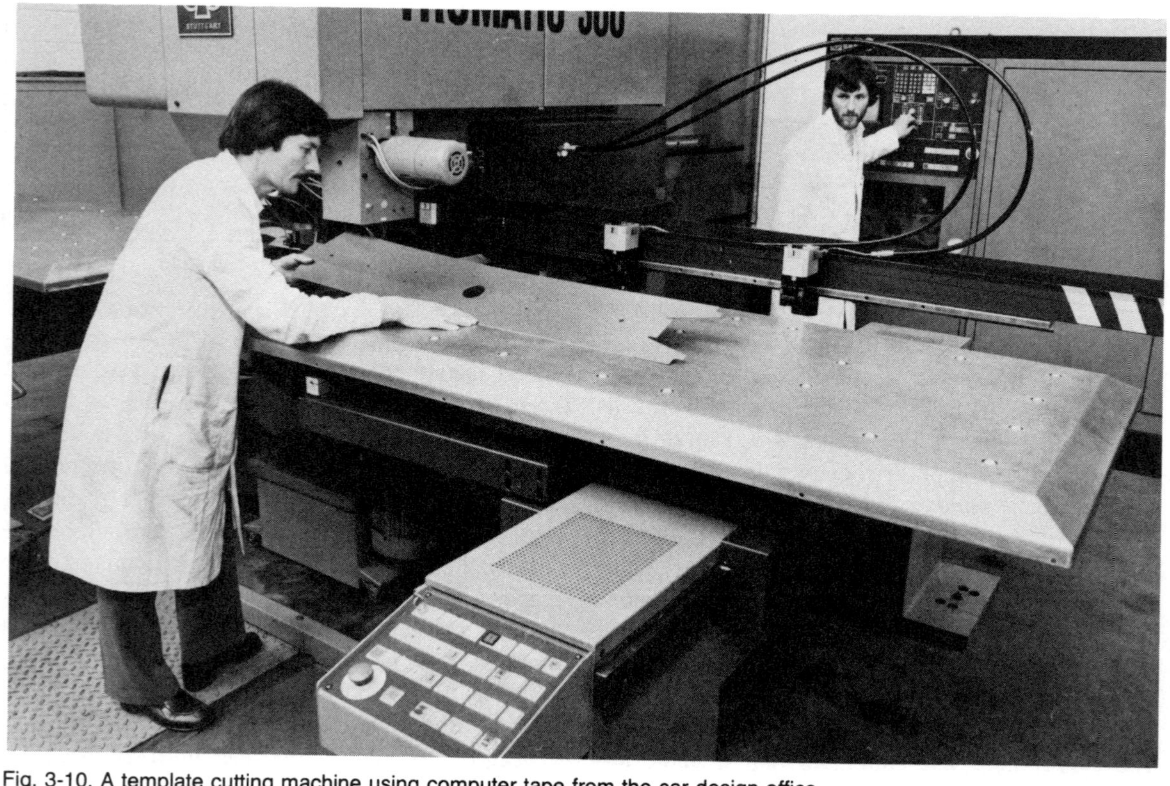

Fig. 3-10. A template cutting machine using computer tape from the car design office.

produce all sheet-metal parts of the completely redesigned 1968 Fairlane and Montego lines. The Ford Maverick was the first completely new car entry ever to have the major parts of its sheet-metal produced on NC dies, according to H. N. Bogart, numerical control director on the Ford manufacturing staff. But Ford was not alone.

GM'S MULTIROUTE APPROACH

"General Motors developed three electronic processes simultaneously during the early sixties," according to Harold G. Warner, executive vice president in charge of the operations staff. "The first of our three individual programs was the DAC-1, perfected by the General Motors Research Laboratories. The second was photogrammetry, and the third Integrated Numerical Control Approach, called INCA, developed by the Manufacturing Development Section of the Manufacturing Staff."

Why three systems? Warner explained: "Each of the three methods proved a valuable asset in designing and transmitting information into draftings. During recent years, we have used the three programs in concert, with each segment producing a vital portion of the overall process. Each has its specific value in the method of going from the clay model to eventual production dies."

The full scope of the advantages offered by NC was not apparent at first, but became clear only after some experience had been gained. Application of the DAC-1 project demonstrated that the productivity of a designer could be greatly increased as he learned to control the powerful medium of an interactive graphic console.

When final approval of design sketches and the clay model of the proposed vehicle had been obtained by the styling staff, the DAC-1 process began with scanning of the clay surface with an electronic surface recorder. The probe of the surface recorder glided lightly over the clay and automatically drove a recording stylus to reproduce the surface as a two-dimensional drawing on a pantograph.

Meanwhile the second phase of activity began. That meant acquiring data from other areas of the object through a process known as photogrammetry. Stereo photographs were taken to show the depth of an object rather than the flat surface of normal photography. These pictures were placed on a plotting device that electronically produced a tape representing three-dimensional coordinate clay information needed to produce a master drawing on a numerically controlled drafting machine.

The INCA system showed that the basic mathematical models could be derived from available graphic data and used effectively as a base for numerical control operations to machine the complex shapes required in wood or metal. From the die model—modified to tooling conditions—the hard model scanner converted the surface measurement into tape for use on a numerically controlled kellering machine to produce the production dies.

The first application of the three systems resulted in exterior body sheet metal for the 1965 Corvair. The following year, Chevy II major body components were produced in the same fashion. Since that time, use of the system for moving from clay models to prototype dies, and in many cases production dies, has produced both interior and exterior metal stampings for many GM automobiles.

From the electronics industry came new technology needed to integrate the various functions of the three GM systems, and the machine-tool industry began offering ESEM (Elasto-Static Elements Method) equipment (Fig. 3-11). Ford, General Motors, Chrysler, and Mercedes-Benz were quick to put it to use.

More recently, Ford adopted the SAMIS (Structural Analysis and Matrix Interpretive System) developed by the Jet Propulsion Laboratory in California. General Motors switched to FORMAT (Force-Oriented Matrix System), a development

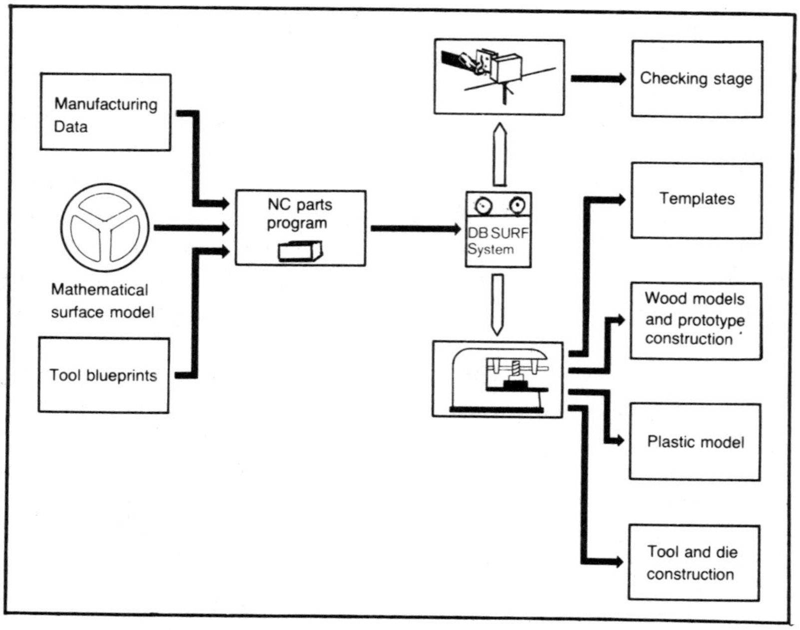

Fig. 3-11. A method for using body design tape to make tool and die drawings for use in numerically controlled production machinery.

from the research laboratories at Wright-Patterson Air Force Base. For some years, Chrysler used SAP (Structural Analysis Program), developed by the computer center at the University of California at Berkeley.

Fiat developed its own AMSA 20 system in the early 1970s (Analisi Matriciale Strutture Automobilistiche), that, by 1975, led to AMSTRAN, borrowing certain elements from NASTRAN methodology. The method called NASTRAN was developed by NASA (National Aeronautics and Space Administration) for space-vehicle design and declassified for industrial use in 1971. It has since come to play a considerable role in car design and tool making.

The letters mean NASA Structural Analysis Computer Program. By incorporating NASTRAN predictions into its design process, Ford's Advanced Analytical Technology Department reported a 60-percent improvement in predicting the behavior of components under stress and a time saving of two-thirds in achieving such calculations.

Chapter 4

Car Size and Vehicle Architecture

SOME OF THE BASIC QUESTIONS CONCERNING the layout of the car-to-be must be solved in the concept stage. For instance, how many seats? Where to put the engine? For many car companies, the answers are foregone conclusions. They are preparing a new model to replace an old one, the engine will stay where it was, and the seating capacity will remain the same.

Others are more questioning of their own previous practice of engine location and seating configuration. To them, it is not to be taken for granted that the seats in a four-door sedan must be aligned in pairs, one pair behind the other. Or that a six-seater must be the same, only with wider seats. Some ask why the wheels must follow the same arrangement (two front wheels on the same axis and two rear ones, side by side, on a shared axis).

SEATING PLAN AND ENGINE LOCATION

Some of the answers do not depend strictly on layout and architecture, but on dynamic principles affecting the vehicle's behavior on the road. This applies in particular to the number and positions of the road wheels. How many wheels, and where you put them, are matters that will greatly affect the balance of the complete car. The wheel arrangement is in turn tied in with the seating configuration. It would be difficult to keep a classic seating pattern if you deviate from the conventional positioning of the four wheels. It can be done with three wheels, but it's not a popular solution. We shall soon see why.

The engine location has a big influence on weight distribution. This is a main criterion in road-holding and directional stability. The bigger the engine the more critically its position must be examined. Because of the importance of this question for vehicle-control and safety purposes, it must be an engineering decision and not a design decision.

If a car manufacturer really wanted to bring up the most basic questions for reexamination, where would he start? Engine location? Number and position of road wheels? Or the seating configuration? The problem is that the three are inseparable; they must be considered together.

One line of reasoning might go something like this. To make a four-seater car, it's fairly obvious that you don't put all the seats in line. That would make a long, thin car with maneuverability and parking problems. Also, it is not a good way to save weight or cost. Just think of all those doors, and the reinforcements needed to keep the floorpan from sagging.

Nor is it practical to place four seats side by side. The car would be too wide. The legal limit in the United States is 80 inches. This gives ample room for three, with door thickness allowing for wide guard beams, but not four, including a driver, who needs a certain amount of elbow room.

Seats in a lozenge formation have been considered. That would permit interesting new body shapes and wheel configurations. It would allow the driver to sit centrally in front, all alone. Behind him there would be a bench-seat for two or a single seat along each side of the body, separated by a passage. There is no proof that the driver gets any actual benefit from sitting in the middle. But what really kills the lozenge-seating plan is the door arrangement. It is very difficult to provide convenient access to all seats. Power train complications also argue against this type of layout.

SOME HISTORICAL ODDITIES

Sunbeam began automobile production in 1901 with a curious contraption called the Sunbeam-Mabley; it was designed by inventor Mabberley Smith. It had two wheels on a narrow-track driving axle in the middle, with a single wheel on the center line front and rear, in true lozenge formation. The engine was mounted in the rear and the driver sat in the rear. The passengers' seat had a side-facing bench for two, with curbside entrance at the front end. The Mabley was soon replaced by more conventional Sunbeam cars.

Gabriel Voisin was toying with lozenge wheel formation and novel seating arrangements in the 1930-33 period. And in 1947, Professor Alberto Morelli, trying to reconcile a new concept of an aerodynamic body shape with an automobile chassis, also went to lozenge formation wheels. Professor Morelli moved the center pair of wheels far back, however, so as to permit the fitting of all doors ahead of them.

Another Morelli-designed prototype, PfX (Fig. 4-1), built by Pininfarina in 1960, applied the same layout to a subcompact four-seater that attracted wide attention. The Morelli-Pininfarina collaboration continued in 1961, with the creation of the PfY using the Fiat 600 chassis, having a conventional wheel configuration and a rear-mounted engine.

Without straying from the conventional carriage concept with front and real axles, Dr Frederick W. Lanchester, in Coventry, created a novel layout for seating and engine location in 1903. Odd as it might seem today, the Lanchester was an extremely rational design.

Dr. Lanchester rejected all previous patterns for vehicle architecture, British and Continental,

Fig. 4-1. The PfX four-door sedan with four wheels in lozenge formation.

and pursued his own reasoning, without wavering, to its logical conclusions. His car designs were free of influence from bicycles and carriages, and show us one of the first pure automobile configurations.

He moved the axles as far apart as possible, and came up with a 94-inch wheelbase for a five-seat touring car with an overall length of 137 inches. He placed the engine so that its center of gravity coincided with the midpoint in the wheelbase (in the interest of even weight distribution).

Dr. Lanchester also had a theory that the driver would best be able to judge the position of his vehicle if his eyes were at the same height from the ground as when walking. This made for an uncommonly low chassis for its time. In order to provide an unobstructed view of the roadway ahead, Dr. Lanchester moved the front seat as far forward as practical for legroom requirements (with no hood).

One passenger could sit alongside the driver. The other three had a forward-facing seat in a rear-entrance tonneau. The engine was easily accessible; it occupied the whole section between the tonneau and the front seat.

Another theory of Dr. Lanchester's led to the use of long cantilever leaf springs. Spring rates were calculated to give a natural frequency corresponding to the normal gait of a man walking.

Lanchester built cars of this type until 1914. His following was small, however, and some adopted parts of his architecture without accepting his theories. In 1903, Standard of Coventry built a car with a vertical engine mounted between the driver's seat and the passenger compartment. The Lanchester had a specially balanced flat-twin. NEC (New Engine Company) and James & Browne built cars with pancake engines, installed centrally, below the driver's seat.

MIXED GOALS AND MIXED RESULTS

Then there was the Rolls-Royce in 1905. Henry Royce's business managers, Claude Johnson and C. S. Rolls, were not pleased at the number of electric carriages being sold in London. They were losing a lot of business with wealthy customers to battery-driven machines. Early in 1905 they decided to ask Royce to make a car that would offer some competition. Their request went into specifics with regard to styling (it must resemble the best electric carriages), layout (the car should not have a hood, and the engine should be as far removed from the passenger compartment as practical), and operational characteristics (car must be silent, need little or no gear changing and the engine must be vibrationless, smokeless, and odorless).

Royce chose a large-displacement, slow-running V-8 engine that was accommodated in the front end of the frame. The engine fit deep down behind the front axle, raising the floorboards enough to cover the accessories.

Several versions of the car were designed. First came the Landaulet Par Excellence, then the Motor Landau, and third the Touring Car; all were advertised with the slogan "invisible V-8 engine." A variant was designed with the front seat moved toward the middle of the chassis carried coachwork, it had a low, wide hood.

All models with the forward-control driving position were built on the same 90-inch wheelbase. The legal limit was a wheelbase of 106 inches. The chassis weighed just over 2200 pounds. Not many were built. Factory records show that three cars were completed and a fourth had been under construction. All are thought to have been destroyed.

Experimental cars with odd seating arrangements and wheel configurations appeared from time to time in the following 50 years, but none got into production. Worth mentioning are the movable lounge chairs in William B. Stout's 1935 Scarab, and the buslike Dymaxion streamliners of R. Buckminster Fuller.

Jules Brandt exhibited his Reine (Queen) prototype in Paris in 1950. It was a light car, of remarkable proportions and seating configuration, with the four wheels mounted on in each corner. It had four bucket seats and no side doors. Rear passengers entered through a side-hinged tailgate and walked up a central corridor to their seats.

The driver could enter by opening the front panel and stepping over the engine. There was no hood in the ordinary sense; a panel extended downward to bumper level in the slope of the windshield. The prototype had no roof (so that a lot of acrobatics were avoided), but provision was made for a canvas top. The lack of side doors contributed to lightness, rigidity, clean airflow, and low cost. But the public would probably not have accepted that seating arrangement in a production car.

ASYMMETRICAL VERSUS CLASSIC LAYOUTS

The attractions of the centrally mounted engine, even when small and light, that stem from the ideal weight distribution it affords continue to occupy the minds of many designers and engineers. Chris Humberstone, a well-known British designer, holds patents for an interesting seating configuration for a six-seater sedan with central engine (Fig. 4-2). His Quartette was exhibited at the London motor show in October, 1970. Built on an original chassis with the twin-rotor Wankel rotary-combustion engine from the NSU Ro-80 mounted behind the front seats, the seats were placed close together in the tapering nose section.

Two bucket seats were placed on each side of the engine, with open legroom between the front seats and the door sills. Two rear seats were placed in line behind the front seats (separated only by a tunnel for the propeller shaft) with legroom behind the engine cover. It would not work with a bigger, taller engine. Some engine-cover length could be saved by mounting the engine transversely, but that would give rise to drive-line complications.

Asymmetrical four-wheel configurations have

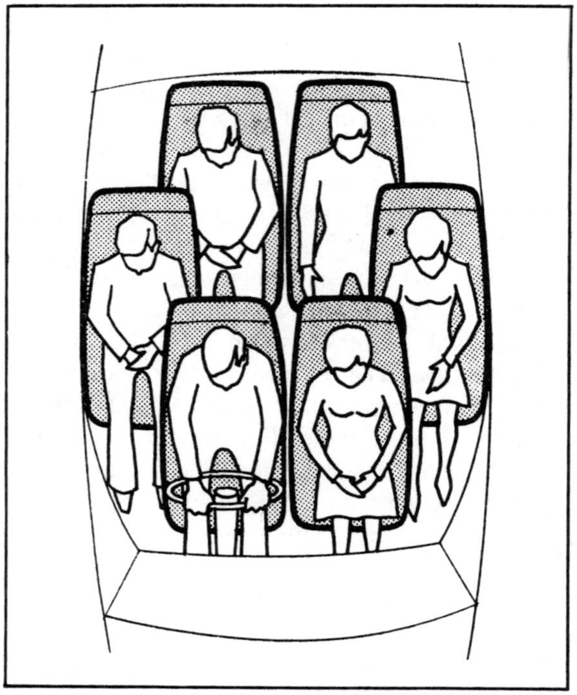

Fig. 4-2. A patent drawing for seating plan in sedan with a centrally mounted engine.

existed, but the variations in track or wheelbase have been minor and caused by mechanical solutions in the chassis design. On the Renault 16, one rear wheel was brought forward to avoid collision between the transverse torsion bars on the rear-suspension trailing arms.

There is nothing further to gain by mounting one front wheel further forward or backward than the other architecturally, technically, or in terms of cost. This applies whether the wheels are driven or not and it goes equally for the rear wheels.

Variations in track? A streamlined body that tapers towards the tail end can dictate a narrower rear track (as on the Citroen DS-19). But in general, the chassis engineers see it in their interest to give the car the widest possible track (consistent with body shape both front and rear).

There are excellent reasons why today's family cars (regardless of size and price) continue the classic layout with seating for two or three persons side by side in front, and a similar arrangement right behind. The use of four wheels, has become almost universal. It would be universal if it were not for the existence of three-wheeled vehicles of various types.

WHY THREE WHEELS ISN'T ENOUGH

Three wheels is the minimum required for static balance. Three-wheel cars enjoy some popularity in Great Britain, and three-wheeled light commercial vehicles are common in Italy and Japan. But their popularity is based on fiscal, not technical, advantages. Three-wheelers are taxed at a lower rate in those countries.

Any photographer or land surveyor will tell you that there is nothing with more static stability than a tripod. The car is a dynamic machine that is subject to outside forces acting on it in three planes as it moves along. The stability of a three-wheeled car depends on the location of the single wheel (front or rear), track and wheelbase, weight distribution, center of gravity height, and several secondary parameters.

A single front wheel almost automatically locks the design into rear wheel drive (unless the engine is so small and light that it can conveniently be mounted on the front wheel or its suspension fork). This layout permits exceptional maneuverability thanks to the theoretically unlimited steering angle sweep.

A single rear wheel can easily be driven via chain or shaft. This layout is also suited for front-wheel drive. Morgan chose the former and BSA chose the latter for their small three-wheeled sports car.

For a practical car, however, the three-wheel solution has too many drawbacks. With a single front wheel, there is no obvious place for the engine. In a light commercial vehicle, you could put it under the driver's seat. That's hardly possible in a car with a rear seat (due to drive line requirements). Under the back seat? Technically feasible, but it runs into noise-and-vibration problems, difficult access for service and maintenance, and could force the designer to raise the seat and end up with an unattractive roofline.

Consequently the engine tends to get pushed back further, and can be neatly accommodated in the tail. But there it steals trunk space. With a single front wheel, it is very difficult to design for much luggage space at the front end.

The three-wheeled car with a single rear wheel also runs into a trunk-space problem. The rear wheel is, almost by definition, located on the car's center line. It cuts the trunk in two, and that restricts the size of individual luggage items that can be carried.

There is also a high risk of overloading the rear wheel. That can be counteracted by fitting a bigger wheel and tire, but that way you lose the standardization of tire size and end up having to carry two spare wheels. Where can you find space for them? If you standardize the bigger wheel size, the front wheels are over-tired and rolling resistance goes up. That wastes fuel. It also adds to the cost and means that an ordinary four-wheel arrangement would be less expensive to build than a technically satisfactory three-wheeler with a single rear wheel.

MULTIWHEELERS

Six-wheel cars have been made, with two rear axles, mainly for military or other special purposes. There was even an eight-wheeled car, the Reeves Octo-Auto, with two front axles and two rear axles. It was created in 1912 in an attempt to improve ride comfort. The highway system has evolved since then; there would be no advantage today.

Any car design with more than four wheels will

be heavier, usually longer, and certainly more expensive. Four wheels have proved capable of meeting all normal requirements of a passenger car in terms of load capacity, traction, and safety.

ENGINE LOCATION

For many years, most designers have thought that the narrow space between the front wheels made a logical bed for the power unit. It is a space that is not easily adaptable for other uses.

The current trend is towards smaller and lighter engines, and therefore it is easier to accommodate them inside the vehicle structure. The engine has less influence on weight distribution, for instance, and its reduced space requirements opens new opportunities in vehicle architecture.

In big cars that demand a powerful engine, it is natural to place the power unit between the front wheels. In any other position, it would give rise to problems with weight distribution, space allocation, and possible mechanical complications.

In such cars it is also natural to drive the rear wheels. Why? Because of the engine weight, driving the front wheels would set up a forward bias in cornering force distribution that could cause odd behavior on the road. Because such cars often carry a full load of passengers and luggage, the rear wheels carry more than half of the total weight. Then it makes sense to drive the rear wheels. Such cars are also often used for trailer towing, and the trailer's tongue weight adds to the load on the rear wheels. Therefore, it's important to drive the rear wheels whenever trailers above a certain weight are involved

Gains in space allocation and weight distribution can be made in front-engine/rear-wheel-drive cars by moving the transmission away from the engine and attaching it to the final drive unit (forming what is called a transaxle).

Just how far forward it is advisable to push the engine is something that chassis engineers and vehicle-dynamics experts are still arguing about. Let's just say for now that there is a risk of seriously compromising the vehicle's handling response by trying for radical solutions with heavy engines.

Front-engine and rear-wheel drive necessitates a propeller shaft running under the whole length of the floor. Few modern cars of this type have a flat floor. The shaft runs through a tunnel that intrudes on legroom.

Tunnels are biggest on cars that have rear axles as opposed to cars with independent rear suspension. The axle must have room to bounce around; the final drive unit in a car with independent rear suspension is bolted to the chassis frame or body shell. The propeller shaft undergoes only minute angular changes, and the tunnel can be lowered.

VEHICLE SIZE

The general rule about engine location is not the same for small cars and big cars. Front engine and rear drive might be maintained indefinitely as the best-balanced solution for heavy cars. For small cars, the rule is that the engine must be located at the end where the driving wheels are. Front-wheel drive demands a front engine and rear-wheel drive demands a rear engine.

Won't all cars be small in the future? It's not as clear-cut as that. For rational design, car size must be related to the intended payload. Because it is expected that there will be a continued need to carry more than four people and more than a small amount of luggage by car over more than local distances, it is hard to see how the large car can be condemned to the rank of an endangered species (Fig. 4-3).

Dr. Friedrich Goes of Volkswagen has an enlightening way of explaining his view: "Obviously, it is nice to have a large vehicle which meets all or practically all one's requirements. However, it is seldom that the whole potential is used at the same

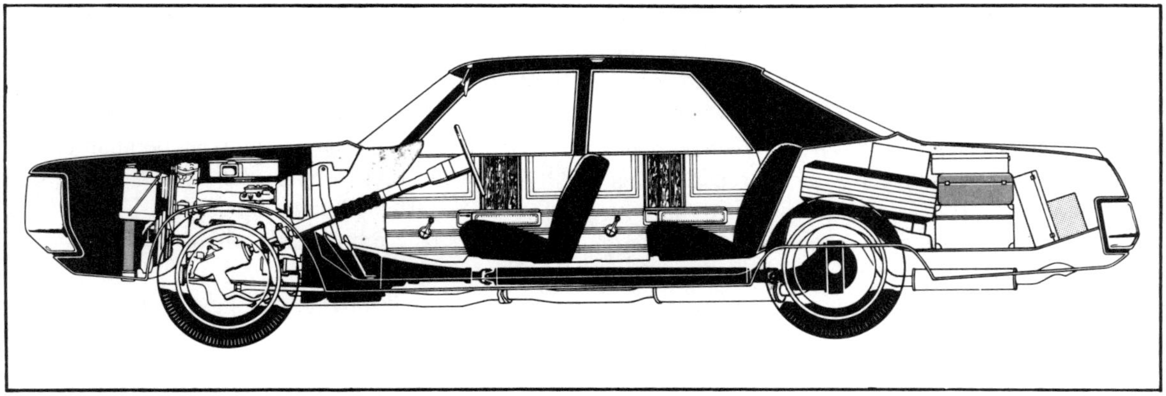

Fig. 4-3. Vast amounts of useless space add to the weight of the car. This increases production cost as well as operating costs.

time. Many drivers never make full use of the cars they buy, so that the capital they invest is never properly written off. For example, with an average of two occupants, four seats usually remain unused. The same applies to trunk space which is used to capacity even less often. High engine performance is always an expensive luxury and seems rather pointless when the maximum speed exceeds that which is permitted by law.

"What the big car has in effect is extra engine weight, extra body weight to house the big engine, and extra fuel consumption which is always greater than necessary. Such a car cannot be economical.

"Excessive size, engine power and weight do not merely prevent economy. The excessive size of the most popular vehicles in the United States, on the crowded streets of today, is threatening to rob the automobile of its very reason for being—its mobility. It is well known that the number of vehicles on the roads (and therefore the road space requirement) is growing at a much faster rate than new road construction. Yet for the foreseeable future there seems no hope of even maintaining the road space per vehicle we now have, let alone increasing it.

"All the same, it would be quite simple to master the problem or at least to put off the time when the roads become completely choked. A simple lumberjack's trick should be remembered. If a stream tends to become blocked by logs, the logs are cut up into shorter sections. The lumberjacks know that in this way a lot more timber can be floated down the same stream."

There is no doubt that smaller cars would make room for more vehicles on city streets. But I do not think we should see that statement as a condemnation of the big car as such. We can interpret it, instead, as an indictment of the man who uses a big car to move light loads that could well be carried in a smaller and more economical vehicle.

NO COMFORT WITHOUT LIVING SPACE

No one can win an argument stating that to carry six persons and their luggage, one saves energy by using two four-cylinder, 50-hp cars rather than one V-8, 100-hp car.

One company that is convinced of the big car's future is Mercedes-Benz. The Stuttgart-based organization has long traditions in rational design of roomy cars, and will not give up the big car. The company has ample statistics to prove that owners of their big cars carry more people—and luggage—and drive longer distances than the average. Because of the higher payload, larger size is only logical. Because of the longer distances, greater comfort must be provided, and that can't be done within small-car dimensions.

"For many of our customers, his work place is at the wheel of his car," says Werner Breitschwerdt, technical director of Mercedes-Benz. "Labor unions insist on a certain minimum space for factory workers, and for truck cabs, some minimum dimensions are laid down by law. Roominess is a safety factor, for it helps reduce driver fatigue. We cannot offer the man who buys our car less safety and comfort than truck drivers and factory workers have."

Designing a big car is not so difficult, I am often told by engineers who work mainly on small-car projects. It is true. There is a bigger margin for wasted space and a lesser premium on economy in both materials and cost. Because of space and weight restrictions in the design objectives, small cars are necessarily more specialized. Each type is tailor-made to the requirements of a specific category of buyers.

The biggest group consists of people who need transport for professional use and local journeys. Dr. Friedrich Goes reveals some details about the basic needs. "For these short journeys, there are usually one or two occupants per car. For shopping, a large, easily accessible trunk is once again of importance. The small number of passengers means that, if necessary, the rear seat space can be utilized to increase the luggage room. The smallest possible external dimensions and unimpeded visibility for driving in heavy traffic and parking in busy zones are essential, especially for commuting drivers. A two-seater with a small trunk of about 3 cubic feet could be adequate."

PRACTICALITY OF SMALL CARS

Cars that small are not yet the most popular—even in Europe. Market analysis observe that two-car ownership still has not progressed far enough do away with one vehicle being called upon to perform a lot of family-car duties. If current trends continue, they say, the need for increased product specialization will force a rise in multicar ownership.

As it stands today, the most popular car in Europe is the lightweight, 900-1100cc, two-door, four-seater with a 10-cubic-foot trunk. Usually it is a hatchback with individually folding rear seat backs (Fig. 4-4).

Doing something new to show a clear advance in product design and gaining a competitive advantage in that class is an extremely difficult task. But the reward for the manufacturer who succeeds will be ample.

In a short (lightweight), economy four-seater, the worst conflict between styling demands and practical needs occurs in the area of headroom for the rear passengers. The conflict can be resolved the way Austin did it with the Mini and what Renault did with the R-4 (for example, with the wagon-type rear end. The more stylish Renault R-5 shows a slant-back, wagon-type solution. In both cases (and others like them), there is no conventional trunk. Consequently, luggage capacity is restricted and loading is often inconvenient.

The 1961 Ami-6 solution by Citroen is practical, safe, and aerodynamically unimpeachable, but its esthetic principles have been called dubious—at best. See Fig. 4-5. The reverse-slanted rear window provides ample headroom in the rear and provides convenient hinge mountings for the trunk lid (a trunk lid of conventional size and operation). Trunk arrangements are not nearly as important for the everyday user as the question of where the engine is located. Not because he cares particularly about the engine, but because of what the engine location dictates for the overall car design.

INFLUENCE OF ENGINE TYPES

Where the engine should be placed depends to a great extent upon its size, shape, and weight.

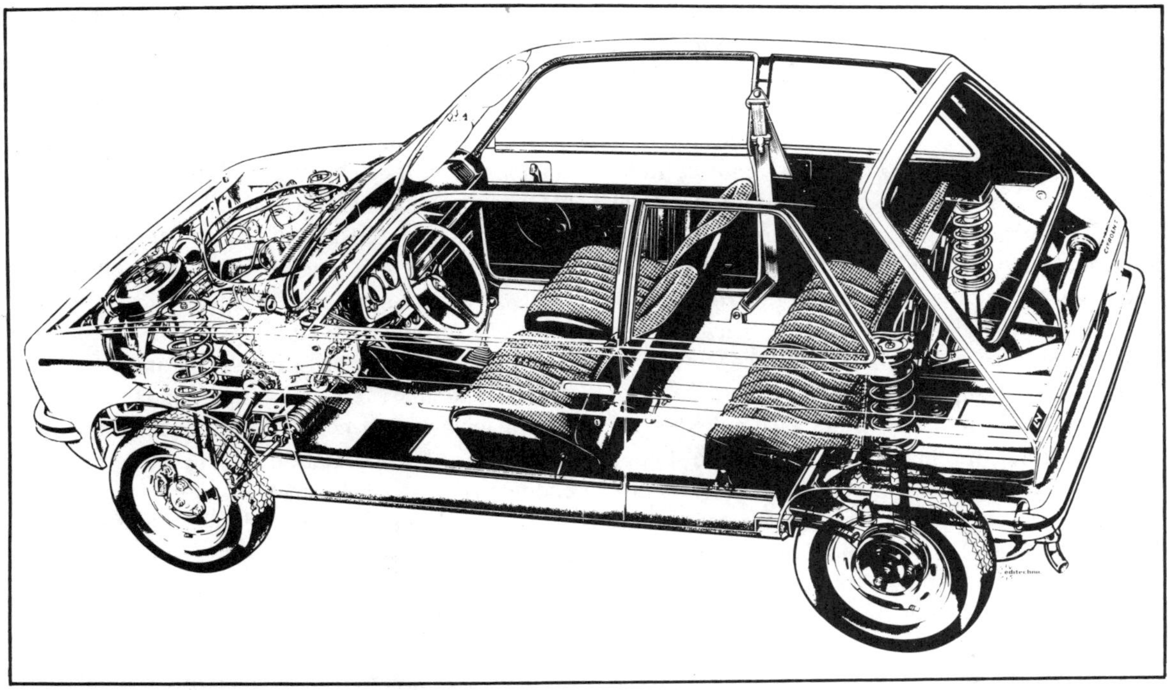

Fig. 4-4. The typical European small-car layout features transverse engine and front-wheel drive, with semiwagon hatchback.

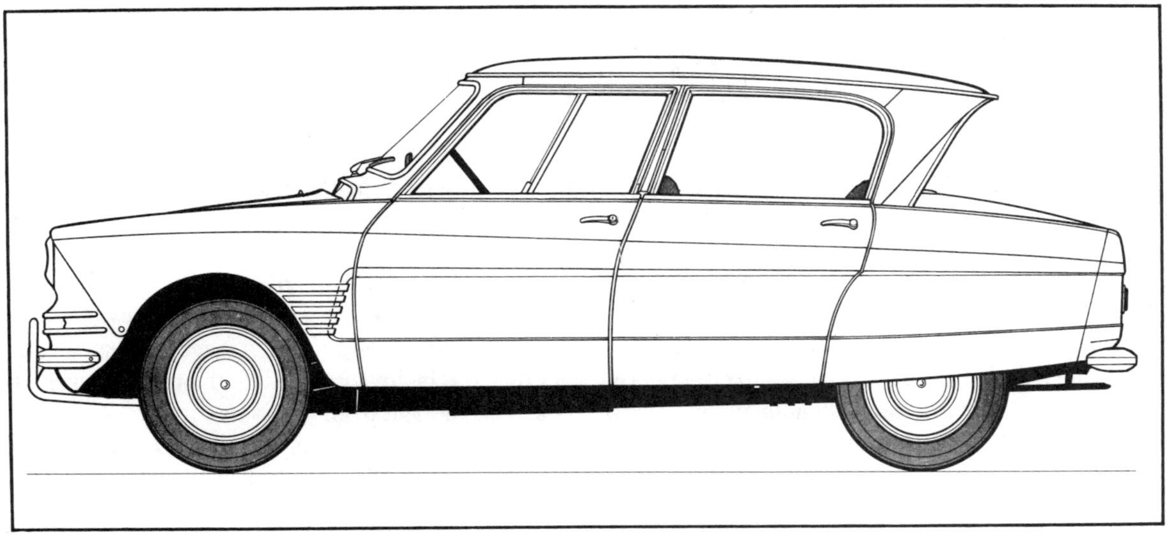

Fig. 4-5. An example of a short sedan with a large trunk, standard hinged trunk lid, and generous back seat headroom (Citroen Ami-6).

Horesepower and torque are to some extent proportional to engine displacement. Assuming an "ideal" cylinder size of 250cc to 400cc, we find that the car's power requirement dictates the number of cylinders.

Single-Cylinder Engines. One-cylinder engines are used only in minicars, and are not likely to find application in cars with horsepower requirements beyond 15-20 hp. Single-cylinder engines can be installed vertically, horizontally, or slanting at any angle as long as the lubrication system, carburetion and exhaust, and accessories are arranged accordingly.

Two-Cylinder Engines. Two-cylinder engines come in three popular types: parallel-twins, V-twins, and flat-twins (usually from 15 to 30 hp). The parallel twin has two cylinders in one compact block, while the flat-twin has two cylinders extending to opposite sides of the central crankcase. The parallel-twin is usually less expensive to produce, and fits better into a modular design system (two, three, and four cylinders of identical size, for instance).

The flat-twin has better balance and is not imcompatible with modular design systems (the Citroen GSA flat-four for instance, can be regarded as two 2 CV flat-twins put together). The flat-twin has no installation problems, but is regarded as more suitable for front-wheel drive layouts. The parallel-twin is generally preferred for a rear-engine installation (Fiat 126). The parallel twin can be installed vertically, horizontally, or at any angle. The flat-twin is invariably mounted horizontally in order to give the lowest center of gravity.

The flat-twin is usually installed as front overhang (with a longitudinal crankshaft). Consequently it demands a more expensive final drive unit (with pinion and ring gear rather than simple spur gears, as used with transversely mounted engines).

The V-twin solution is common for motorcycles but has practically disappeared from use in cars. The V-twin takes up more space than the parallel-twin, costs more to make, and is not ideally suited to a modular design system.

Three-Cylinder Engines. A three cylinder engine is usually derived from a four-cylinder in-line design, and can therefore be produced economically. It can be installed transversely or on the longitudinal axis, and be tilted at any angle, from vertical to horizontal. A radial-type three-cylinder engine can well be visualized, but would have no advantage over the in-line three in any aspect of automotive use.

Four-Cylinder Engines. The in-line four is the most popular. It is short, light, and has reasonable manufacturing costs. It has equal versatility with regard to installation as the three-cylinder engine. The V-4 can be made shorter, but has more complex balance problems. It has not been successfully combined with a modular design system, as a partner of V-6 or V-twin units.

The flat-four engine is the only type of four-cylinder unit that is inherently balanced, with two horizontal cylinders on each side of the crankcase. For reasons of installation, the crankshaft must run longitudinally. That restricts the car designer's freedom in drive train layout.

Five-Cylinder Engines. Five-cylinder engines are either shortened in-line sixes or lengthened in-line four-cylinder units. By their very nature, they belong in a modular design system. Transverse installation is possible, but not popular due to lack of width in the car for the full length of the engine.

Six-Cylinder Engines. Three types of six-cylinder engines exist, but the third is rare. Most common is the in-line six. It is inherently balanced, but tends to be heavy. The length of the crankshaft is a disadvantage from a noise-and-vibration viewpoint. Transverse installations have been built (Austin 1800, 2200, and Princess), but the in-line six is almost invariably mounted longitudinally. Most in-line, six-cylinder cars have rear-wheel drive.

The V-6 can be made shorter and lighter, but has a balance disadvantage even when equipped with a crankshaft that provides equal firing intervals. It is eminently suitable for transverse installation and is extensively used in longitudinal installations.

The flat-six was standard on Chevrolet's Corvair 1960-69 and is still used by Porsche for the air-cooled 911 (installed in the tail end). The flat-six, unless too wide, is also suitable for front-end installation (and front-, rear- or four-wheel-drive). It fits well onto a modular design family.

Eight-Cylinder Engines. For passenger cars, only one type of eight-cylinder engine remains in production: the V-8. The typical V-8 is like two in-line, four cylinder engines mounted at right angles to each other on a common crankcase. It is compact and has no serious balance problems. Like the in-line four, it is equally suited for transverse and longitudinal installations.

PACKAGING AND WEIGHT DISTRIBUTION

The other large masses that must be accommodated in the car are the fuel tank, spare wheel, radiator (for water-cooled engines), and battery. The fuel tank should be positioned for maximum protection—preferably within the wheelbase, but at least not extending behind the rear wheels—so that it is not damaged in a rear-end collision. It should be so positioned that its being empty or full has the least possible influence on weight distribution.

Such rules have been widely disregarded in the past. For instance, in 1950, Jaguar put a fuel tank in each rear fender on the Mark VII sedan.

The radiator must be placed for ideal airflow conditions. This usually means placing it in the nose of the car.

The spare wheel can go anywhere it remains accessible. Some cars carry it on a separate rack or shelf below the trunk floor (accessible without opening the trunk lid). Others have it stored below the trunk floor. That is inconvenient in case of having a flat tire when you are carrying luggage. On many cars, the spare is stored vertically in the side of the trunk.

Several small cars with front-mounted engines carry the spare wheel strapped to a bracket under the hood. The temperature is high there but apparently no threat to the rubber. If the engine is not mounted in front, the spare wheel, fuel tank, and battery usually go in the space between the front wheels. That steals trunk space, but is often necessary for reasons of weight distribution.

REAR ENGINE CUTS COST

The temptation to mount the engine in the tail end stems from a combination of reasons. A rear-mounted engine (Fig. 4-6) gives the shortest overall length of the body for a given passenger space. The front-wheel drive car comes close, but due to the conflict between front legroom and space needed for the mechanical components and accessories it ends up with a slightly longer body. Thanks to the simplicity of the body construction when the engine is mounted in the rear, the structure is also very light.

The rear-mounted engine encourages the use of a short wheelbase, and the freedom from power train elements in the front end opens the way for very wide steering angles on the front wheels. That combination guarantees a tight turning circle.

With a rear engine, there are no heat leaks into the passenger compartment from the power unit or its cooling system (provided the engine is air-cooled or the engine is also mounted in the rear. The absence of exhaust pipes under the passenger-compartment floor makes it possible to put the floor lower, and thus obtain a lower center of gravity.

What has perhaps counted most to many makers of rear-engined cars, however, is the reduction in cost. The small rear-engined car costs less to

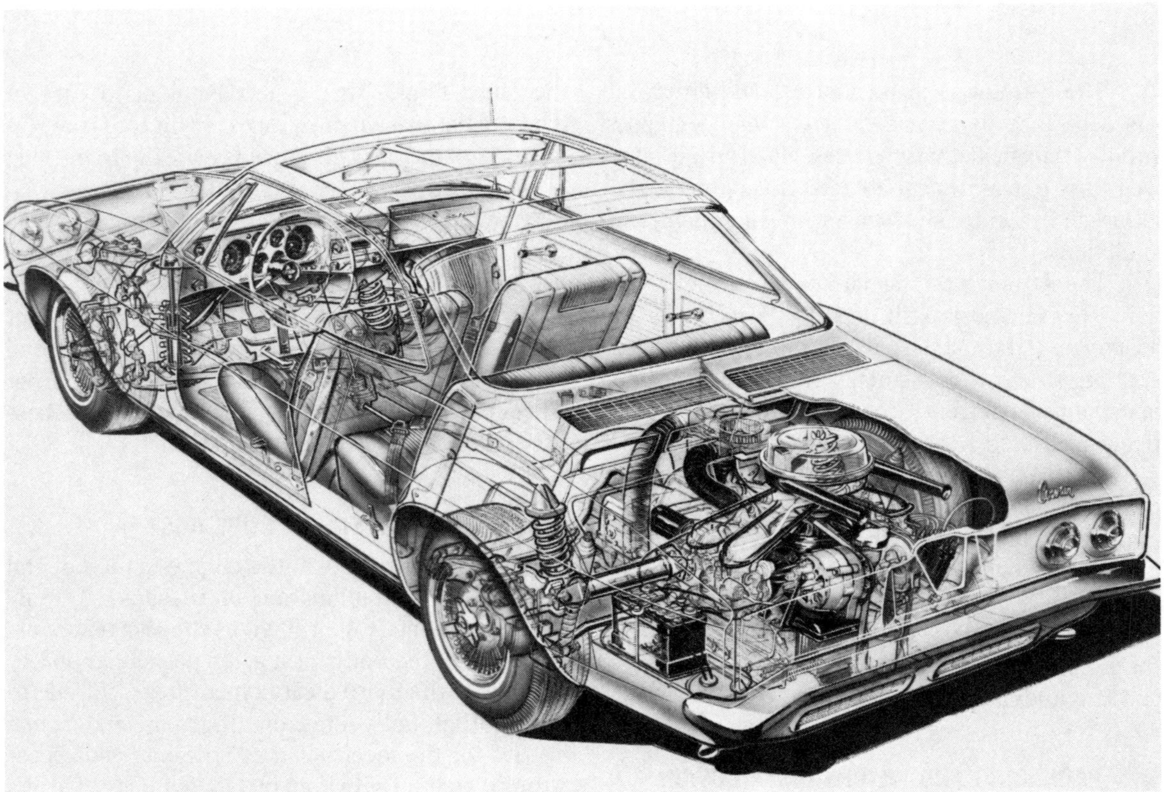

Fig. 4-6. A flat-six rear engine eliminates rear luggage compartment (Chevrolet Corvair).

build than a front-wheel drive car of the same size and power.

Compared with front-wheel drive, the rear-engine solution avoids the use of expensive constant-velocity universal joints (as required with driven front wheels because they also have to steer the car). The front suspension can be made lighter and cheaper. Compared with a front-engine/rear-wheel drive car, the cost of the propeller shaft and its universal joints is eliminated.

REAR ENGINE DRAWBACKS

The drawbacks of rear-engine construction are a rear bias in weight distribution, limited luggage capacity, lack of versatility in the body design (poorly adaptable to station wagon, for instance), long and delicate control linkages, difficulties in providing adequate engine cooling, and loss of efficiency in interior heating and windshield defrosting.

The engine occupies the space normally used for a trunk. Finding equal trunk capacity at the front end is difficult due to the need for the front wheels to steer. They demand much wider wheel housings than the rear wheels, and leave a narrower area in the middle.

The engine and transmission control links are long and therefore tend to lack precision in their action. It is not so serious for the throttle control, for instance, but it is a severe handicap in the gearshift.

The rear engine limits the practicality and versatility of a car. A station wagon with rear-mounted engine suffers a high floor at the loading end, the engine environment is a hot zone that demands insulation, and it robs space exactly where other wagons offer advantages. For a large-scale car manufacturer, a rear-engine sedan limits the market for a derivative wagon. No new sedans with rear-mounted engines have been designed since the 1950s.

Then there is the weight-distribution problem. The rear-mounted engine constitutes—for a station wagon—a preloading of the cargo-carrying end. This limits payload capacity and works against dynamic stability in the vehicle. In a light car, a heavy rear-mounted engine can cause dynamic instability, with consequent handling and control difficulties. Technical solutions exist (bigger wheels, fatter tires, altered suspension geometry), but tend to be expensive and thereby defeat the purpose of having chosen a rear-mounted engine in the first place.

The rear engine also gives rise to styling problems. The engine package is difficult to integrate in certain car shapes. Air intakes for engine cooling are difficult to accommodate without running into conflict with aerodynamic or esthetic ideals.

THE URBAN CAR PUZZLE

In 1957, a year after launching the Dauphine into mass production, Renault began working on a forward-control derivative with a very roomy back seat. It was intended mainly for city use (taxi, airport limousine) where the up-front driver's seat was preceived as an advantage. Due to the use of the Dauphine chassis with its rear-mounted engine,

Fig. 4-7. The Renault city-car project based on the Dauphine chassis.

Fig. 4-8. The revised city-car project with improved entry and exit.

there was no trunk at all, and luggage had to be carried inside. Figure 4-7 shows that entry to the front seat was difficult. The long front overhang, needed for front seat legroom, combined with the long tail to make the car needlessly heavy and expensive.

But the formula was not given up. Renault's Project 900, from 1959, had a similar seating arrangement, but the engine was moved inboard so as to make room for a trunk in the tail (Fig. 4-8). The engine did not go under the back seat, but behind it; this pushed the rear wheel axis back and made the car longer in a badly chosen section. A prototype was built by Ghia, but the project was rejected for production.

Ford's Bloomfield Squire styling prototype (Fig. 4-9), from 1964, reveals similar thinking. There is a rear engine, a short hood, raised-level front seats above the front suspension, and two wide bench seats on the low floor inside the 80-inch wheelbase. Thus it would give seating for eight (seven plus driver) within extremely compact dimensions.

A sharp contrast is provided by a proposed taxicab built, in 1972, at the invitation of the New York Museum of Modern Art for its show: "the Taxi Project—Realistic Solutions for today." The aim of the show was to stimulate builders to look for a city transport vehicle that would improve the environment by burning less fuel and polluting the atmosphere less, while at the same time offering greater comfort to both driver and passengers. The result was a car that would be more practical, shorter, lighter, and less expensive to build than the Ford and Renault design exercises.

It came into being when Alfa Romeo commis-

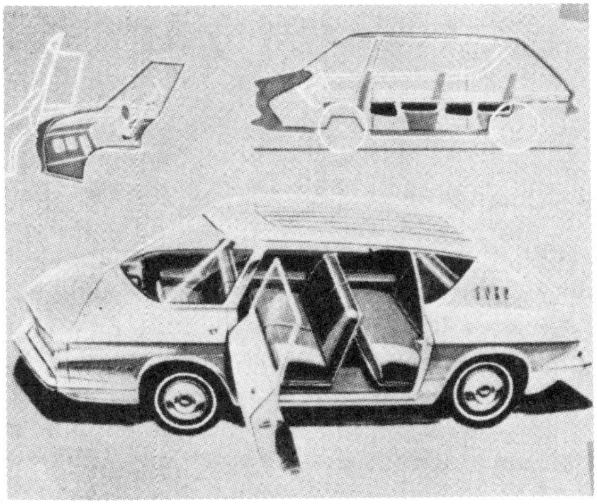

Fig. 4-9. The Ford Bloomfield Squire.

sioned Ital Design to make a taxicab proposal on the basis of its front-wheel drive F-12 minibus chassis (Fig. 4-10). The water-cooled, four-cylinder engine is carried in the front overhang, and the driver sits above the front suspension. It was built on a very short (90.5-inch) wheelbase and was 160 inches long overall.

The 160 inches are enough to comfortably seat five passengers and a driver, and give ample and practical luggage space. Furthermore, it is easy to get into the passenger compartment. Last but not least, from the taximan's point of view, the vehicle is easy to handle.

The body is protected by a wide rubber strip running around it that helps to soften the impact in crashes and protects the vehicle from those slight,

Fig. 4-10. An Ital Design taxicab proposal, built on Alfa Romeo F-12 light van chassis.

everyday scrapes so common in urban transport; in this way, it cuts downs maintenance costs.

The passenger door slides along so it doesn't jut out into the street when opened and makes it easier to get into the car. A retractable ramp in the floor of the vehicle can be used to get wheel chairs and baby carriages into the car. Passenger and luggage space is considerable; there is almost 20 cubic feet beneath the seats where up to 5 people can be accommodated (Fig. 4-11) with headrests and safety belts. Wide windows ensure perfect visibility.

THE STATION WAGON

The layout and shape of a station wagon are based on design objectives that differ in many ways from those that apply to sedans. These differences are mainly due to the demand for greater carrying capacity. This means passengers as well as cargo (not necessarily just luggage, but objects of all kinds). Yet most car manufacturers build station wagons that are adapted from four-door family sedans. Why? Because a station wagon designed from the ground up, without any attempt to share existing components with the sedan, would prove to cost far too much to produce and end up not being competitive in the market.

Consequently this discussion of station wagon design is reduced to considering modifications to be carried out on an existing sedan. With a conventional layout, little or no change is needed ahead of the C-post. But behind, both body and chassis design must be specific to the wagon. Body modifications are dictated almost by definition. The wagon adopts the profile of a panel van due to the desire for obtaining the greatest useful interior volume within the allotted length. That led to the adoption of the full-length roof and the vertical or near-vertical tailgate.

The next question concerns the division between passenger space and cargo space. In this area, wagon designers have shown great ingenuity over the years. They have devised easily transformable interiors that can be quickly altered without any use of tools to suit the transport needs of the moment.

Why chassis modifications? First, because of the additional payload capacity. Second, because of the peculiar requirements of station-wagon interiors.

Reinforcement of the rear suspension to cater for the increased load capacity is common. This is most simply done by beefing up the springs (whether they are coils, leaves, or torsion bars.)

Fig. 4-11. The seating arrangement in the Ital Design taxicab proposal.

Some car manufacturers prefer other, perhaps more expensive, solutions calculated to give a better compromise between ride comfort when running with light load and vehicle attitude (rear squat, nose up) with a full load. For instance, Peugeot went to dual coil springs for its 404 and 504 wagons.

Complete rear suspension systems have been redesigned for station wagons in many cases. Coil springs that stand up in towers between the rear wheel housings cannot be tolerated. General Motors ran into such a situation with its 1972-model B-body and C-body wagons. They developed an entirely new leaf-spring suspension system that was used by Chevrolet, Pontiac, Oldsmobile, and Buick. Peugeot had a similar problem with the 305, but solved it by mounting the coil springs horizontally, with a bell-crank arrangement from the trailing arms. The Peugeot 504 and 505 sedans have independent rear suspension, but the 504 and 505 wagons have sturdy rear axles.

GREATER SEATING CAPACITY

Certain items might need to be relocated when a sedan is converted into a station wagon. That might apply to such items as the fuel tank and the spare wheel, jack and toolkit—depending on how they were positioned in the sedan. If the standard fuel tank stands up against the rear seat back in the sedan, it must be redesigned with a lower profile so as to fit below the floor. If the spare wheel is carried atop the sheet metal above the rear suspension, it must be moved to a position below the floor or accommodated in a standing position in one of the rear fenders.

To increase the number of passengers, it is common to add a third seat. In some cases, this can be accomplished without going to a longer wheelbase, but most manufacturers prefer to move the rear axle backwards—for two main reasons. One is weight distribution because repositioning the rear axle backwards reduces the risk of overloading the rear suspension. The second reason is access. With a longer wheelbase, bigger rear doors can be fitted. Specific doors make for higher cost, but it is generally a modest penalty compared with the benefits.

In the beginning, third seats were usually installed facing the tailgate, in a dos-à-dos arrangement with the rear (second) seat. Because of wheel housing intrusion, it was usually narrower than the rear seat. Third-seat passengers had to enter through the tailgate—which most adults found undignified. With this type of layout, the third seat was usually reserved for children or pets.

Naturally, the rear-facing third seat could be folded flat with the floor of the cargo platform. The existing back seat was also provided with a folding mechanism. This allowed the seat and backrest to be stored, packed tightly together, against the back of the front seat. This gave maximum floor space for cargo.

In 1958, Citroen introduced its DS-19 station wagon. In it folding seats were attached to the body sides behind the back seat. Two individual seats on the right faced identical ones on the left. There was a high door sill, but the one-piece tailgate was hinged at the roof edge—making for relatively easy access.

About 1965, Ford picked up the idea and began offering the "dual-facing" third seat. Small folding seats were installed along each wall in the rear overhang. With this design, the number of seats doubled, but the drawback of difficult entrance would not have been alleviated unless tailgate modifications also were undertaken. Because of legroom shortage, the dual-facing rear seat was still best suited for children. The forward-facing third seat is theoretically ideal. Its width, however, is compromised by the practical obligation to position it between the rear wheel housings.

That fact also gives rise to access problems. With rear wheels on each side, it is impossible to fit

Fig. 4-12. The Peugeot 505; an eight-seater compact station wagon.

side doors for the third seat. Access must be made available from the front; that means through the rear doors (or at least, one rear door). Of course, the back seat blocks the way. The simplest way to create a passage is to split the back seat (half and half), or one-third two-thirds, according to width) and fit a folding linkage on the curbside portion. That obliges third-seat passengers to enter before the back seat is fully occupied (and exit later), but that is a minor inconvenience that is acceptable in most circumstances.

In 1972, General Motors adopted this system for its biggest wagons. Peugeot now has improved version on the 505 station wagon (Fig. 4-12).

TAILGATE TRIBULATIONS

The first tailgates were patterned on the rear doors of delivery vans. The rear panel was split vertically to give two doors hinged at the rear corner edges.

Next came the more versatile horizontal-split tailgate. The upper part contained the rear window and was hinged at the top end, swinging upwards to roof level. The lower part was hinged at the bottom, and folded down, outwards, to length the load platform beyond the rear bumper (which could on occasion be useful). But this design made loading more difficult, and access to a third seat from the back was made unacceptably difficult.

Ford invented the two-way tailgate when the dual-facing third seat was adopted. The lower tailgate had hinges in the body side as well as at floor level; each set was engaged alternatively depending on the way the handle was twisted. That was an important inprovement, and it was copied throughout the American auto industry.

When E. N. Cole was president of General Motors, he instructed the car divisions and Fisher Body to make use of his patent for the clamshell tailgate. It was a horizontal-split design of sharper curvature than normal. The top end was frameless glass that retracted (electrically operated) into the roof. The lower part was steel, and it retracted

Fig. 4-13. Four versions of the G-series Mercedes-Benz. Top to bottom: open-bodied recreational vehicle; two-door wagon; panel van; and four door wagon.

(electrically operated) into a shelf below the floor. It was expensive and, to many users, its proclaimed convenience was not evident. It was abandoned after five years.

On 1973 A-body cars, General Motors switched to the one-piece, roof-hinged tailgate. This type had been common in Europe; there station wagons, like other cars, were smaller, for upwards of 20 years. The tailgate swings out of the way to offer unhindered access. Because of the simplicity inherent in the principles of hinging and locking, it can be made rattle-free at a relatively low cost. This type is now used almost universally.

FOUR-WHEEL DRIVE UTILITY VEHICLES

Body design for four-wheel drive utility vehicles must take full account of the particular functions of such machinery. High ground clearance is preferred in order to maximize the off-road mobility. That puts a premium on the designers' ability to put the doors where it is possible to provide steps at convenient height.

Despite the existence of several unconventional solutions for military vehicles (some of which also exist in civilian versions), most manufacturers of four-wheel drive utility vehicles have chosen to follow the basic layout of the World War II Jeep (at least as far as the mechanical elements are concerned). Seating arrangements and luggage accommodation are laid out more or less in accordance with the same considerations and principles that apply to sedan-based stations wagons.

Demands for off-road mobility also precludes body designs with a big front or a rear overhang that can get caught on irregularities in the terrain. To allow the highest angles of approach and departure, the front and rear bumpers must be kept as close to the tires as practical (without risking interference with wheel rotation if the bumper is slightly bent in accidental contact with rocks or other obstructions) and raised to the highest legal height from ground level.

A number of body style variations are possible without deviating from the basic structure of a 4-×-4 utility vehicle such as the G-series Mercedes-Benz. In this example, the maker offers two wheelbases, with two-door and four-door station

Fig. 4-14. A two-door version of the Range-Rover. A four-door model is also in production.

wagon bodies (Fig. 4-13), a canvas-top version that can double as pickup truck, and a panel van.

Further variants could easily be added at modest tooling cost, such as chassis-and-cowl only for special bodywork, chassis-and-cab, club-cab pickup, and a long wheelbase, canvas-top, safari trekker.

The Mercedes-Benz G-series, in common with practically all other 4-×-4 utility vehicles, is built with a truck-type frame. That type of construction is favored by the Land-Rover, Range-Rover (Fig. 4-14), Ford Bronco, Chevrolet Blazer, Jeep Cherokee and Wagoneer, Nissan Patrol, Toyota Land-Cruiser, the ARO from Rumania, and the UMM from Portugal.

The exceptions are the Fiat Campagnola (Fig. 4-15) and the Lada Niva. They have unit-construction bodies. The Lada Niva—made in the Soviet Union with engineering help from Fiat and Porsche—has a passenger-car type of body shell. Fiat's Campagnola, which is available in roofless and canvas-top versions as well as in all-steel station wagon form, has a lower structure that reaches from floor to belt line only and is built up on two

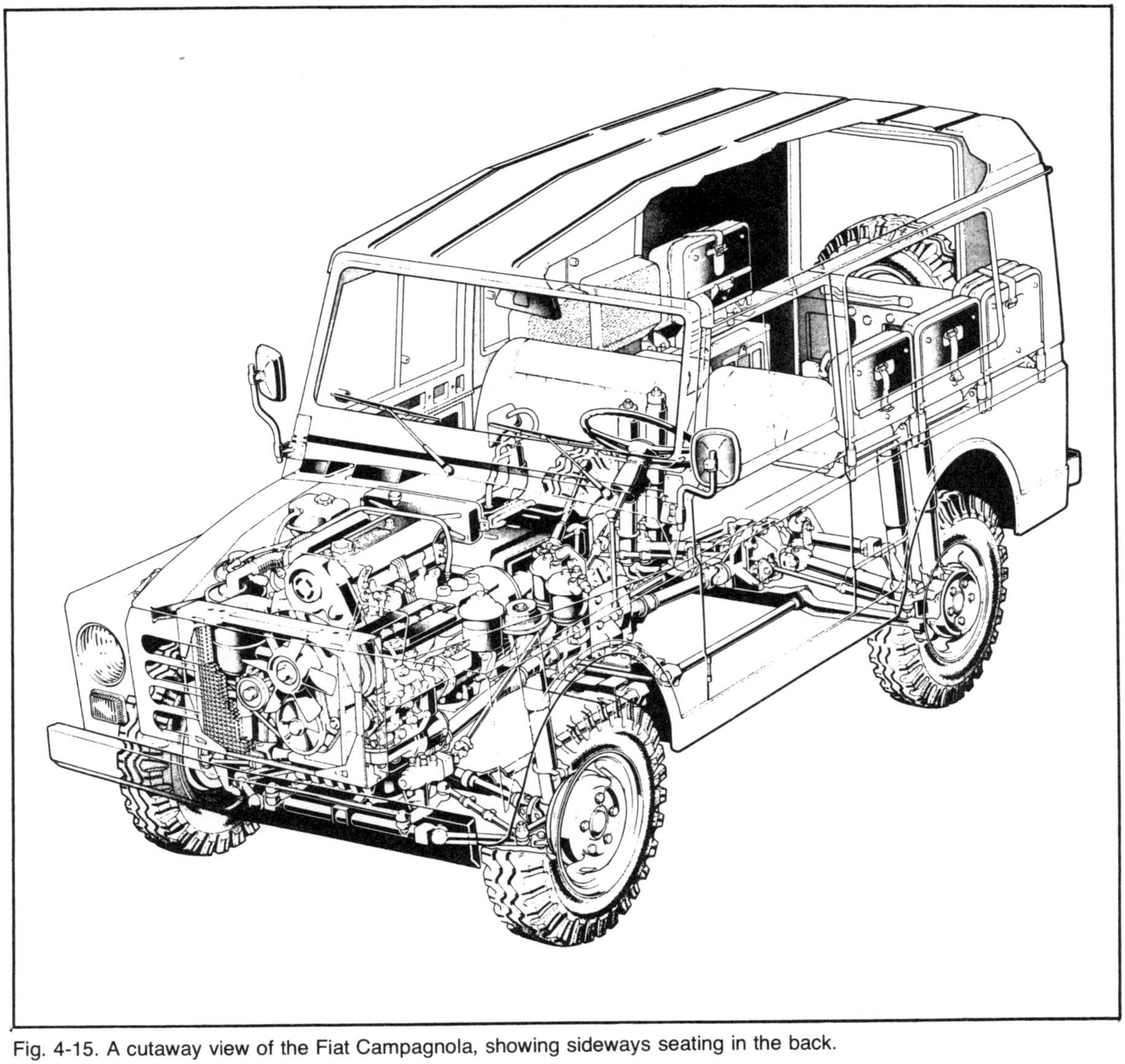

Fig. 4-15. A cutaway view of the Fiat Campagnola, showing sideways seating in the back.

full-length I-beam side members. See Fig. 4-16.

Three French manufacturers offer utility and recreational vehicles on two-wheel drive chassis, styled to resemble the all-terrain 4-×-4 machines. Citroen's Mehari is available either way (4-×-2 and 4-×-4). Renault's Rodeo is based on the front-wheel drive, R-4 chassis with a plastic body by Teilhol, and the Talbot Matra Rancho is based on the front-wheel drive Simca 1100 (still in production as a light van).

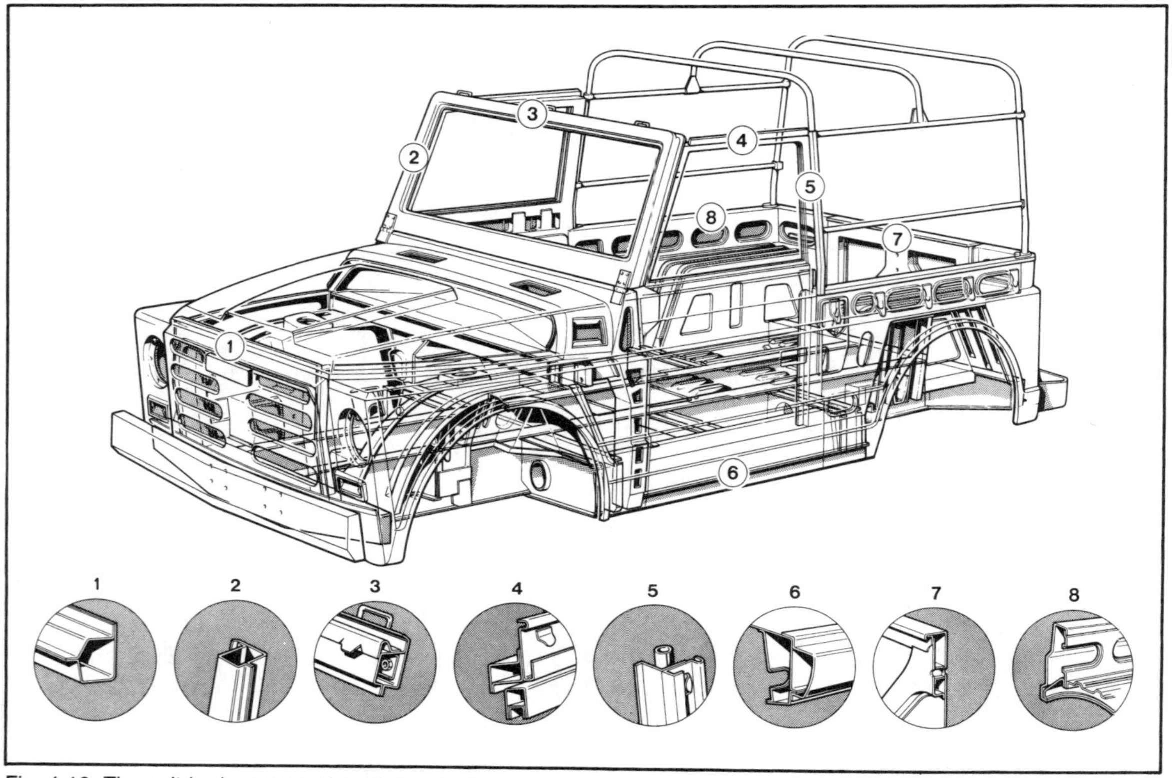

Fig. 4-16. The unit body construction of the Fiat Campagnola.

Chapter 5
Ergometrics

CARS MUST BE DESIGNED FOR COMFORT AND ease of driving. There is no scientific evidence to support Sir Alec Issigonis's contention that a slightly uncomfortable seat helps keep the driver alert. On the contrary, a lot of proof exists to the effect that an easing of the driving tasks and more comfortable seating help reduce driver fatigue. Because fatique is a significant safety hazard, it becomes important to make conditions better for the driver.

That is a large part of the car designers' job. The seating buck is made up before any real work is done on the body exterior. In all car companies, the designers work with a two-dimensional anthropometric model called "Oscar" for making full-scale seating drawings (Fig. 5-1). A simulated human metal structure, dimensionally adjustable, is used for correlating the seating dimensions with the compression of the seats under the person's weight.

BIOMECHANICAL FACTORS

"For years, a major handicap to good auto design has been the ill-conceived and widely accepted idea that the lower a car's roof, the better it is," wrote Anderson Ashburn in the September 14, 1970, issue of *Product Engineering*. "Certainly tall cars don't sell well, or not as well as low ones. But anybody who thinks this makes good design would do well to talk with a doctor about what it can do to the coccyx. Scuttlebutt has it that one type of back ailment is identified by doctors with the name of the most popular of the ground-hugging sports cars."

Seats giving a lean-back posture are preferred. Orthopedic surgeons say that the minimum angle between torso and thigh is 105 degrees, and that 110 to 115 degrees is ideal (Fig. 5-2). Angles of knee, shoulder, and elbow joints must also be selected with full regard for comfort and fatigue. Recommended knee-joint angles are in the 112-degree to 118-degree range.

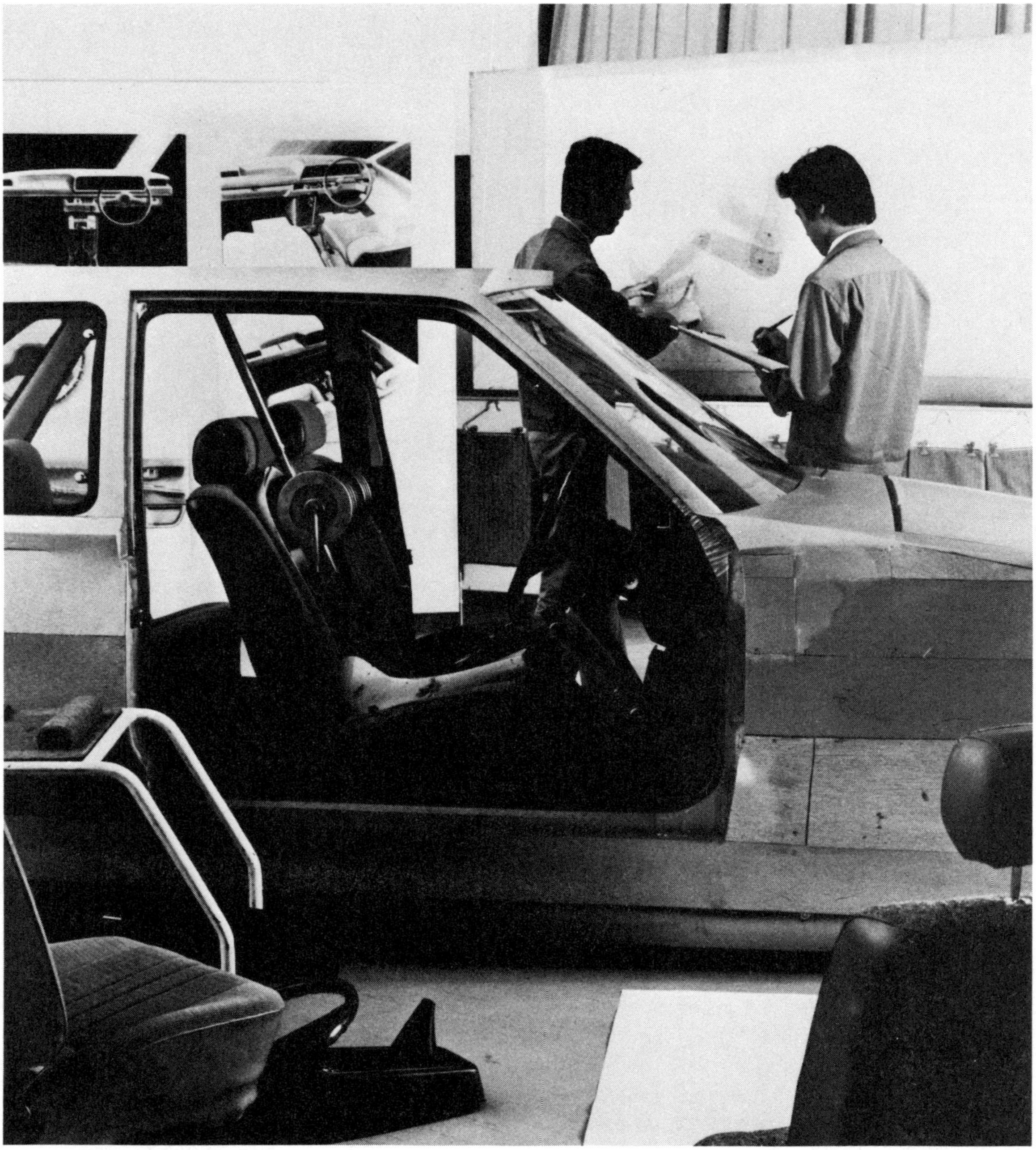

Fig. 5-1. Anthropometric model "Oscar" is used for full-scale seating drawings. A metal structure in the passenger's seat simulates a person's weight; it is dimensionally adjustable.

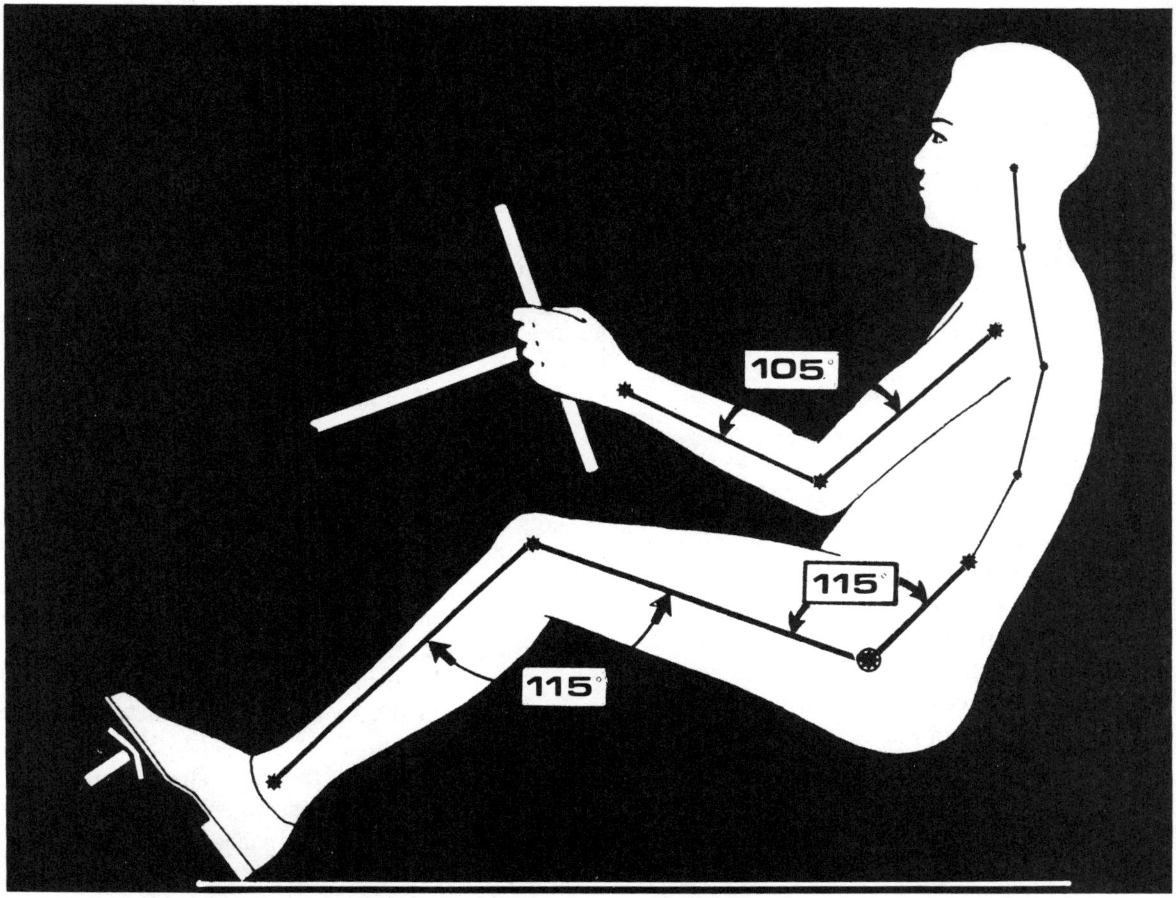

Fig. 5-2. Angles of hip, knee, shoulder and elbow joints must be chosen with full regard for comfort and fatigue.

As a general rule, the seat should be 16 inches above floor level. For proper under-thigh support, the seat should have a cushion length of 16 inches, with a raised front edge, giving a general slope of 5 to 10 degrees towards the backrest.

BETWEEN MAN AND MACHINE

As Peter Kyropoulos—long-time technical director of GM Styling—called it, the ergosphere is not really a human dimension; it describes a relationship between man and machine. The term ergometrics means "work measurement." In a car, the science of ergometrics is concerned with the distances needed for the driver's limbs to reach the controls, their ease of operation, and functionality. All controls must be within reach of the driver, without his having to lean forward from the seat back (Fig. 5-3).

But it's more than just distances. It's also the effort required to operate a lever or a pedal. It is also the relative ease with which an instrument can be read and understood.

Drivers vary in the height of their eye level, their arm's length, thigh length, and lower leg length. The car must provide adequate adjustment

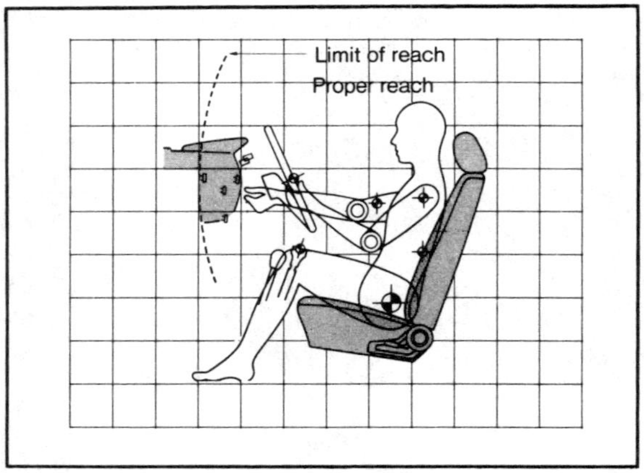

Fig. 5-3. All controls must be within reach of the driver, without his having to lean forward from the seat back.

in the seating to compensate for these variations, and the engineers must find the means to make these adjustments mechanically sound as well as easily operable with the car in motion.

Some people have short legs but a long torso. Others have a high waist and long legs. Consequently it is necessary to provide height adjustment as well as fore-and-aft adjustment for the driver's seat.

TAILORING THE SHORT AND THE TALL

Professor Joachim Elsholz, chief body development engineer of BMW, states succinctly that, "all seating positions in our cars must have adjustment possibilities to satisfy all persons, from the 5-percentile woman to the 95-percentile man." The 5-percentile woman is smaller than 95 percent of the female population, and the 95-percentile man is taller than all but 5 percent of the male population. The average man or woman is in the 50-percentile.

The designer is not bound to put all adjustment possibilities into the seat. For reasons that range from safety to manufacturing considerations, it might be preferable to have the seats in a fixed fore-and-aft position and provide adjustable pedals and steering column. The industry as a whole prefers to keep the pedal mountings fixed and put the seats on runners. Often the seats are in combination with steering columns that are adjustable for both reach and height.

The steering wheel should be the right size and set at the right angle. The effort required should be light, without encouraging overshoot, with a self-centering tendency that increases proportionally with cornering force, without ever rising so high that the driver is rendered physically incapable of going from an initial input to one giving a tighter turn.

Power steering has made that easy for the engineers. But power steering has been abused in the sense that it has been made optional (at extra cost) for certain light cars that had a front-end alignment problem. The cars did not need power steering. I know heavier cars that have had lighter and more precise steering. Instead of solving the problem, the engineers recommended power steering. This merely masked their mistakes.

Dished steering wheels were introduced in the late 1950s in the belief that they offered better crash protection. Now, flat-faced wheels with well-padded hubs are the rule. The horn ring, located halfway between the wheel hub and the rim, has many adherents, but it has practically disappeared from use. Most new cars have the horn on a stalk. That's not foolproof, and it can sometimes be on the wrong hand.

FOR FANCY FOOTWORK

Pedal positions must be coordinated with the layout of mechanical elements as well as anthropometrical requirements (Fig. 5-4). Pedals should not be too large or too small. Their spacing is a crucial point with many drivers. A brake-to-accelerator relationship that makes for easy heel-and-toe downshifting while braking is a safety factor as well as a source of enjoyment to those who use the technique, not to go faster into the curve, but to make the car's motion smoother without wasting time—or fuel.

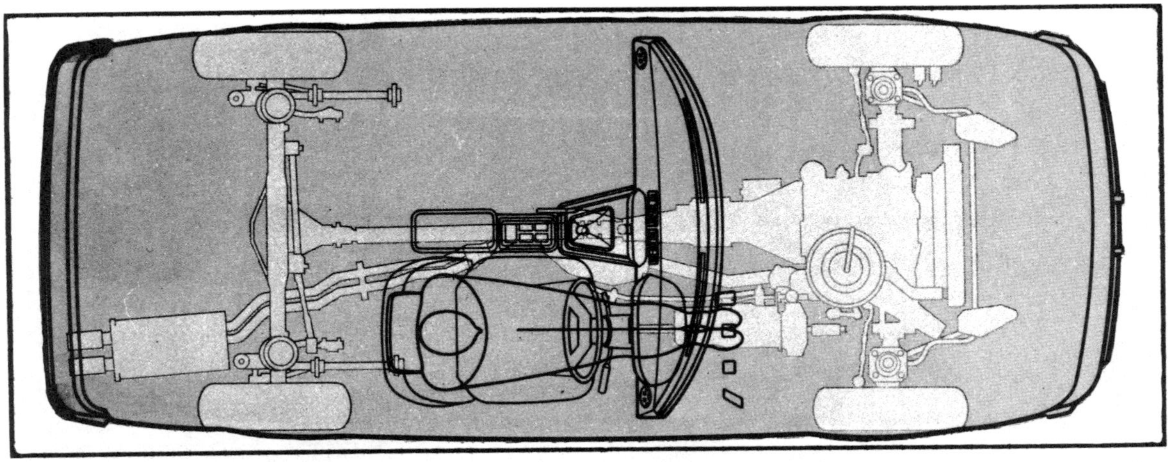

Fig. 5-4. Pedal positions must be coordinated with the layout of mechanical elements as well as anthropometrical requirements.

Enthusiastic inventors have proposed to combine the accelerator and brake pedals into one: push for go, release to stop. Their theory is that eliminating one pedal in some way makes the driving easier. I fail to see the advantage. Many engineers also believe that modulation of the braking is better achieved by stepping down on a pedal than by easing the pressure on it. In any case, it would cause severe retraining problems for millions of drivers.

There is no uniformity in interior design principles (Fig. 5-5). A console can be practical for mounting instruments, accessories, and for providing small storage cabinets. In small cars, the dimensions forbid the use of big consoles. Sometimes they are put in simply to cover up gearshift and handbrake linkages.

Seeing the instruments is important, and it is a good idea to tilt dials and gauges that are mounted the farthest away at such an angle so as to give a head-on view of the faces (Fig. 5-5). Even more important is the driver's view of the world outside the car.

DEPENDENT ON YOUR EYES

Seeing is believing says an old proverb. *La vue c'est la vie* say the French highway safety posters (sight is life, meaning survival). For the driver, vision and visibility are all-important requisites.

Vision provides 95 percent of the information required in normal driving; this gives an idea of the importance of the field of vision of the future driver. Man's field of vision is well known. It is used in drawing up a map of the areas of utilization of the driver's space, taking into consideration easy movements of the eyes, of the head, of the hands on the steering wheel, and of a number of duties or prohibitions specific to driving or to safety.

The car design must offer the driver the most unhindered field of vision in all directions. The car designer must take into account both vertical (Fig. 5-6) and horizontal visibility (Fig. 5-7).

A-posts should be slim or non-existent. Karl Wilfert of Mercedes-Benz tried to keep all eyelevel body components down to a width not exceeding the eye spacing so that the driver's view would be unbroken. But due to structural requirements, it was not always possible.

The hood should be short, but its extent should be clearly visible to the driver. The driver should always be able to see the tops of both front fenders.

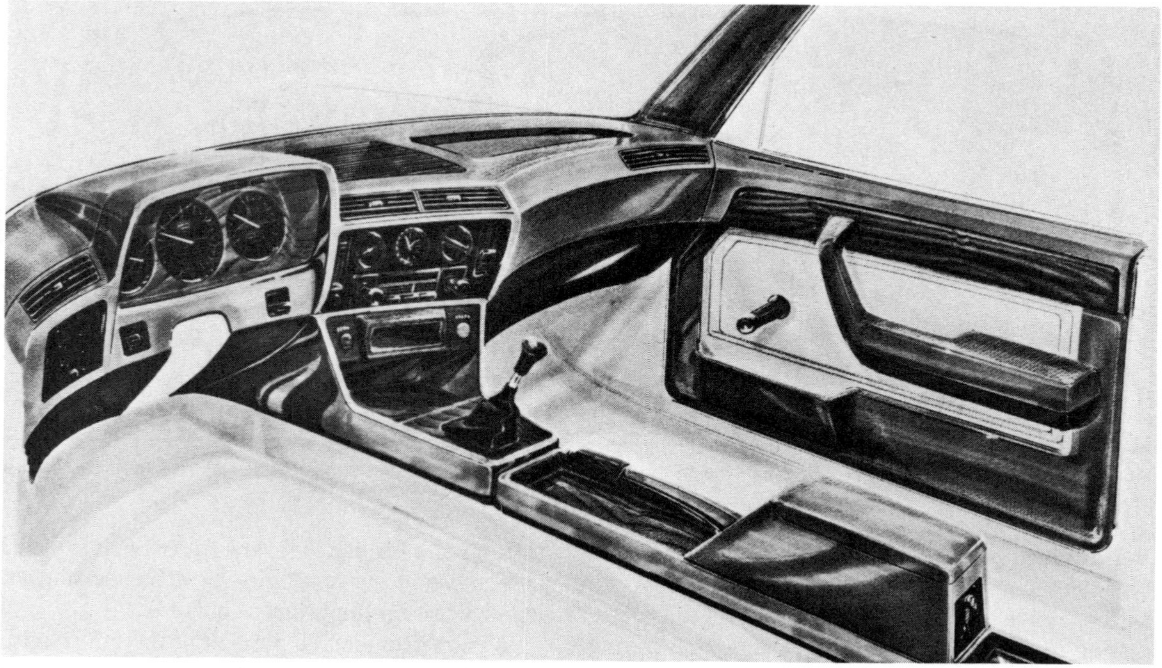

Fig. 5-5. Interior design principles. The cockpit layout with the console between the front bucket seats. Instruments not in front of the driver are mounted at an angle to provide a head-on view of the dials.

Fig. 5-6. Vertical visibility depends on hood and windshield design, rear window dimensions, and angle.

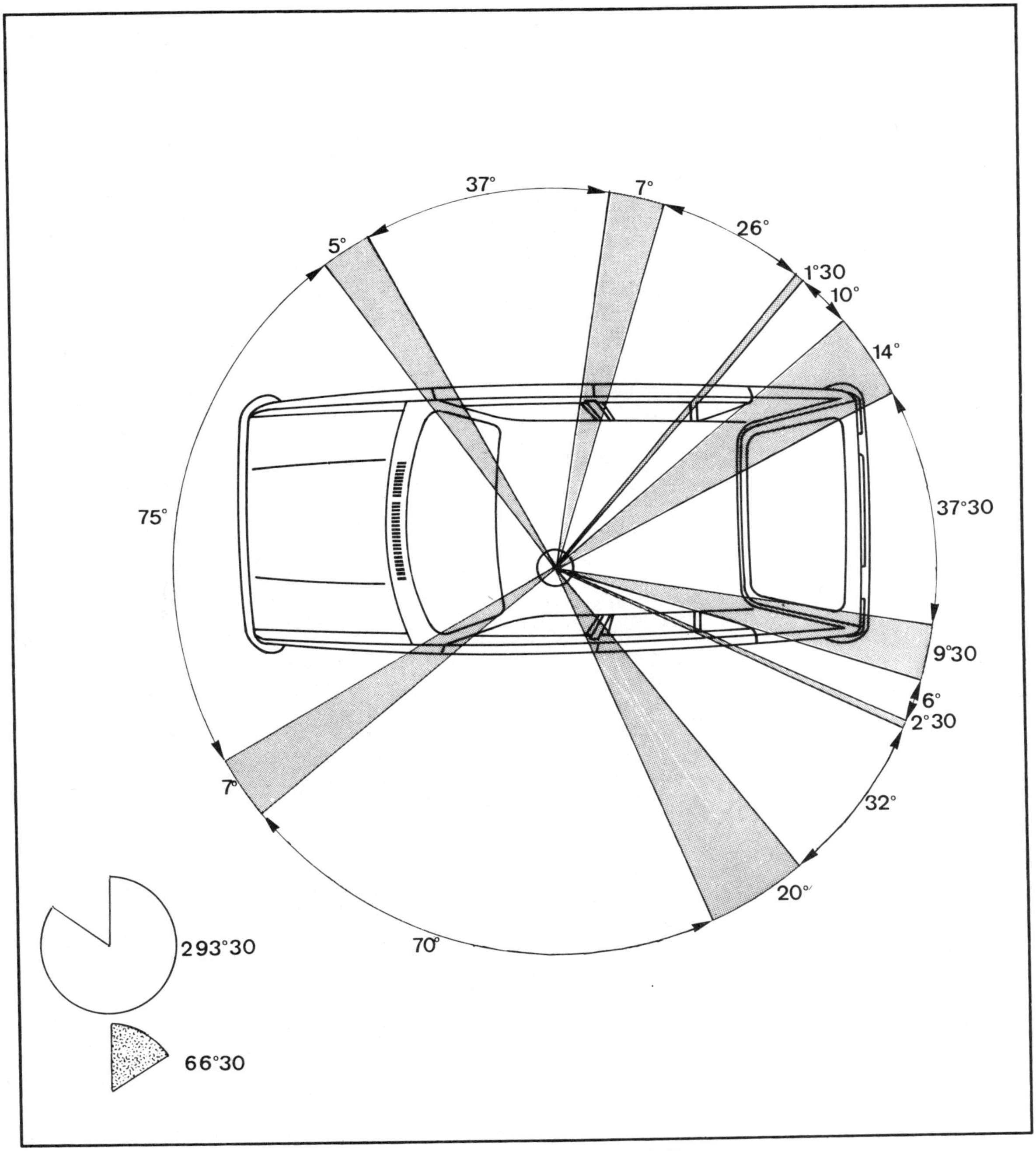

Fig. 5-7. Horizontal visibility is restricted mainly by obstruction from the roof pillars.

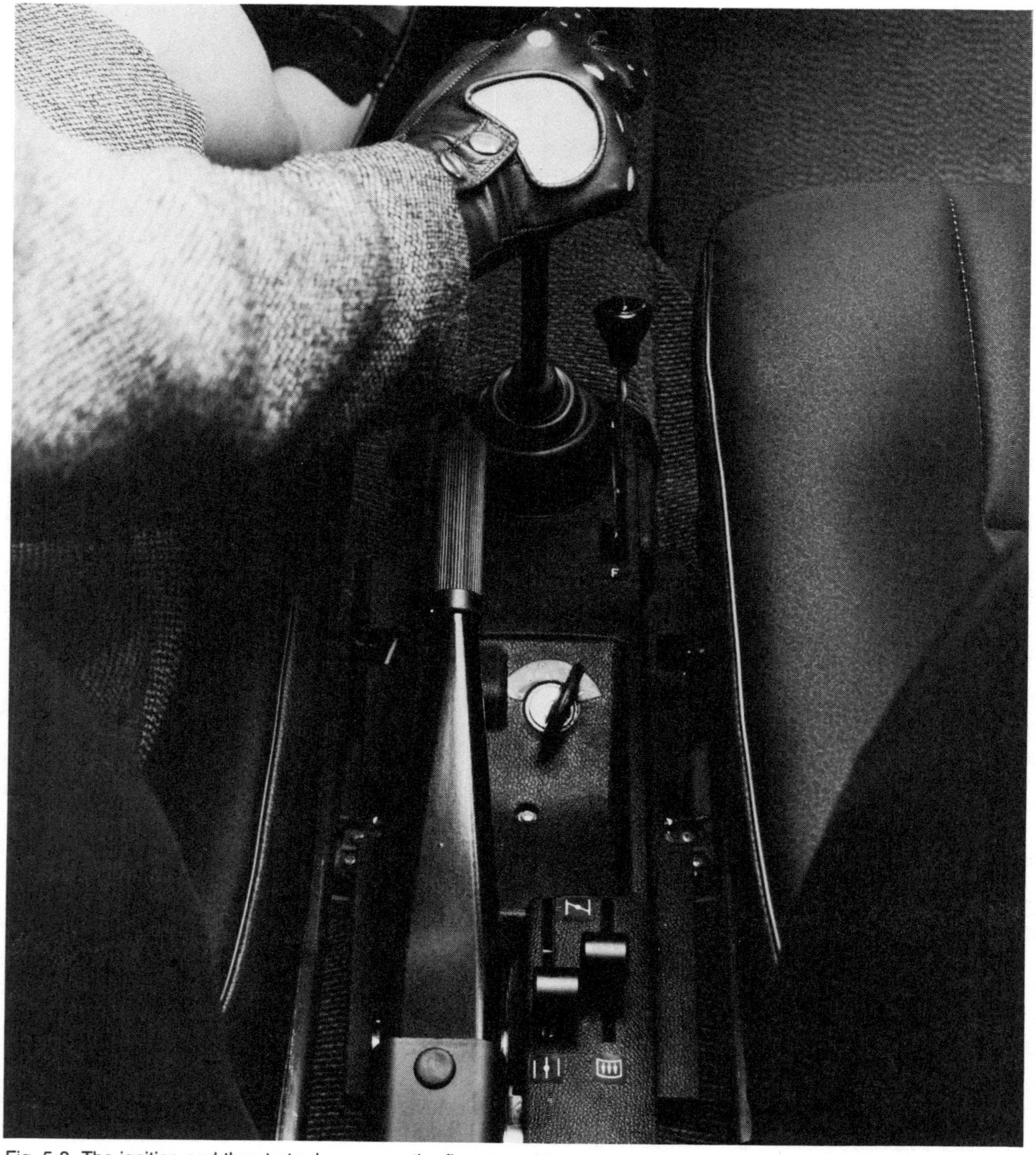

Fig. 5-8. The ignition and the starter key are on the floor tunnel (next to the handbrake lever) in the Saab-99.

Rear windows should be wide and have a low base.

Fashion has its role in the picture. A sharply raked windshield makes the car look speedier or more powerful. It tends to be aerodynamically advantageous, too, but there is an immediate conflict between aerodynamics and safety in choice of windshield angle. The highest angle tends to offer the least interruption of smooth airflow, but it also throws higher demands on the driver's vision. Optical distortion increases rapidly as the angle is made "faster."

The visibility problem is aggravated by use of tinted glass. The risks associated with the higher angle become more pronounced if the driver is wearing sunglasses or driving at night (dazzle from the headlights of oncoming cars, especially). There are ways of curving windshields to obtain excellent airflow with fairly low angles. The Saab 900 is a good example and the S-class Mercedes-Benz is another.

The windshield angle is measured between the center line of the glass and the vertical line from the windshield base. The higher the number the sharper the rake. On modern production cars the angle falls between 45 degrees for economy models and about 60 degrees for sports cars.

According to Karl Wilfert, the rear-window rake should not be exaggerated either. In 1970, he said, that "acutely tilted and curved windows can create optical illusions like double reflections and distortions and for instance impede the view through the rear window for the driver of a following vehicle." Rear-view mirror size and position are important in this connection. Outside mirrors are legal requirements. And that's that.

TACTILE IDENTIFICATION TRICKS

On cars with manual transmissions, the driver never has to look at the shift. It is all done by feel. He knows at all times what gear he is in, and needs no indicator panel to tell him. The situation is not the same with an automatic transmission whether the control stick is mounted on the steering column, on the floor, or in a console. Automatic-transmission shift patterns must be notched so that the driver can move the selector without taking his eyes off the road and not be surprised by getting another gear than desired.

Controls associated with driving safety should be mounted close to the wheel and right at the driver's fingertips at all times. That is particularly important for the wipers, washers, headlight dim-

Fig. 5-9. There are 12 different handles shaped for recognition by touch.

mer, horn, and turn signals. Nonessential controls for the radio, heater, vents, and so on should be removed from the driver's immediate surroundings.

Where the starter key is located might be less significant. Most car makers combine it with the steering lock, below the wheel, for right-hand operation (although left-hand keys are not uncommon). The most unusual car in this regard is the Saab-99. It has the ignition and starter key on the floor tunnel next to the handbrake lever (Fig. 5-8).

Control buttons, for whatever purpose, should be recognizable by touch. The shapes of the knobs and their range of movement can be immensely varied (Fig. 5-9).

While there is a theoretical advantage on standardizing controls for people who often drive different cars (rent-a-car clients, parking-lot attendants, and so on, it should be borne in mind that any standardization locks out the possibility of progress towards a better system. Experimentation must be allowed to continue. Consider that Citroen has offered several interesting solutions such as the multiswitch cylinder on the Visa (Fig. 5-10) and the more radical arrangement on the Karin show car (Fig. 5-11).

INFORMATION BY ELECTRONICS

Instrument panel design is undergoing an electronic revolution (Fig. 5-12). Almost every new car that's

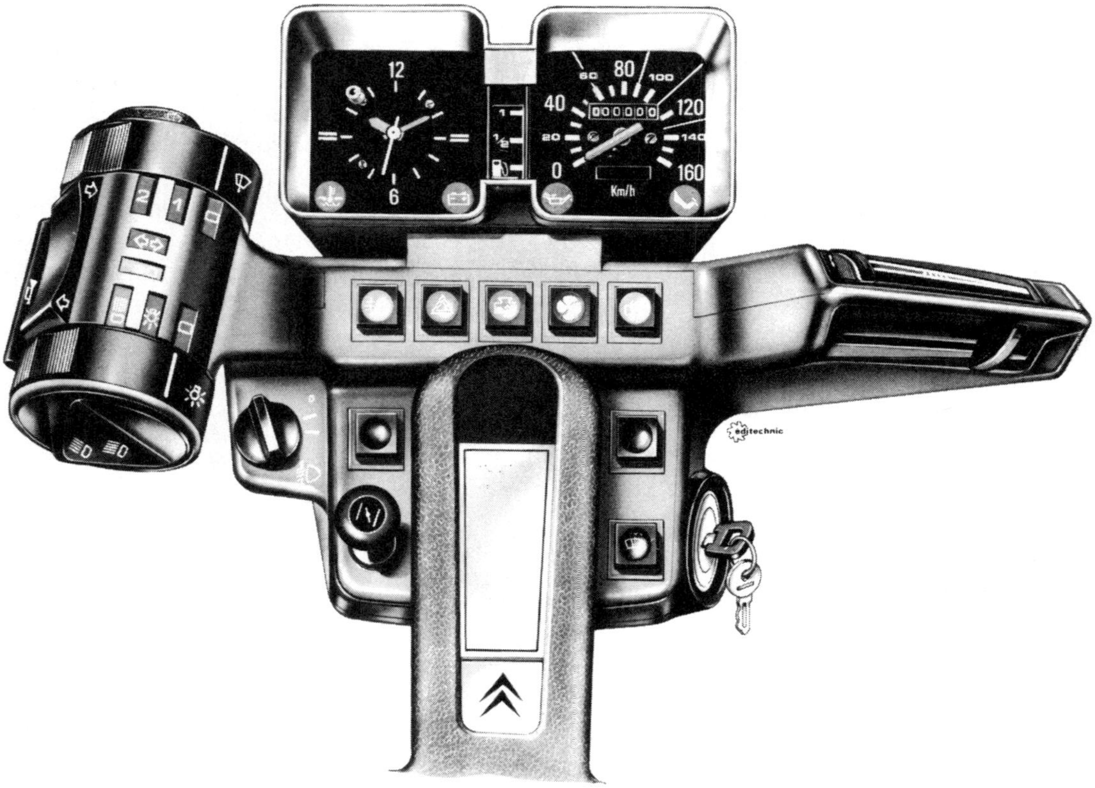

Fig. 5-10. Instrumentation and small controls of the Citroen Visa.

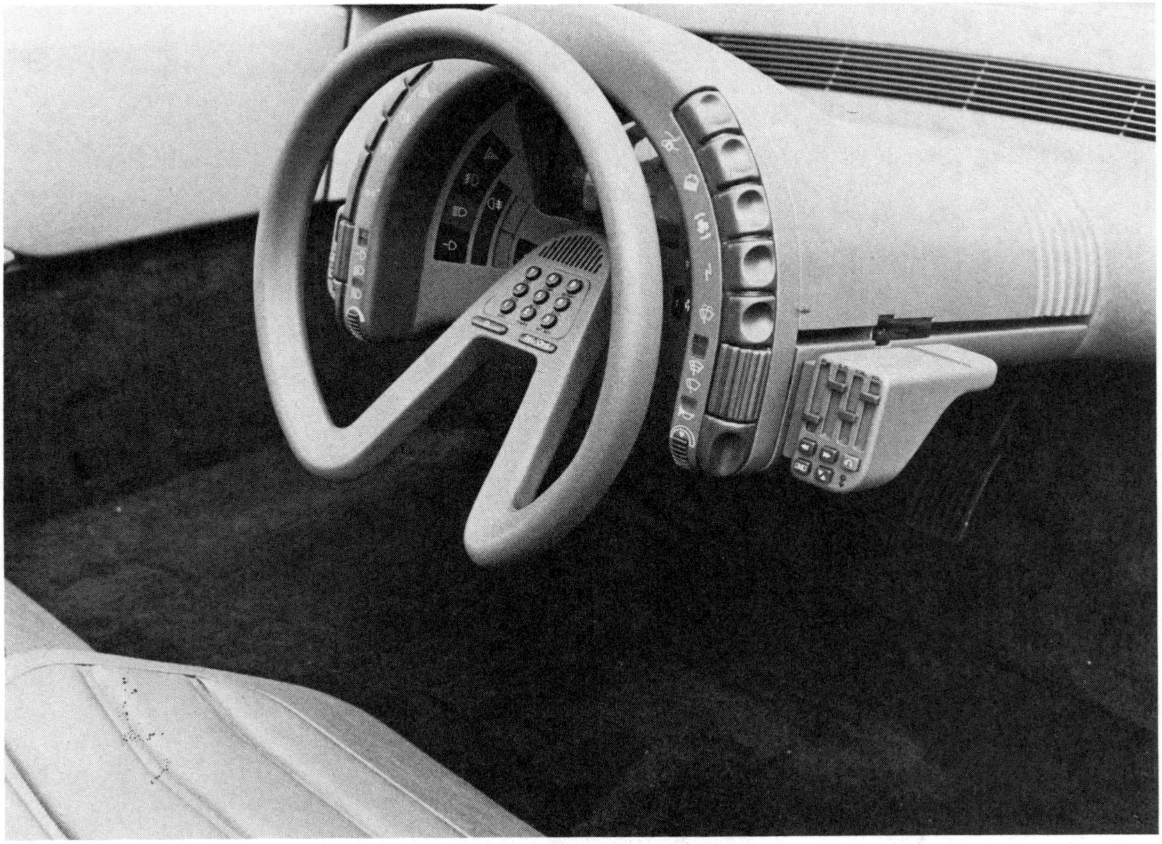

Fig. 5-11. Instrumentation and small controls for the Citroen Karin show car.

introduced in Europe has something new in the way of electronics. From on-board computers to new display systems and sensors and instrumentation to handle functions that are still largely manual (such as checking engine-oil level), you see digital speedometers and hear voice-synthesis warning signals. Is this really coming into the average car of tomorrow or are we caught up in a blind race to electronize wherever possible without due assessment of its worth?

It is my contention that the most important thing about the design of the key instruments is that they should be big. Perhaps I am a slave to old habits, but I do not believe in digital displays for speed or rpm when we can have dials where a glimpse of the hand is sufficient to tell all.

Sports-car enthusiasts who like the precision of oil-pressure gauges, vacuum gauges, ammeters—and consult them continually while driving—speak disparagingly of warning lights as "idiot" lights. In my opinion, however, warning lights are indispensable. It's not enough to have an oil-pressure gauge. If ever the oil-pressure drops below a certain limit, you need immediate visual warning to prevent damage to your engine.

Colored lights can help ease the driver's in-

Fig. 5-12. The instrument panel of the Ford Avant-Garde.

strument-reading task. Reserve red lights for danger signals—brake fluid low and seat belts not fastened are examples. Green can be used for lights on and blue can be used for high beam. Flashing amber can signal for low-fuel warning, engine overheating, handbrake not fully released, or choke on.

Fully electronic display technologies based on liquid-crystal diodes and vacuum fluorescence have now reached a level of dependability that makes them suitable for use in cars. The advantages of the new generation of instruments are reduced depth, savings in weight, and increased possibilities for the display of information. Such features are in keeping with the auto concepts of tomorrow (Fig. 5-13).

The stylist is offered a freedom of design previously unknown. These new "intelligent" systems show the driver in his primary field of vision only that information which is continuously needed (such as speed and information on how to save fuel).

Outputs from the automatic monitoring system integrated in the display on the condition of the vehicle and the engine are given only when faults occur or when there is a danger to driver and vehicle. If several events occur at once, the computer indicates them in the display in accordance with their priority and, if necessary, also reports the fault audibly in plain text.

CRADLING THE DRIVER'S CENTER OF GRAVITY

Keeping the driver informed is great, but he must also be provided with the best means of acting quickly and correctly on the basis of information received. That brings us back to the seat. In Amer-

Fig. 5-13. Instrumentation for the Renault Eve centers on a digital speedometer. On the right, a graphic readout of fuel economy. On the left, a diagram that shows the specific fuel consumption.

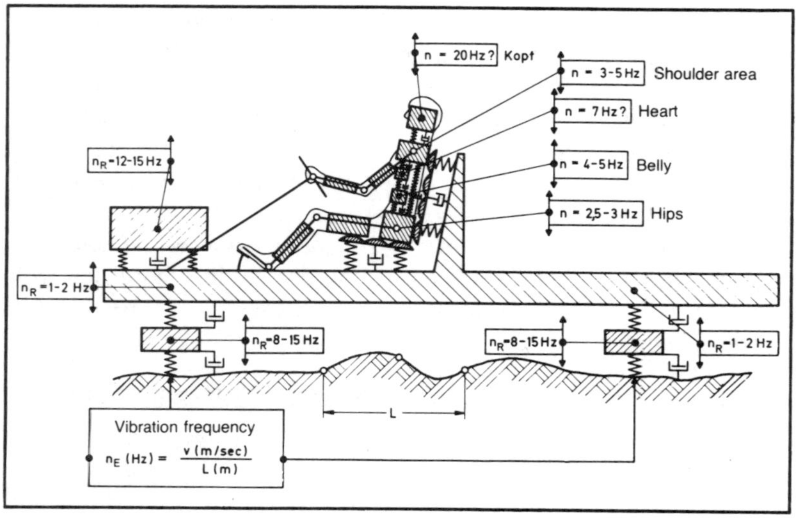

Fig. 5-14. A schematic presentation of the role the seat plays in assuring ride comfort.

ica, seat engineering used to be sacrificed on the alter of seat styling. The seat's functions are not merely static but also dynamic, and the seat construction must take that into account. The driver is subject to the same acceleration forces that act on the car, and the seat must be designed accordingly. See Fig. 5-14.

The seat must provide a firm base for the

Fig. 5-15. The Volvo seat with adjustable lumbar support.

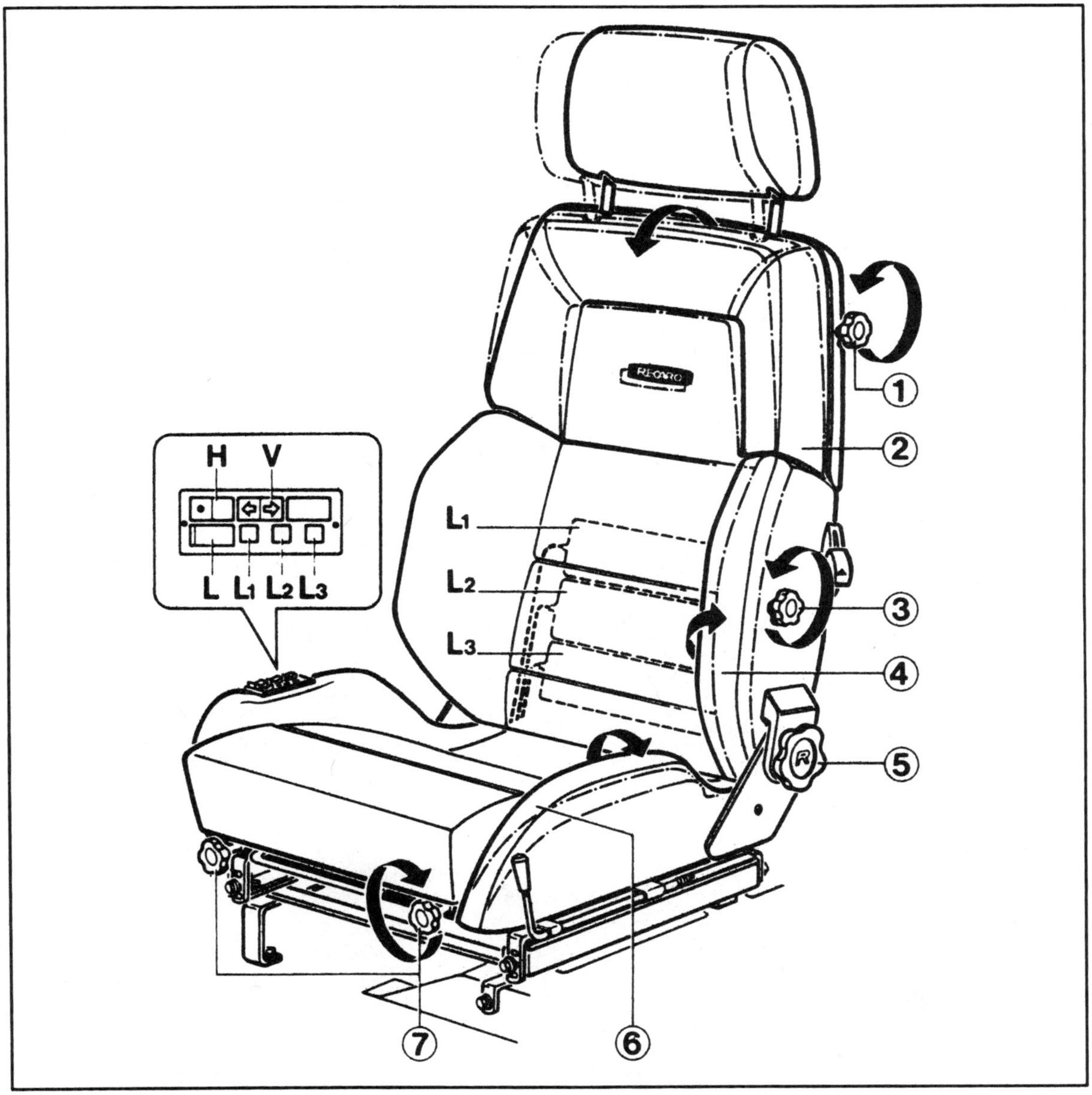

Fig. 5-16. Recaro's concept of the ideal seat.

performance of all driving tasks under all conditions, and for long periods at a time. It must be shaped to minimize the threat to balanced seating imposed by high lateral g-loads (during hard cornering), but only a compromise solution is feasible.

Too little lateral support in the seat cushion

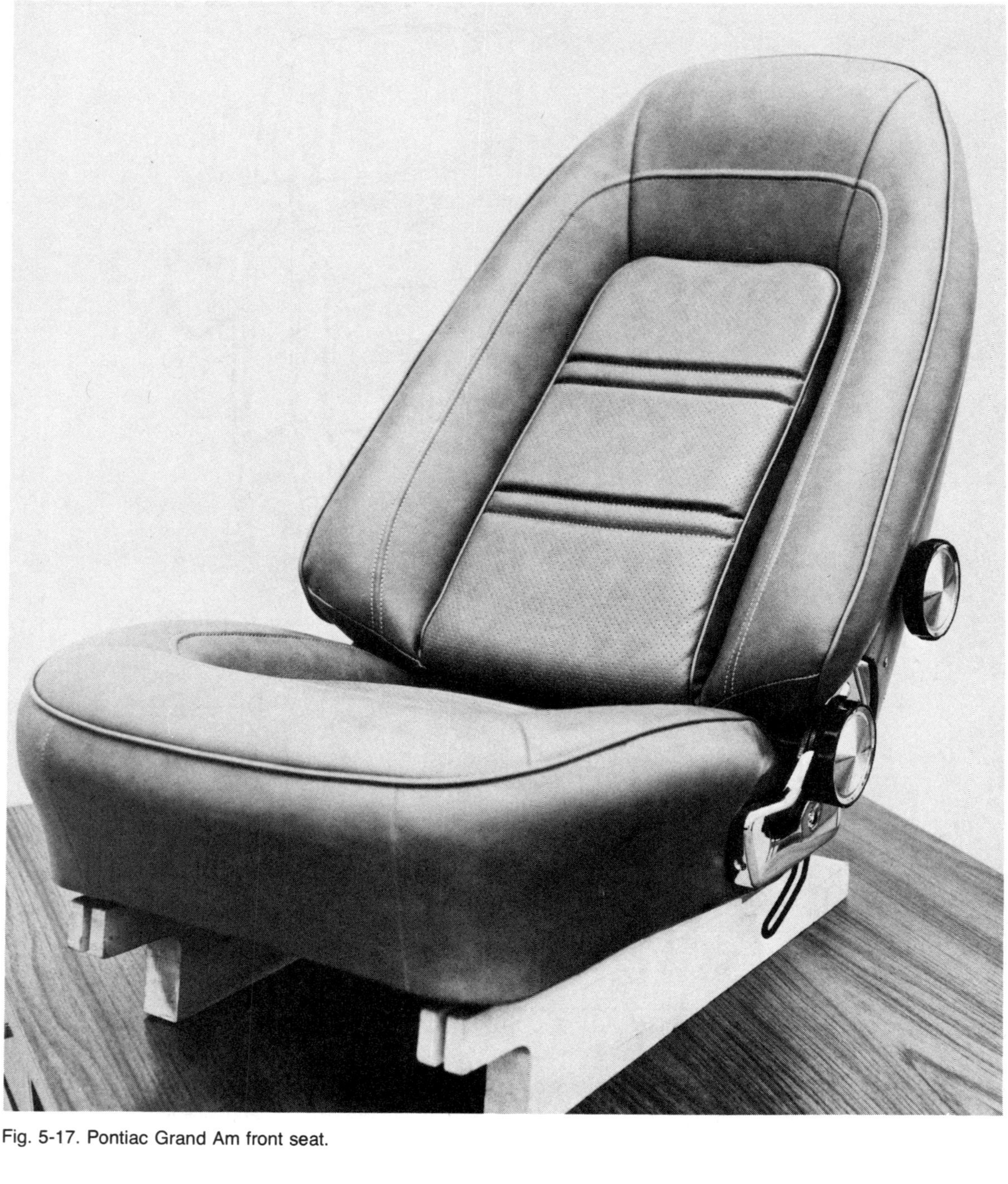

Fig. 5-17. Pontiac Grand Am front seat.

and backrest will jeopardize the driver's physical stability on curves. Too much support will tend to prevent the driver from altering his position from time to time, and therefore threatens to accelerate fatigue in the driver. See Fig. 5-20.

Lack of elbow room can also contribute to fatigue and failure to carry out crucial maneuvers. Many drivers take a skeptical look at the three-abreast seating of the Matra-Simca Bagheera (Fig. 5-19).

Seat designers and engineers must be conscious, above all, of the fact that drivers and passengers alike tend to get uncomfortable when forced to sit in the same position for any length of time. For the driver, seating comfort affects his safety as well as his physical well being.

The seat cushion should preferably be covered by a porous material. There is a risk of making cushions too soft in the quest for extra ride comfort. Soft cushions are compressed under the weight of the occupant. This lowers the effective seat height, tightens the angle between his torso and thighs, and flattens the lumbar curve.

That's why Mercedes-Benz, for example, has always used fairly hard seats. According to Karl Wilfert, "vibration theory would generally favor relatively firm seats so that the natural frequency of the system "seat plus body" does not come critically close to the natural frequency of the vehicle (1 to 2 Hz). The upholstery of the seat must support the body weight at the pelvic bones to avoid constriction of the blood vessels leading to fatigue and cramps." See Fig. 5-21.

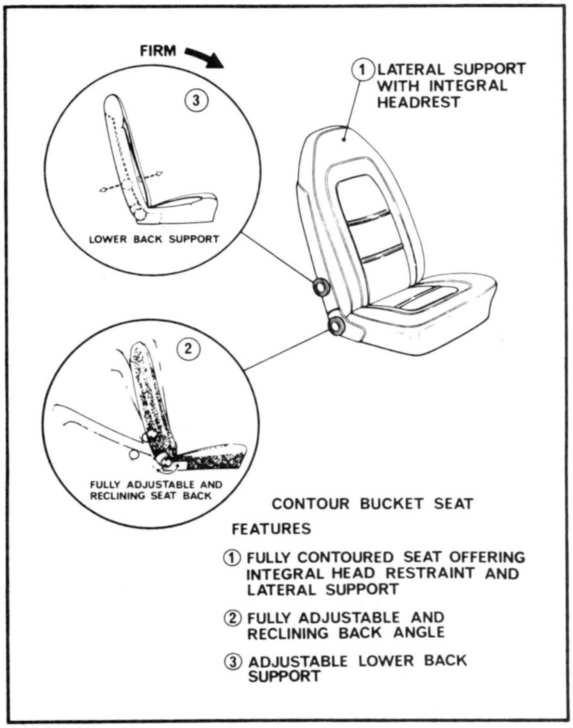

Fig. 5-18. Mechanism of the Pontiac Grand Am seat.

HARDWARE AND SOFTWARE

Until the 1940s, seats usually had coil springs just like most house furniture. Some, like American Motors, persisted in the use of coil-spring seats right up into the late 1960s.

Around 1950, sinusoidal zig-zag springs providing a flat bottom for the seat cushion came into use. They saved a lot of cost. By 1955, the still lower-cost boxlike wire spring grid replaced the zig-zag spring in many cars. Variations on these two metal-spring configurations have dominated until recent times. The current trend is to eliminate metal parts from the basic seat by adopting such materials as plastic for the seat frame and polyurethane foam as the springing element.

The conventional full-width, bench-type front seat was placed on rails to provide fore-and-aft adjustment. A handle on the side released a locking lever from a rack on the rail to let the seat be moved. When released, the locking lever held the seat firmly in its new position. Individual "bucket" seats move on the same principle, but each one is mounted on its own set of rails.

In former days, the seat rails were usually placed at a backwards slope. The theory was that a

Fig. 5-19. The triple bucket seat arrangement in the Matra-Simca Bagheera.

driver who would move the seat back (due to long legs) also had a long torso and had to sit lower in order to conserve reasonable head clearance from the ceiling. Because most people's proportions do not always conform to the norm, this was at best an oversimplification. It caused permanent discomfort for drivers of nonaverage build.

There is general agreement among designers that the backrest angle should be adjustable in closely spaced notches or, better still, continuously variable.

Say that the average person finds his elbow angle in a given car to be 105 degrees. That might or might not be acceptable. The short person, with the seat further forward, is closer to the steering wheel so that the elbow angle becomes 81 degrees. Most people will not be comfortable in that position. This short person might find a higher hip angle more tolerable so that the elbow angle can be relieved. But that can't be done with a fixed-angle backrest. That's where the argument for the recliner mechanism comes in.

In the same car, a tall driver, seated with a 105-degree hip joint angle, will find his elbow angle at 113 degrees. The angle of tolerance depends greatly upon steering wheel height (but not on angle). It has been proposed that the most restful driving position allows a person to hold the wheel at a quarter to three (3 o'clock and 9 o'clock) with the hands below the level of the heart. Medical studies indicate that holding the hands higher than the heart can cause fatigue.

The advantages of a reclining backrest can be lost, however, if the hinge points are put in the wrong place. It happened to a GM option that was made available in 1964, but quickly abandoned because people did not like it. The hinge points should have the same instantaneous center as the pivot point of the occupant's hip joints. This instantaneous center is an imaginary point usually located in the air around the steering wheel.

Many experts also recommend an adjustable lower lumbar support, controlled independently of the backrest rake, as first offered by Volvo in 1967. See Fig. 5-15.

A company that tries to make the ideal seat, with multiple adjustments for all possible requirements, is Recaro (an outgrowth of Reutter in Stuttgart). See Fig. 5-16. In addition to the variable backrest angle adjustment, it offers a lumbar support insert that can be repositioned, an adjustable thigh support, and a shoulder support adjustment. The stamped steel frame, with its contoured, wide seat bottom, also has vertical adjustment. The Recaro people point out that the driver has 20 different ways of fine-tuning his comfort.

The first of the properly designed seat options for American cars was the bucket seat for the 1973-74 Pontiac Grand Am (Fig. 5-17). Design objectives included lower back support, recliner, and lateral restraint capability. Normal back angle was 26.5 degrees; the recliner mechanism added a 20-degree range for a maximum of 46.5 degrees. The adjustment knob turned an eccentric that caused an

Fig. 5-20. Deep bucket seats in the Simca 1000 Rallye-2.

Fig. 5-21. The latest Mercedes-Benz bucket seat design (380 SEC and 500 SEC).

inner gear, with 29 teeth, to walk around the inside of an outer ring gear, with 30 teeth, increasing or reducing the slope of the seat back (Fig. 5-18). The lower back support consisted of a 3-inch-wide canvas band attached to the seat frame on one side and attached to a movable plate on the other side. Turning a knob on the side rotated a threaded rod inside the seat back; this slid the plate along its track and it tightened or loosened the band accordingly.

Chapter 6
Safety

SAFETY IS A RELATIVE THING. WHEREVER YOU are—in or out of your car, your home, your place of work—you are never immune to danger. Perhaps we can be made secure from a specific hazard or set of hazards at a given time, but trying to keep oneself absolutely safe against all dangers all the time would be a full-time occupation.

PRIMARY AND SECONDARY SAFETY

It follows that total safety in a car is an unattainable goal. The car designers and engineers do what they can to guard against certain known dangers. Wherever there is traffic, accidents would occur continuously unless the cars were equipped with brakes. Brakes are the most basic automotive safety factor. But it is up to the driver to use them. Driver error or misjudgment are to blame for a majority of avoidable accidents.

Avoiding accidents must be the first commandment for all road users. That is called primary or active safety. See Fig. 6-1. Good brakes, accurate steering, and balanced and predictable handling are the main ingredients on the car's side.

Not all accidents are avoidable. To protect the car occupants against the consequences of a collision is called secondary or passive safety. There are three principal approaches:

☐ Reducing the forces imposed on occupants in a collision, by controlled crumpling of the front of the vehicle.

☐ Distributing these forces over each passenger's body by means of a protective device (safety belt or any other similar device).

☐ Limiting the deformation of the passenger compartment so that crumpling and the retention devices of the front end can be fully effective in providing protection.

NONAUTOMOBILE FACTORS

The road is the government's responsibility. It is up

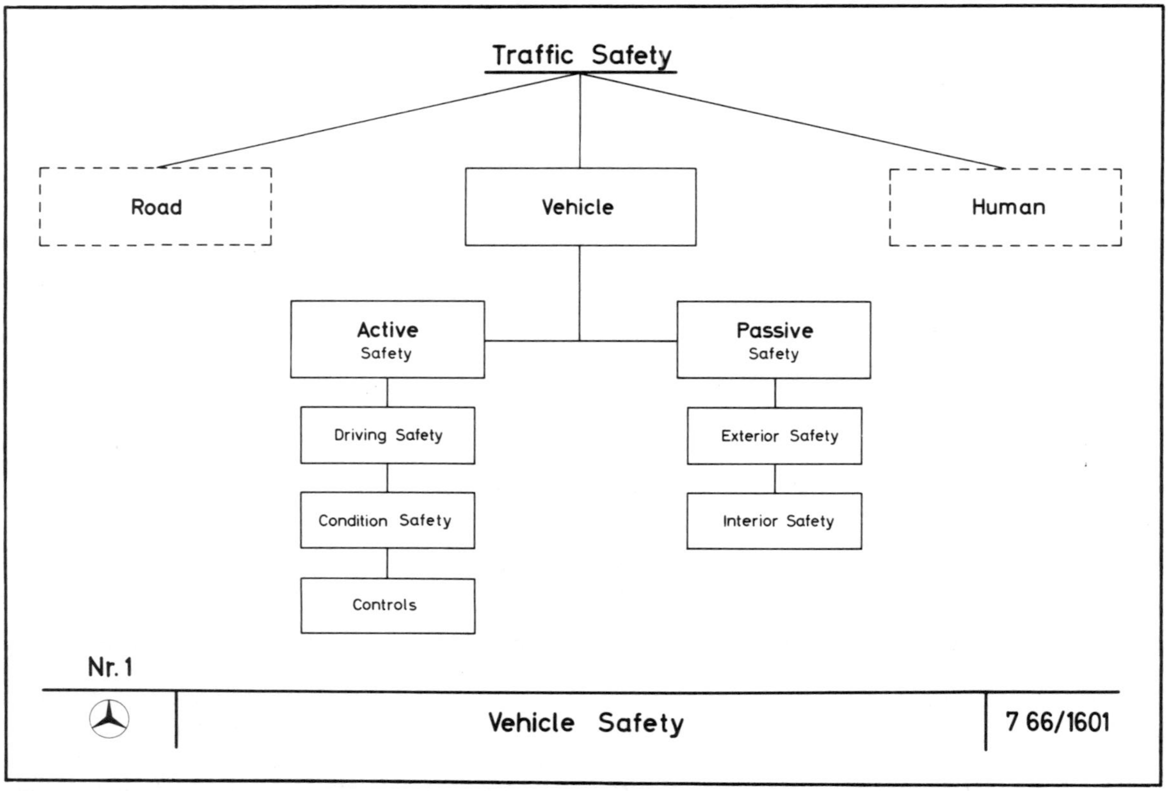

Fig. 6-1. What traffic safety is all about, according to Karl Wilfert of Mercedes-Benz.

to the public authorities to provide safe roads. Highway engineers, lawmakers, and law enforcement agencies have major roles to play.

There are several things that can be done to prevent accidents beyond simply lowering the speed limits (which has limited effectiveness and carries a heavy price in terms of wasted time). One method is to separate the various categories of highway users so as to keep the motorcycles away from the cars and the cars away from the trucks. Pedestrians should be kept entirely separate from wheeled traffic.

The authorities have stopped short of any such solutions by hiding behind arguments that the proposals are "impractical" or too costly. When the auto industry mentions cost/benefit ratio, it is widely accused of being callous.

A VIEW FROM DEARBORN

Bill Innes, vice president of Ford, told the press in 1973: "Obviously, there is a lot of sense in the notion that we will never prevent all accidents and that the safe packaging of occupants is essential to reducing accident tolls. But I don't want to leave anyone with the impression that we can solve the accident problem simply by building safer cars.

"We know, for example, that eight of ten drivers in fatal crashes last year were in violation of a traffic law at the time of the accident. We also know that one-third of all deaths occur in single car accidents, and that some of those lives could be saved by removing hazards from the roadside. Finally, we know that half the fatal accidents involves a driver or a pedestrian who was legally drunk."

"With driver error a major factor in accidents, it would make sense to me to toughen the test for driver's licenses, and to require periodic re-examination of drivers. Cars should be subject to a periodic safety inspection, too. In short, when the federal safety acts were drawn up in 1966, it was generally agreed that accident prevention would involve attention to three factors; the car, the driver and the roadway. To date, the car has received most of the attention."

The automobile was singled out for action because the car industry is an easy target, and all costs are passed on to the consumer. A lot of legislation was passed in a great hurry without obtaining reliable data on how and why traffic accidents occur, and exactly what it is that injures or kills drivers and passengers. Now, 15 years and countless millions of dollars later, cooler heads are again prevailing, and design for safety is making progress subject to the cost/benefit ratio the market demands and can afford.

NO STANDARD ACCIDENT

In 1966, the United States government wanted to set rules that would assure survivability in frontal collisions, at progressively higher speeds, whenever a target was met. The government happily set performance standards for crashworthiness while completely overlooking the fact that there is no standard accident.

Detroit countered with "you can't legislate new technology into existence. Congress can't alter the laws of physics." The authorities stared themselves blind at barrier-crash test performance data while totally disregarding the fact that a barrier crash has little or no relationship to actual road accidents. The key to the problem is energy absorption. The nonresilient barrier is equivalent to an impact, central and full width, with another car coming in the opposite direction at the same speed as the impacting car.

ENERGY MANAGEMENT

A vehicle of mass M traveling at velocity V has a kinetic energy $E = 1/2\ MV^2$ (drag and friction are overlooked). See Fig. 6-2. In a collision, the kinetic

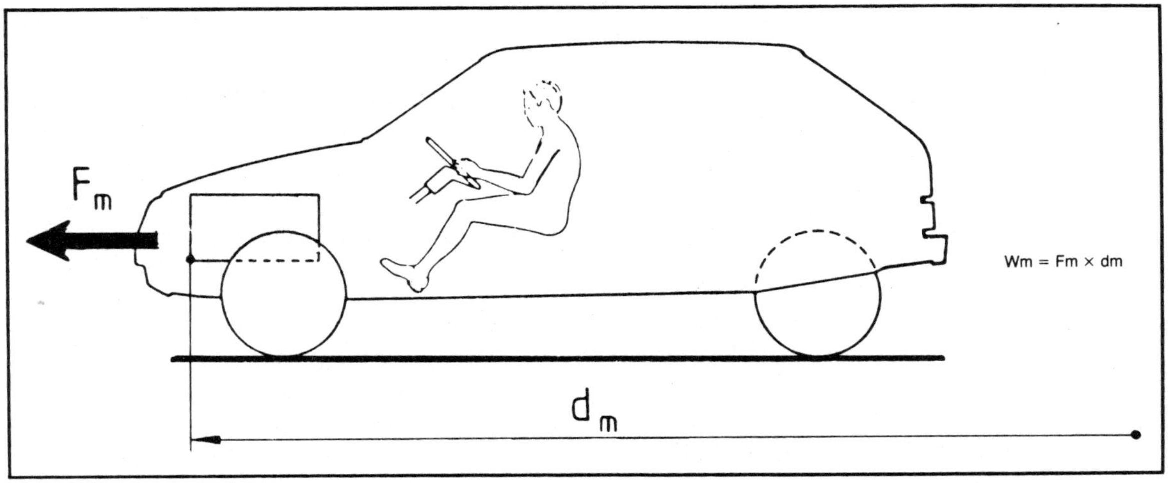

Fig. 6-2. A schematic of the energy-absorption problem in a head-on collision.

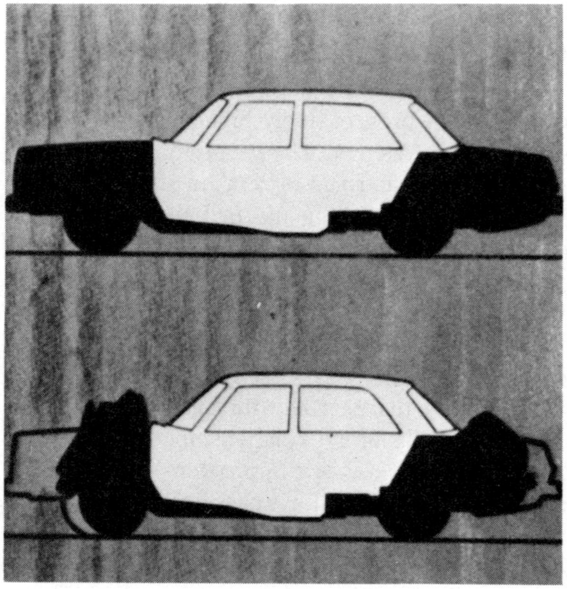

Fig. 6-3. The principle of localized crush zones is to leave an undeformed survival space after the impact.

energy 1/2 MV^2 of the vehicle is dissipated by the work of deformation or crumpling of the front end. The stopping distance of the vehicle is this deformation. See Fig. 6-3. Making the vehicle more rigid would reduce deformation, but at the cost of increasing the force applied to the vehicle and to its passengers.

Making it longer—to increase stopping distance and reduce the deceleration rate to g-loads the human body can withstand—would lead to cumbersome dimensions and increased weight.

The American auto industry came under strong attack for its low spending on safety research. The accusers conveniently neglected to count primary safety and concentrated entirely on passive safety. And even here, the industry's record was not so bad.

SUSTAINED SAFETY RESEARCH EFFORT

General Motors, Studebaker, Hudson and Packard began crash-testing programs about 1933. They rolled cars off cliffs, built ramps to test the bodies for rollover strength, and even made barrier-impact tests. Europe started later but with a more scientific approach.

Mercedes-Benz built its first special crash-test vehicle in 1939. It had an extra-rigid floor, side protection in the doors, and a three-piece collapsible steering column.

Academic circles in America began taking an interest in traffic safety prior to the start of World War II. Cornell University at Ithaca, New York, was first with a crash-injury study program, in 1940, aimed at trying to determine the causes of traffic accidents and how to prevent them.

In 1949, the Institute of Transportation and Traffic Engineering at the University of California at Los Angeles (UCLA) undertook an auto-crash injury research program. And Colonel John Paul Stapp started the U.S. Air Force automobile crash research project in 1954. Ford introduced a "safety package" as an option for the 1956 models. It included interior padding, deep-dish steering wheel, seat belts, lift-bolt door locks (Fig. 6-12) and many other features.

The controversy over belt use and air bags is outside the scope of this book and is not discussed.

FROM PATENTS TO PRODUCTION

The first type of safety door lock was a cone-bolt design that became standard on the Mercedes-Benz 170 S and 170 D models in 1949 (Fig. 6-13). The same company was a leader in developing safe arrangements of interior door locks and handles (Fig. 6-14).

Bela Barenyi's patent for the Mercedes-Benz safety body shell—with a virtually nondeformable passenger compartment in the middle and impact-absorbing "crush" zones front and rear—had been granted in 1952 (Fig. 6-4). Its principles were built into the new 220 series launched in October, 1958.

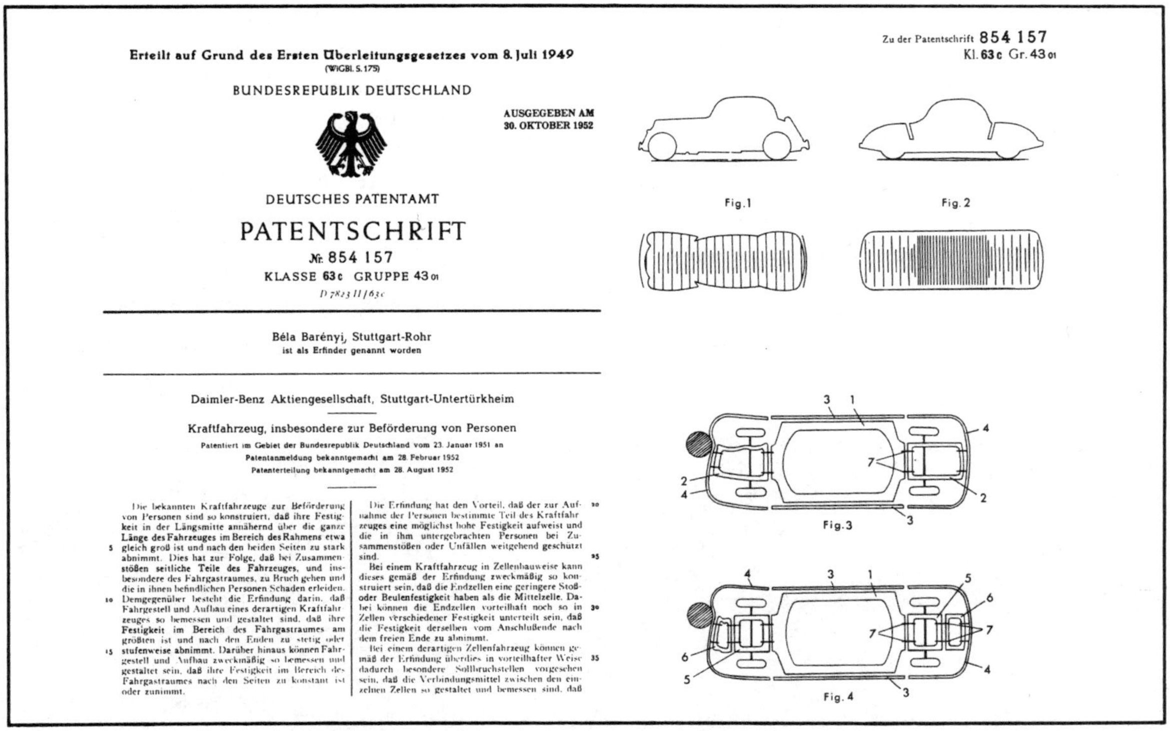

Erteilt auf Grund des Ersten Überleitungsgesetzes vom 8. Juli 1949

(WiGBl. S. 175)

Zu der Patentschrift 854 157

Kl. 63c Gr. 43 01

BUNDESREPUBLIK DEUTSCHLAND

AUSGEGEBEN AM 30. OKTOBER 1952

DEUTSCHES PATENTAMT

PATENTSCHRIFT

Nr. 854 157

KLASSE 63c GRUPPE 43 01

D 7823 II / 63c

Béla Barényi, Stuttgart-Rohr

ist als Erfinder genannt worden

Daimler-Benz Aktiengesellschaft, Stuttgart-Untertürkheim

Kraftfahrzeug, insbesondere zur Beförderung von Personen

Patentiert im Gebiet der Bundesrepublik Deutschland vom 23. Januar 1951 an

Patentanmeldung bekanntgemacht am 28. Februar 1952

Patenterteilung bekanntgemacht am 28. August 1952

Die bekannten Kraftfahrzeuge zur Beförderung von Personen sind so konstruiert, daß ihre Festigkeit in der Längsmitte annähernd über die ganze Länge des Fahrzeuges im Bereich des Rahmens etwa gleich groß ist und nach den beiden Seiten zu stark abnimmt. Dies hat zur Folge, daß bei Zusammenstößen seitliche Teile des Fahrzeuges, und insbesondere des Fahrgastraumes, zu Bruch gehen und die in ihnen befindlichen Personen Schaden erleiden. Demgegenüber besteht die Erfindung darin, daß Fahrgestell und Aufbau eines derartigen Kraftfahrzeuges so bemessen und gestaltet sind, daß ihre Festigkeit im Bereich des Fahrgastraumes am größten ist und nach den Enden zu stetig oder stufenweise abnimmt. Darüber hinaus können Fahrgestell und Aufbau zweckmäßig so bemessen und gestaltet sein, daß ihre Festigkeit im Bereich des Fahrgastraumes nach den Seiten zu konstant ist oder zunimmt.

Die Erfindung hat den Vorteil, daß der zur Aufnahme der Personen bestimmte Teil des Kraftfahrzeuges eine möglichst hohe Festigkeit aufweist und die in ihm untergebrachten Personen bei Zusammenstößen oder Unfällen weitgehend geschützt sind.

Bei einem Kraftfahrzeug in Zellenbauweise kann dieses gemäß der Erfindung zweckmäßig so konstruiert sein, daß die Endzellen eine geringere Stoß- oder Beulenfestigkeit haben als die Mittelzelle. Dabei können die Endzellen vorteilhaft noch so in Zellen verschiedener Festigkeit unterteilt sein, daß die Festigkeit derselben vom Anschlußende nach dem freien Ende zu abnimmt.

Bei einem derartigen Zellenfahrzeug können gemäß der Erfindung überdies in vorteilhafter Weise dadurch besondere Sollbruchstellen vorgesehen sein, daß die Verbindungsmittel zwischen den einzelnen Zellen so gestaltet und bemessen sind, daß

Fig. 6-4. A Mercedes-Benz patent, from 1952, describing the theory of front and rear crush zones.

A German patent issued to Bela Barenyi, in 1956, shows a small car with its fuel tank located safely near the center line, and deeply inside the body structure, well ahead of the rear axle (Fig. 6-5).

General Motors installed its first impact sled

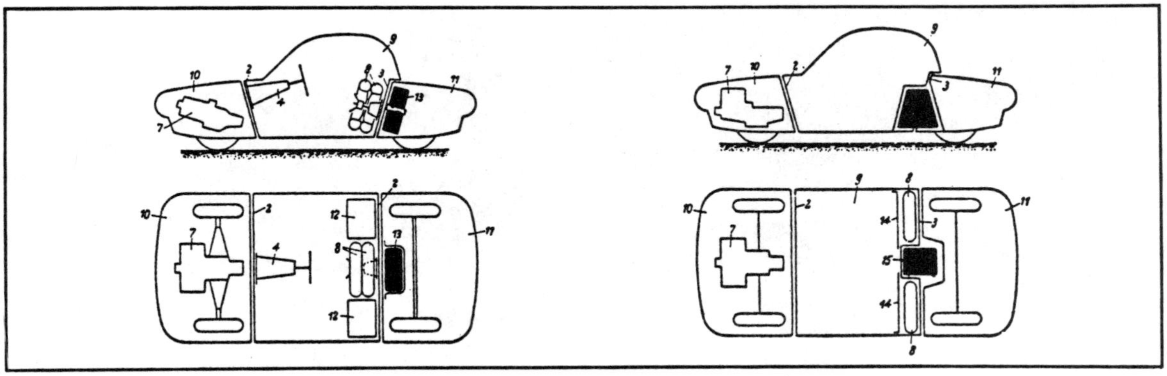

Fig. 6-5. A Mercedes-Benz patent from 1956 shows the protected position of the fuel tank.

in 1963. This costly device for simulating crash conditions in the laboratory subjected complete structures to 40-g acceleration forces.

ARE SMALL CARS LESS SAFE?

Crashing big cars into small ones became a popular pastime at several research establishments (Fig. 6-6). One test involved a collision between a 1969 Volkswagen sedan weighing 1800 pounds and a 1957 Ford sedan weighing 3400 pounds. Each vehicle was traveling at approximately 30 miles per hour. Anthromorphic dummies were placed in the right front seats of the two vehicles to measure the shock loads.

A lot of experts were fooled by the results that proved the greater safety of the big car. They concluded it was because one car was heavier and they attacked small cars because they were light. But what mattered was simply the greater deceleration distance (crash zone) of the bigger car.

TEST METHODS AND INNOVATIONS

American car makers and foreign ones selling cars in the United States had to build barrier-test fa-

Fig. 6-6. Small cars tend to receive more damage in a collision with a bigger car.

Fig. 6-7. The barrier test after the impact.

cilities and submit certified test results to the National Highway Traffic Safety Administration before their cars could be legally sold. After the head-on barrier test (Fig. 6-7) came the side-impact test (Fig. 6-8). Rules setting out limits for body deformation in rollover tests (Fig. 6-9) were to follow. Research on restraint systems were conducted with dummies on test sleds (Fig. 6-10), and a docket was opened for rule-making on pedestrian protection (Fig. 6-11).

General Motors did not wait for a law on collapsible steering columns and developed its own, which was standardized in 1967 (Fig. 6-15). Later, Volkswagen developed a 4-inch-long collapsible hub (Fig. 6-16) that would allow the steering wheel to participate in the energy management in the "second collision" (occupant against the inside of the car).

BUMPERS FOR OR AGAINST SAFETY?

In the mistaken belief that stronger bumpers could help reduce loss of life in traffic accidents, the NHTSA passed a rule demanding front bumpers that would withstand a 5-mph impact without damage to the car. That became effective with the 1973 models. It increased the aggressiveness of the

Fig. 6-8. The side-impact test.

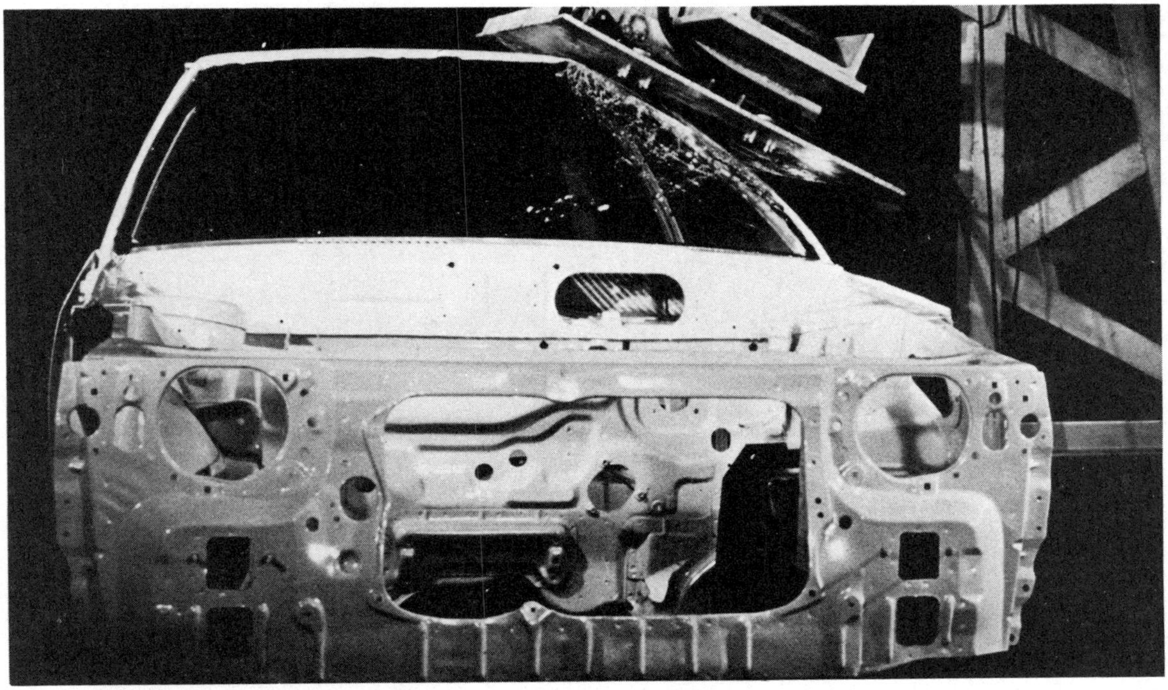

Fig. 6-9. A simulated rollover test.

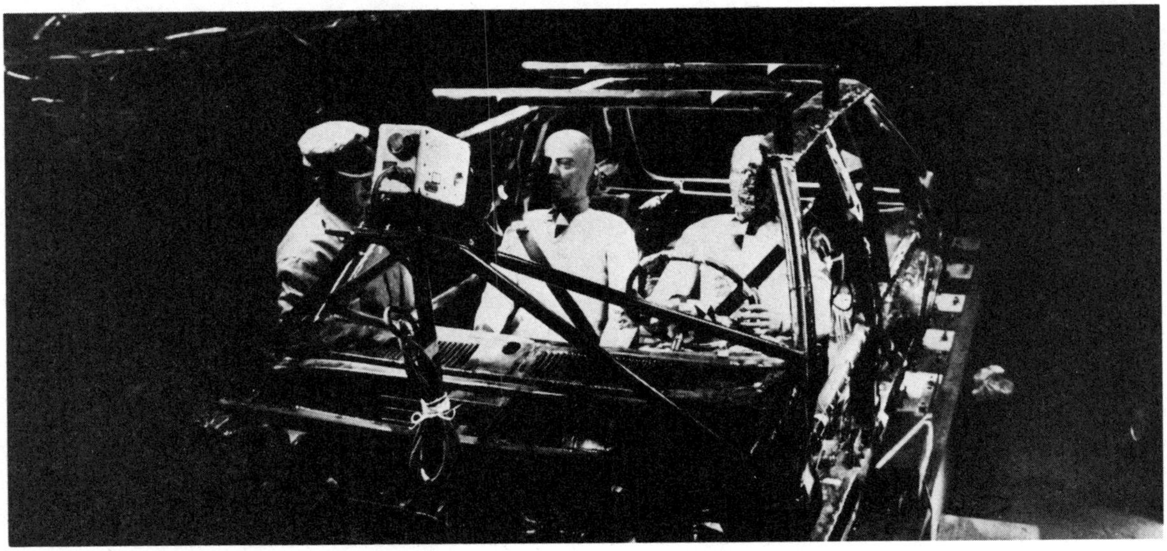

Fig. 6-10. A test sled with anthropometric models ("dummies").

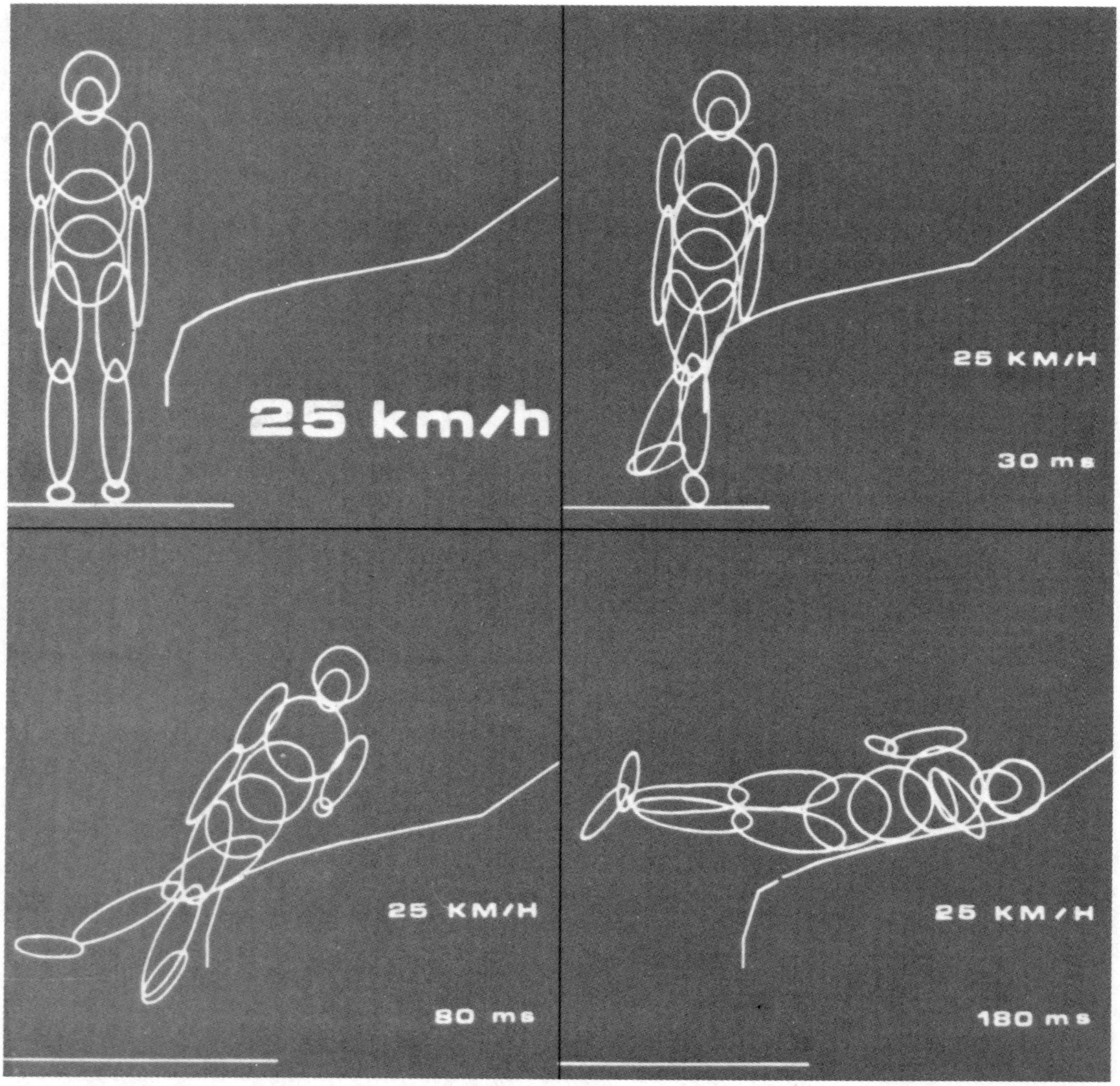

Fig. 6-11. A computer simulation of an adult hit by the front of a car at 15.5 mph impact speed.

owners of the cars, and collision repair costs skyrocketed.

"You know what they should have done?" said a New York-based friend of mine. "No. What?" I obediently took the bait. "Outlaw bumpers," he bellowed. "If nobody had them, they'd be a darn sight more careful not to scratch their own paint, dent their own fenders, break their lamps, bash in their precious grilles or mangle their trunk lids."

He had a point; that's easy to see. With the

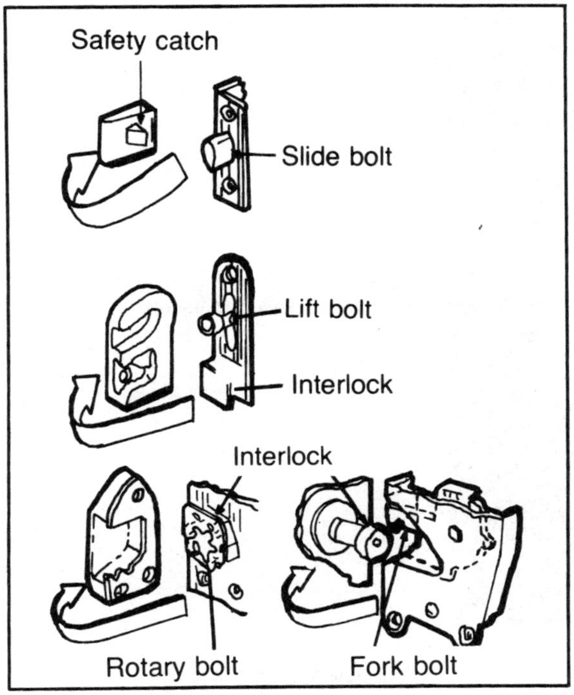

Fig. 6-12. The evolution of GM door locks during the 1931-1958 period.

same kind of logic a police officer once argued that the act of carrying a radar-warning device in your car was proof of intent to break the law. Therefore you could conclude that having bumpers on your car constitutes proof of your will to commit agressions against other cars.

I had visions of a scene from Bob and Ray's Rorschach, their spoof on the popular TV series *Kojak*, with the boss roaring: "Lopez, Steinberg, Caruso! In here on the double! I want you to cover all highways and byways in this city. Lopez, you take the tollgates. Steinberg, block all bridges, and Caruso, the tunnels. If anybody tries to get into town with a bumper on his car, shoot to kill!"

It took the second- and third-generation bumpers to meet the standards without adding greatly to the weight and length of the car, and without causing excessive damage to the impacted vehicle in a collision. About the same time, the uselessness of the barrier crash test was exposed.

VALUE OF REAL-LIFE DATA

Renault's safety experts, Claude Tarriere and Philippe Ventre, set about to look at real accident statistics and found that, in France, head-on colli-

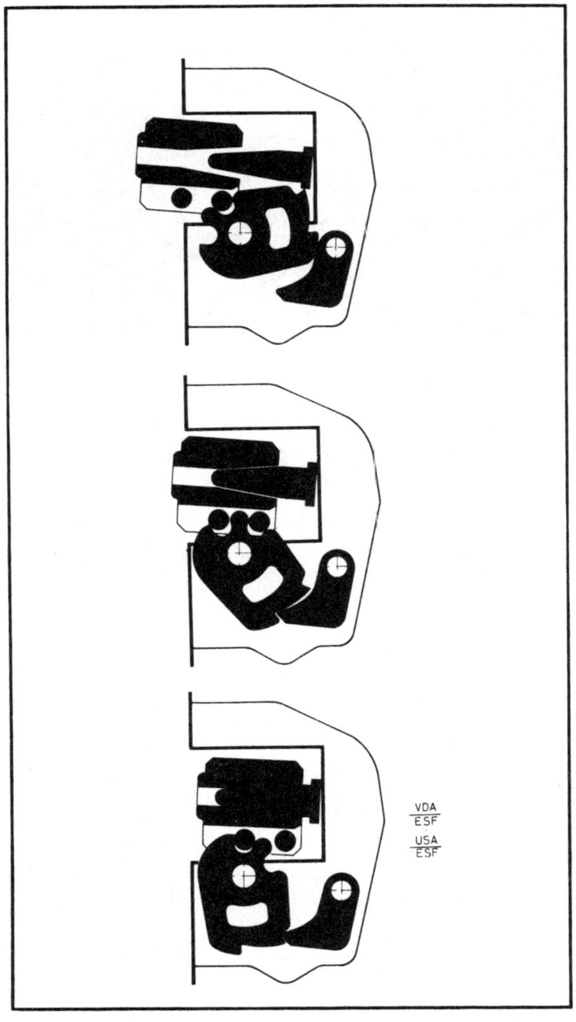

Fig. 6-13. The latest version of the Mercedes-Benz door lock.

Fig. 6-14. The safe arrangement of an interior door lock and handle.

sions accounted for two-thirds of all accidents, causing two-thirds of the reported injuries and fatalities. That was in agreement with American statistics.

But the Renault studies soon revealed basic differences. Full-width frontal impacts (as in a barrier crash) were found to be relatively rare. They acounted for no more than 20 percent of all frontal collisions. Most of the fatal accidents that killed or maimed peopled involved in frontal impacts occurred in asymmetrical vehicle-to-vehicle contact. Two main types of frontal collisions were defined. In one case, the two cars were directly opposed, but offset laterally, so that only a portion of the front end on each car received the full impact load.

In the other case, one car was traveling at an angle to the path of the other so that the full impact force was taken by a corner, in a diagonal direction to the car's center line. These two main types of frontal collision accounted for more than half of all fatalities and injuries.

In a very learned paper loaded with highly scientific data and explanations, Philippe Ventre submitted to the NHTSA that the barrier test would make more sense if the face were oblique to the impacting car's line of travel. Even that does not account for the driver's last split-second efforts to take evasive action. Heavy breaking tilts the car nose-down, to some extent, or front wheels cocked to one side introduces a lateral component into the dynamic energy stored in the car.

The study also repeated what many others had pointed out earlier. In a car-to-car collision, both vehicles participate in impact-absorption. In a barrier crash, the impacting car takes the full load. And that is totally unrealistic. You don't need an engineering degree to understand that the barrier test was worthless in terms of reducing the number and consequences of traffic accidents.

Renault's solutions were built into the structure of the new R-30, introduced in 1975. This led to modifications on the R-5 and other models.

The construction of experimental safety cars had been sponsored by the NHTSA, since 1970, with the main contractors outside the auto industry. This had its background in preceding attempts to build cars with maximum secondary safety.

THE SAFETY-CAR PIONEERS

Frank Crandell was chief engineer of the Liberty Mutual Insurance Company of Boston, Massachusetts in 1952, when he made a proposal to the management for building a "crash-proof" car. The idea was accepted and he had the car ready in 1957. It was built in collaboration with Cornell Aeronautical Laboratory of Buffalo, New York. Among its architectural features was a driver's seat placed in the middle and the car had no steering wheel or column. Handles operated a hydraulic device that was intended to turn the wheels.

An updated version was built, four years later, based on a 1960-model Chevrolet Bel Air. In this Liberty Mutual Survival Car II, all occupants sat in capsule seats, fixed in place, with four-point harness. It had a flexible steering column and a thick interlayer in the laminated windshield. A fire extinguisher was included in the equipment. It had a wraparound windshield with the A-post standing far back, making entry to the front seats awkward.

The first European safety-car proposal was built by private industry (without sponsorship of any kind). It was called Sigma, and it was the creation of Pininfarina (Fig. 6-17). It was not a running car, however, because it had no engine or drive

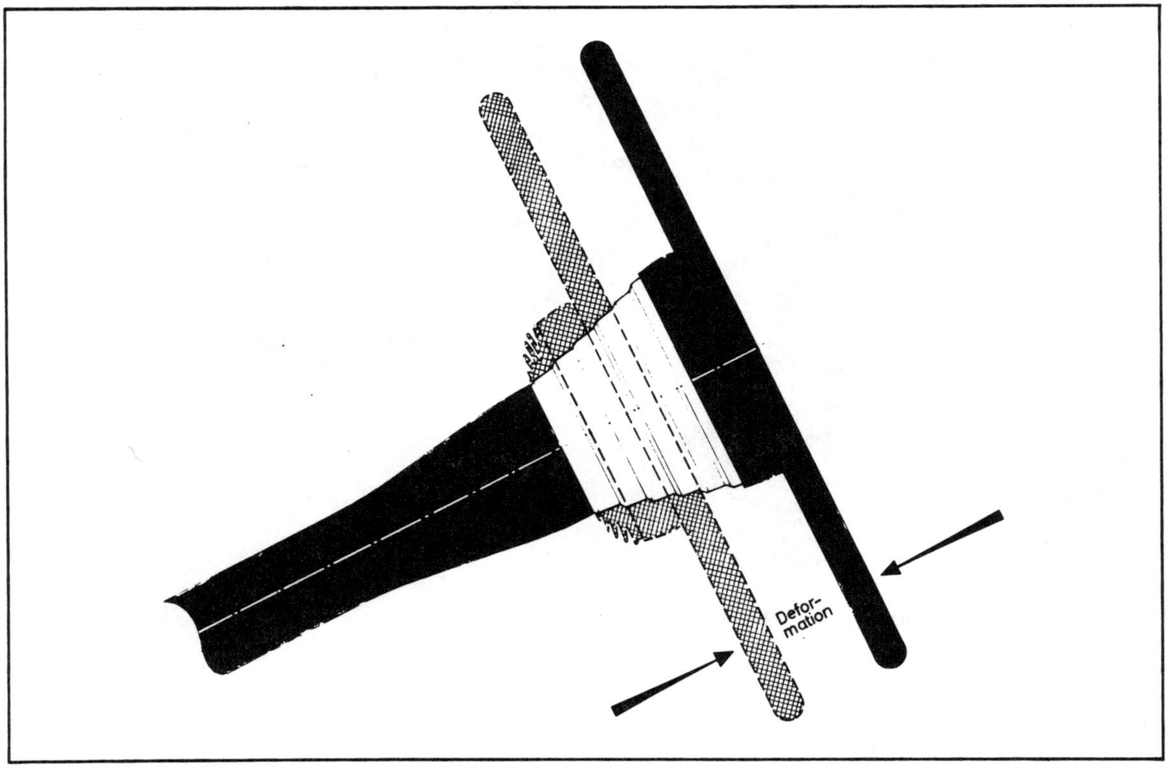

Fig. 6-15. The principle of the collapsible steering column.

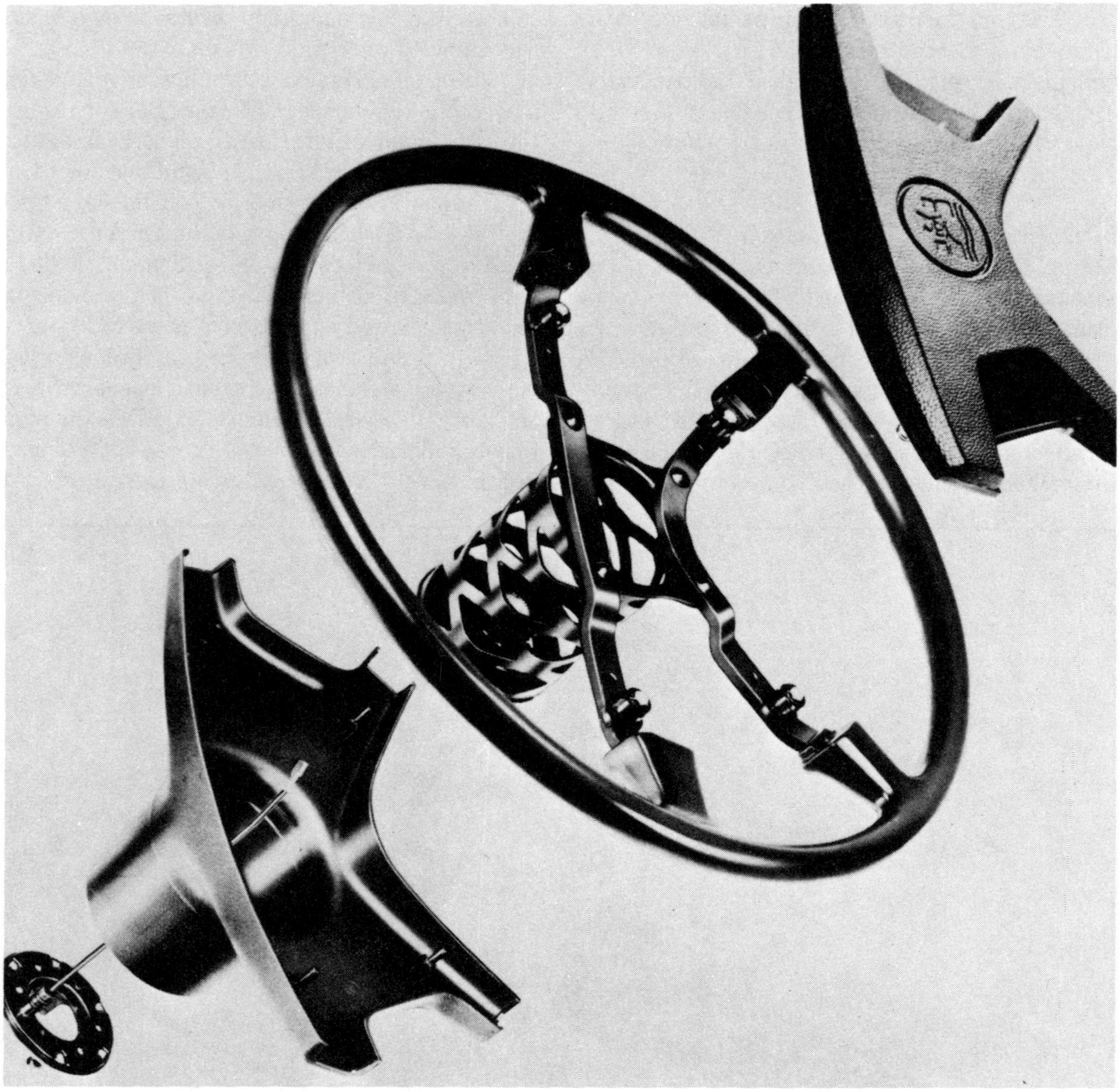

Fig. 6-16. A production version of the VW collapsible steering column.

train. The designer had concentrated on frame and body. Pininfarina pointed out that the design was flexible enough to accommodate almost any kind of engine, and stated that patents held by a number of scientific organizations and such car manufacturers as Rover, Citroen, and Mercedes-Benz were intrinsic to the concept.

The Pf Sigma was a two-door, four-seater

Fig. 6-17. The Pf Sigma safety car.

sedan of average dimensions. The central part was very rigid while the front and rear parts were deformation zones compressed into minimize size. Sliding doors were engineered to stay closed in lateral impacts as well as frontal collisions. Longitudinal impact loads tend to close the doors harder instead of trying to force them open (as is the case with swing-open doors).

The exterior surface was smooth and rounded to avoid any sharpness or angularity that could cause injuries to pedestrians. The bumpers were partly covered with rubber and had an enlarged face in order to reduce specific pressure in impacts.

Seating layout was conventional, but each oc-

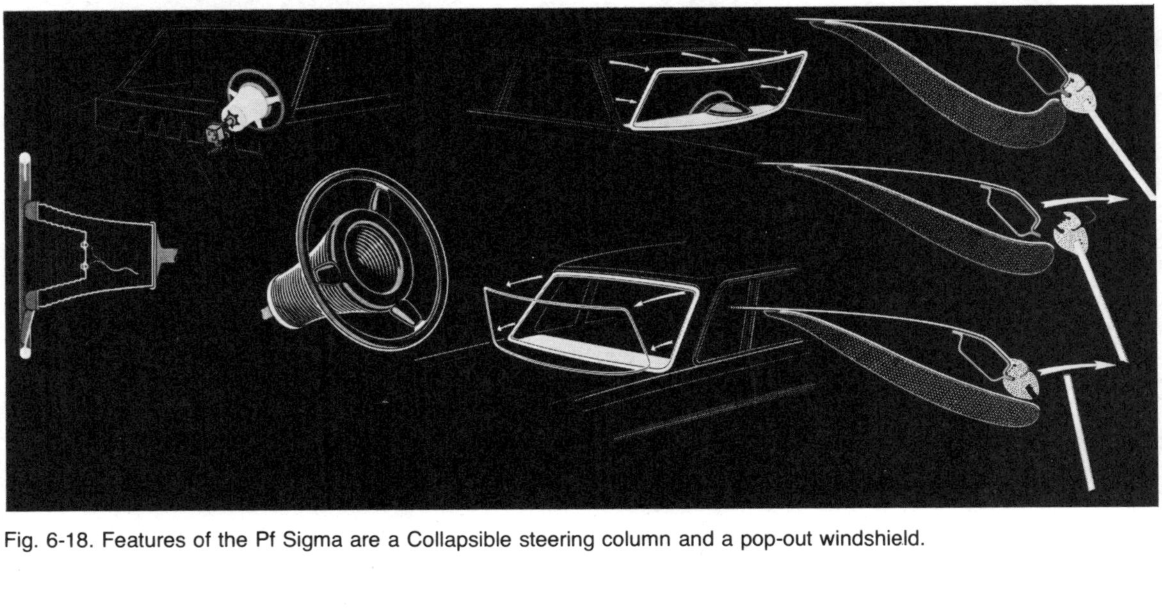

Fig. 6-18. Features of the Pf Sigma are a Collapsible steering column and a pop-out windshield.

cupant had his own bucket seat of anatomic design. All seats except the driver's had an integral headrest. All inner parts of the body were thickly padded. The instrument panel was a padded metal structure mounted with elastic elements on the cowl to let it slide away from the driver in a frontal impact.

The steering system was collapsible (Fig. 6-18), with a steering gear mounted high up and as far from the front end as possible, and a very short column. The steering wheel was mounted on a corrugated sheetsteel cone, capable of absorbing impact energy, while the hub was heavily padded.

Windshield and rear window frames were of the "pop-out" type (Fig. 6-18). At that time, most European experts maintained that it was not the job of the glass to play a role in preventing ejection of the occupants. Their prime concern was mainly to prevent violent contact of their heads with the hard glass, and to provide an exit in case the doors could not be opened after the accident.

The Americans took the opposite view that has come to prevail worldwide since the development of plate glass with high flexibility and body reinforcement techniques that minimize the risk of jammed doors.

THE NEW YORK STATE SAFETY SEDAN

In 1965, Senator Edward J. Speno proposed and the state of New York passed a bill allocating $100,000 to a safety-car program. A contract for the design and construction of a prototype was signed with Republic Aviation in Farmingdale, Long Island.

The design objectives were set out by John O. Moore, a thoughtful engineer, then director of research, Department of Motor Vehicles, New York state. Republic's program manager was George Hildebrand, who had once worked for LeBaron. Stanislaus Dubowski was named project engineer.

The design objectives for the passive safety of the New York Safety Sedan (Fig. 6-19) included a capability to sustain frontal and rear-end impacts at 50 mph, side impacts at 40 mph, and rollovers at 70 mph, with no more than half the severity of injuries

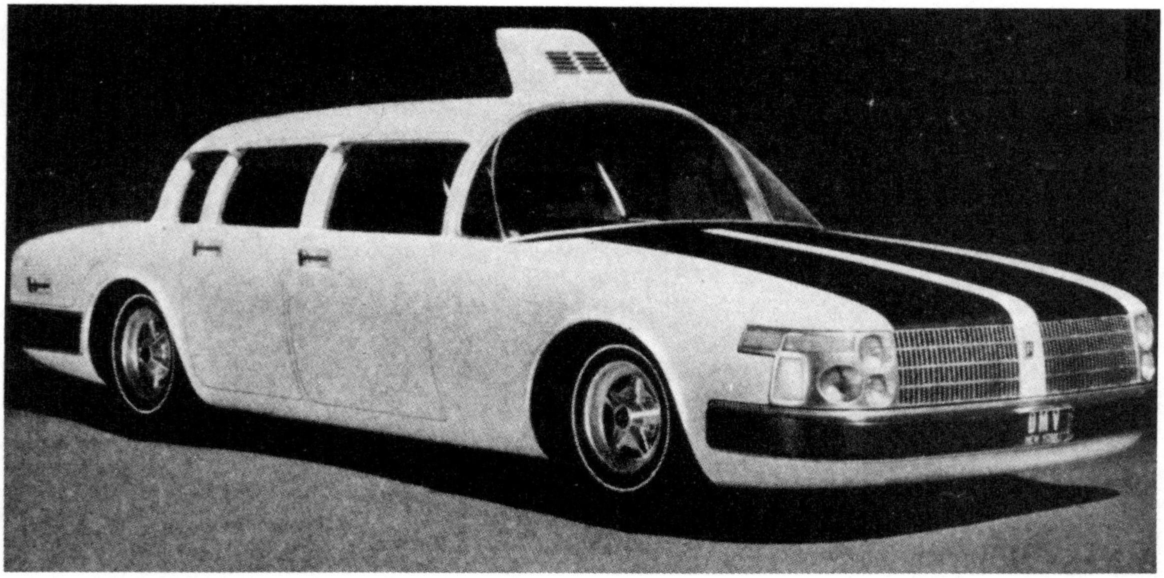

Fig. 6-19. The New York State Safety Sedan.

that would be inflicted on the occupants of a regular production car in the same situations.

The car was built on a 123-inch wheelbase. It was powered by a front-mounted, 260-hp Ford V-8 driving all four wheels via an automatic transmission and a Ferguson self-equalizing drive system.

The body was built on aircraft principles. Therefore all parts of the structure had to be functional. Instead of a conventional frame, the body was built up on a platform that also carried the mechanical elements. The designers called the body a semimonocoque because parts of the skin helped carry stress loads.

THE SURVIVAL CELL

Three fabricated steel hoops representing the A- B- and C-posts, fastened to the platform at the bottom and linked by two longitudinal roofbeams, made up the inner cage.

The windshield was a single piece of laminated glass held in a steel frame extending from the A-post hoop. The car had a conservative windshield slope and relatively flat glass linked with the side glass ahead of the A-post. There was no body structure directly below the windshield so that the 20 inches from the windshield base to the A-post would serve as part of the impact-absorbing zone.

The B-post hoop incorporated a heavy cross-member reaching from the top of the doorsill to the belt line. It was intended to prevent unbelted rear-seat passengers from being thrown into the back of the front seats. The C-post hoop was filled in below beltline height to act as a bulkhead to strengthen the fuel tank and guard against penetration into the passenger compartment of objects carried in the trunk.

The spare wheel was placed at a 30-degree (from horizontal) tilt on a recess in the platform above the rear suspension. The trunk lid was hinged on the side with an opening only to the curb side.

Multiple, tapered dovetails engaged the door structures with the hoops, roof beams, and platform to assure the integrity of the total structure in a collision, and to maintain full load-transfer capability.

The doors were two-piece designs. In addition to the normal door, a section of the roof was hinged at the roof beam and opened to improve the access (partly offsetting the drawback of the high sills).

The dashboard was heavily padded, and the interior surface of the doors was covered with energy-absorbing material. Door handles and window regulators were mounted in niches on the upper surface of the armrests to prevent accidental operation.

The front seat was fixed in place except for vertical adjustment. The car had steering wheel and pedals with adjustable mounting systems. The steering assembly featured a four-spoke wheel with a 10-inch-diameter flexible rim, a large hub with padding, and a collapsible steering column.

The front bumper automatically extended forward a full foot and locked in place when the speed of the Safety Sedan exceeded 37 mph. The bumper was connected to a pair of hydraulic cylinders to absorb some impact energy. This hydraulic absorption of energy is the most efficient known in terms of displacement and force, according to John O. Moore. The front bumper was arranged so as to act as a swinging gate on impact with any off-center object, such as a tree or bridge abutment, and thereby deflect the car to one side.

A large zone ahead of the radiator was taken up by a backup system for the bumper. It consisted mainly of triangulated steel box-section subframe bolted to the front of the platform. Seating capacity was restricted to four occupants; this was primarily due to the desire to provide side protection. High and wide door sills made up the main line of defense against lateral impact.

VISUAL INFORMATION

What most of the people who saw it probably re-

Fig. 6-20. The AMF safety car.

Fig. 6-21. The Ford safety car.

Fig. 6-22. The General Motors safety car.

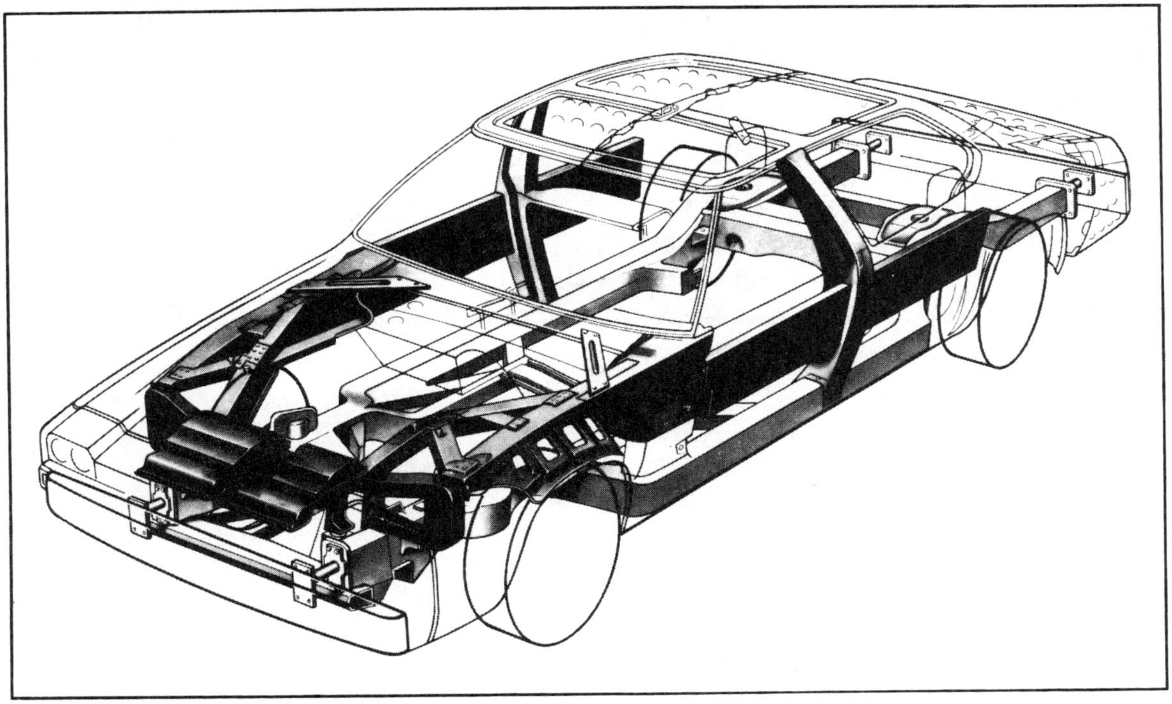

Fig. 6-23. The structure of the GM safety car.

Fig. 6-24. The front seat of the GM safety car.

member best about the car was its use of a rear-view mirror mounted in a streamlined periscope extending from the A-post hoop. The image from the periscope mirror was reflected, without distortion, on a screen mounted in the instrument panel. The periscope mirror was claimed to be less sensitive to glare from the headlights of following cars.

It was intended to give an unobstructed view of the traffic behind, regardless of the number of passengers carried, their positions, and the amount of clutter carried inside the car. The periscope also carried a forward-viewing mirror to enable the driver to spot a slowing down in the traffic flow.

A triple-light signal system adorned the periscope pod to give information to following drivers. A green light indicated steady speed, a yellow light indicated slight deceleration, and a red light indicated heavy braking.

The New York State Safety Sedan was stated to weight 5,100 pounds, but it is likely that the true figure was higher. In the days of inexpensive gasoline, that weight did not seem unreasonable.

Since then, we have found that the cost of setting 5000-pounds-plus in motion to carry a maximum of four people is simply prohibitive. The benefits (on a statistical basis) are not worth the price.

It is now accepted that, in some accidents, survival is not feasible—within our means. A whole generation of experimental safety cars appeared, however, before that realization penetrated the brains of the NHTSA staff.

One safety car was built by Fairchild-Hiller, to designs by George Hildebrand, using a Plymouth body shell and chassis as the base. It had massive bumpers that extended 12 inches when the car reached a speed of 25 mph.

Another safety car was the AMF project (Fig. 6-20). A structural frame made of high-strength steel surrounded the passenger compartment. Aluminum honeycomb composite panels were built into all doors (making the doors thicker and cutting down inside room). Front seats were fixed in posi-

Fig. 6-25. The back seat of the GM safety car.

Fig. 6-26. Fiat's ESV 2500 prototype.

tion and pedals were adjustable. AMF also adopted the periscope mirror.

Not qualifying for federal funds, the auto industry nevertheless built experimental safety cars and presented them to the NHTSA for testing. Ford started its program in 1970. They took the approach of using a standard production car as a base and modifying it rather than to design a car from the ground up. See Fig. 6-21.

The body front end featured corrugated aluminum sheet metal to form a complete hoop after the top section had been bolted on. Additionally, the front end side structure contained longitudinal corrugations toward the firewall for load support. The passenger compartment had major structural components intended to improve compartment integrity under all crash modes.

The Ford used a fixed front seat permitting the installation of a strong lateral beam between the B posts. Its front bumper system incorporated a high-strength steel bar supported by two hydraulic struts attached to the frame.

GM's $250,000 CAR NOBODY WANTED

The Ford ESV ended up with a weight of 5675 pounds, compared with the 4300 pounds of the standard 1973-model Galaxie 500. The General Motors experimental safety car started out at 4700 pounds (Fig. 6-22). It was a five-passenger, four-door family sedan with a 124-inch wheelbase, using the Oldsmobile Delta 88 chassis.

The frame was a massive structure with diagonal bracing. Body reinforcements (Fig. 6-23) were huge and heavy. GM's bumper was a mammoth shock absorber made up of the reinforced bumper car, the car frame, two telescoping absorption units, and an extender system to reset the bumper after a low-speed impact.

The roof was cantilevered on the B-post so that the A-posts could be eliminated and the door glass made to meet the windshield totally without obstruction of the view. All body sheet metal was aluminum; that explains why the GM was so much lighter than Ford's.

Other special features included: a cross-car structure or "credenza" (Fig. 6-25) designed to protect unrestrained rear seat passengers from impacts up to 30 miles per hour; 12-cubic foot rear trunk accessible from either side; single-pivot front seat (Fig. 6-24) with movable pedals and a "message center" instrument panel above the steering wheel; and fixed side glass to reduce the risk of ejection during rollover. Power-operated access sections gave openings for tollgate payments.

FIAT PROVES A POINT

European and Japanese car companies were brought into the safety-car program; most of them selected sizes conforming to their current production models.

Fig. 6-27. Fiat's ESV 2000 prototype.

Fig. 6-28. Fiat's ESV 1500 prototype.

Fiat built three different sizes of safety car. All of them supported the contention that beefing up the structure to guarantee survival space for the occupants would result in deceleration loads beyond human tolerance at the NHTSA crash speeds of 40 and 50 mph.

The ESV 2500 prototype (Fig. 6-26) was based on the Fiat 124, with front-mounted engine and rearaxle drive. Body reinforcements added 8 percent to the length and 43 percent to the weight.

ESV-200 (Fig. 6-27) was based on the Fiat 128, with front-wheel drive. Big front and rear bumpers protruded about 8 inches from the body shell, and the side band thickness was about 2 inches. Three bracing structures were put into the body: one at floor level, a second at the belt line, and a third in the roof. It weighed 2750 pounds; that was 45 percent more than the 128.

Smallest of the three was the ESV 1500 (Fig. 6-28) based on the Fiat 500 (which had a rear-mounted engine). Several versions were built, with slightly different priorities, and weight increases from 35 to 50 percent above the standard Fiat 500 (variations of 175 pounds).

THE STATE OF THE ART

Leadership in safety-car design probably belonged to Mercedes-Benz, but none of its prototypes met the NHTSA demand for survival in a 50-mph barrier crash. In its progression from the ESF-05 of 1971

Fig. 6-29. The Mercedes-Benz ESF 22.

Fig. 6-30. The first VW experimental safety car.

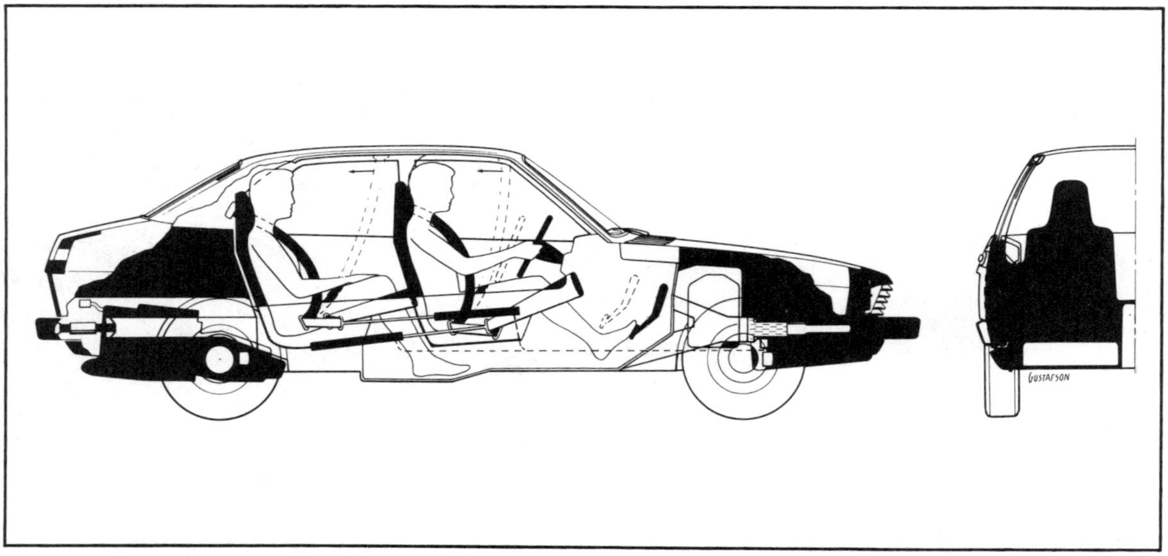

Fig. 6-31. The ESVW-1 had a rear engine and "passive" belts.

Fig. 6-32. The ESVW-2 was closely based on the VW Golf (Rabbit).

via the ESF-13 of 1972 to the ESF-22 of 1974 (Fig. 6-29), the safety car concept came closer and closer to the actual production model that had evolved with full benefit from safety research.

Volkswagen's first entry (Fig. 6-30) was built in April, 1972, and was based on the 412 chassis and structure (with rear-mounted flat-four engine). On its 110-inch wheelbase, it weighed 3200 pounds, and VW engineers stated that it would enable its occupants to survive a 75-mph impact with a parked car of twice its own weight.

The ESVW is notable for having introduced the "passive" belt system (Fig. 6-31). Because Volkswagen was converting its production program to front-wheel drive in great haste, the first ESVW rapidly became a museum piece. A second prototype was built based on the front-wheel drive NSU Ro-80 chassis and body shell.

After the Golf (Rabbit) went into production in 1974, Volkswagen used it also as a basis for a new safety prototype named ESVW-II (Fig. 6-32). Very close to the production model in its dimensions, it weighed only 1978 pounds, and exceeded the 40-mph crashworthiness criteria set for it.

Chapter 7

Body Engineering and Materials

MODERN AUTO BODIES ARE BUILT ON THE unit-construction principle. Such a body is an enormously complicated assembly of steel pressings (Fig. 7-3) forming a base structure for attachment of the exterior surface panels ("skin"). The whole body is built up section by section. These sections are formed in high-power presses by stamping flat sheet steel into shape between a pair of heavy steel dies with corresponding contours.

A sheet of steel, used in the sound-effects room, can be made to resonate like thunder. Yet when it is formed into a well-designed pressing, it assumes a strength and stiffness that is very high for its weight. The thickness of steel body panels is less than a 1/20 of an inch in most cars. It is therefore necessary to stiffen many of the panels.

Unit-body reinforcements at crucial points do not usually take the shape of a heavy steel bar or rod welded onto the sheet metal. Instead, most body engineers use light steel pressings of considerable cross section, boxed or U-profile, thereby providing great bending stiffness in one or more planes for a very modest increase in weight (Fig. 7-4).

When the roof section, floor pan, cowl structure, side panels and other supporting sections are all welded together, the resultant structure is extremely strong and capable of receiving the car's mechanical units and taking the suspension, drive- and brake-stress loads.

Because it uses the boxlike shape of the body itself as the basic structure, it offers ample resistance to deflection. Additional rigidity can be provided by cross members welded under the rear seat area and to the box side rails at the extreme rear end of the body. Reinforced roof bows give greater strength to the roof panel. The radiator support, wheel housings, and cowl panels add structural strength to the forestructure and body assembly.

THE BODY-PRODUCTION CYCLE

The machinery needed for mass-production of such body shells is expensive, takes up a lot of space, and output must be checked after nearly every welding station. Body-shell assembly can be fully automated. A maze of machinery surrounds the body structure being made at Ford's Wixom assembly plant (Fig. 7-1). The massive arms of the pillar "buck" swing inwards, and steel fingers grip the body parts to be joined.

The product of the full sequence of assembly and welding processes is called "body-in-white" because its surfaces are still untreated. Application of antirust coatings and paint come much later in the production cycle (Fig. 7-2).

Citroen's body assembly at the Aulnay plant starts by preparation of the forward unit and assembly of the CX (Fig. 7-5) base plate (forward unit, floor and rear part). The base-plate is assembled on an auto-welding transfer machine 203 feet long

Fig. 7-1. An auto body structure in a welding jig.

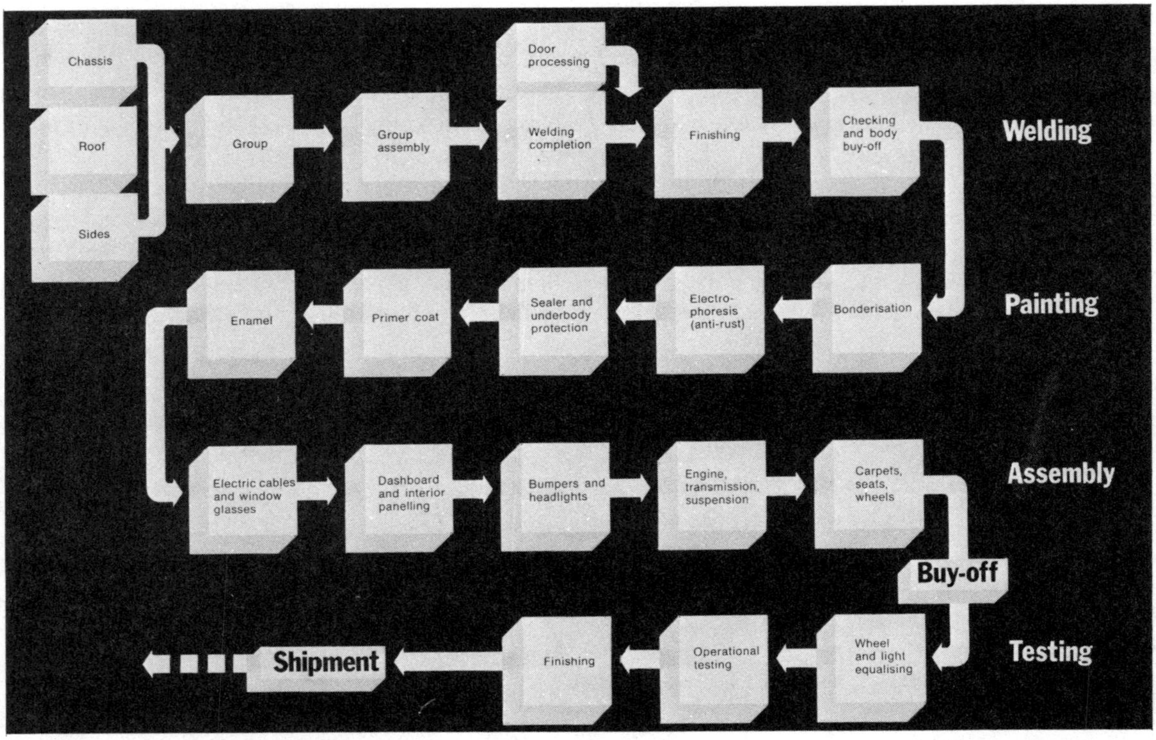

Fig. 7-2. A diagram of the production cycle.

(worked by two operators). It makes 217 spot welds in 1 minute 25 seconds, allowing a production rate of 48 base-plates per hour. Once the body is unloaded, assembly continues in the following order:

☐ Base-plate and side panels; pneumatically controlled robots carry out 44 spot welds.

☐ Windshield-frame assembly; 72 welds.

☐ Roof assembly; an electromechanical feeler detects the type of body concerned and issues an order to the loader, which will then collect and position the corresponding roof parts.

☐ Fender assembly; at this stage, an automatic feeler machine also checks the base-plate for flatness.

☐ Door fitting is automated by means of jig-gantries that position and maintain the doors while the operator fixes the hinges.

☐ Hood and trunk lid are bolted on. Assembly is complete.

BODIES FOR OPEN-AIR MOTORING

Citroen is one of the many mass-producers that do not build convertibles. The European car makers who have convertibles in their model ranges usually farm out their body production and assembly to a specialist coachbuilder. Volkswagen farms out its convertibles to Karmann, Peugeot and Talbot to Pininfarina, and Fiat to Bertone (Figs. 7-6 and 7-7) and Pininfarina.

There are two main reasons for this: Convertible bodies require special reinforcements to compensate for the loss of a roof structure, and they are

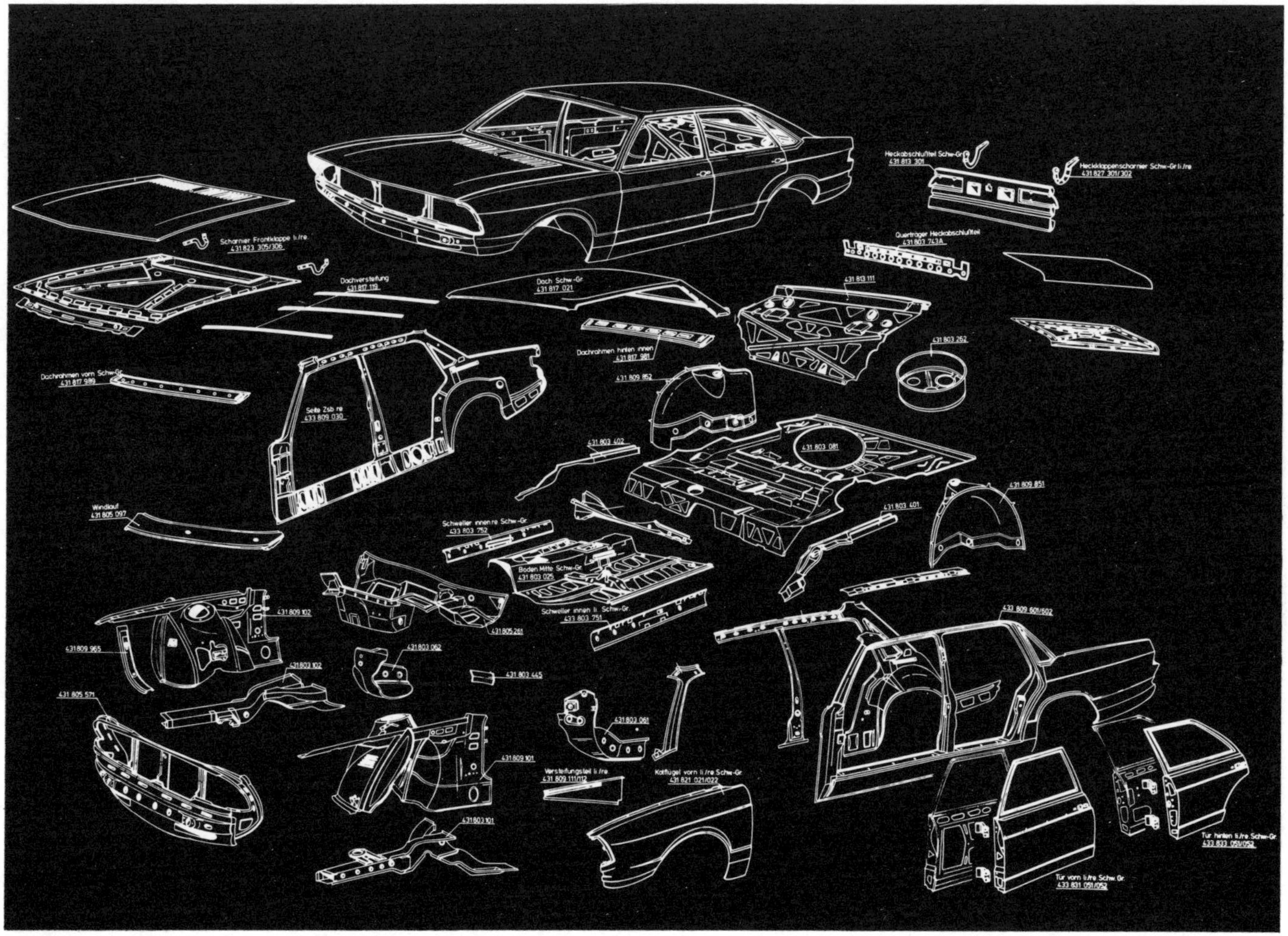

Fig. 7-3. The separate pressings that go into making an Audi body shell.

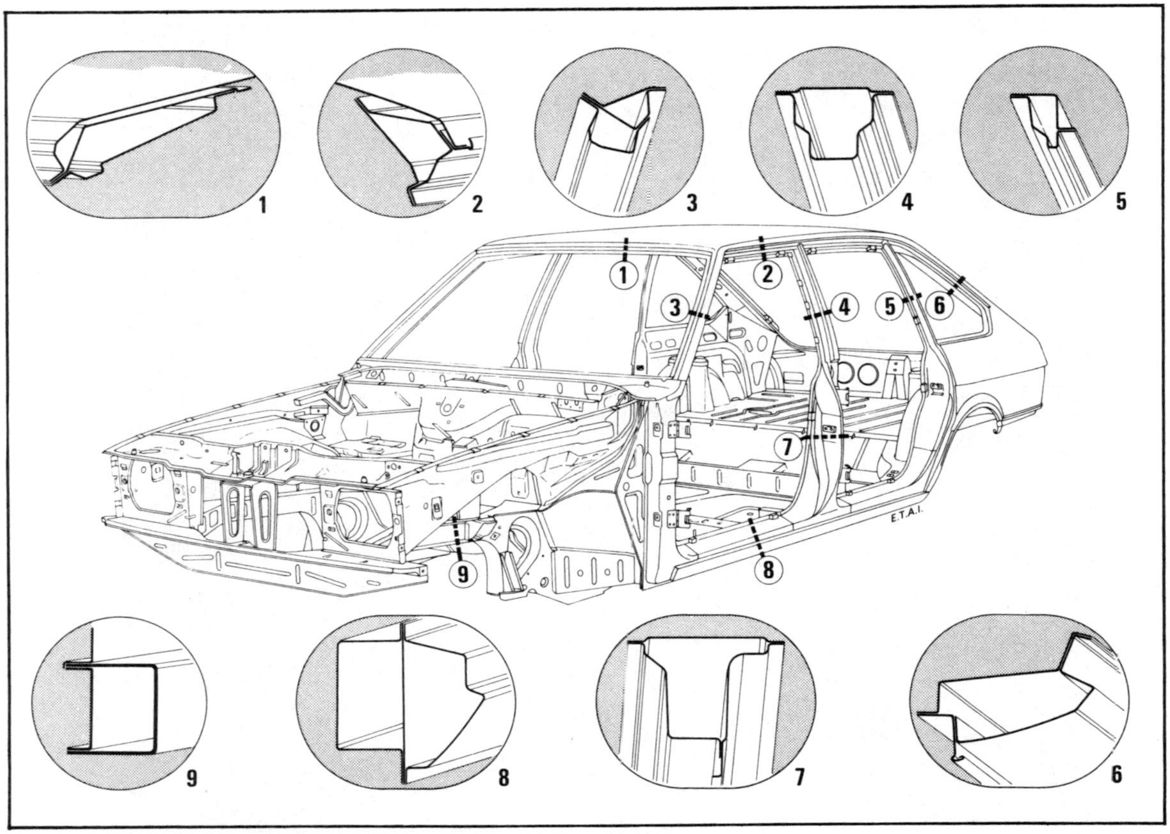

Fig. 7-4. How reinforcements are made without adding massive steel parts.

more expensive to produce (which keeps the market fairly small). Production volume is therefore restricted to low numbers that do not justify mass-production tooling investment.

The soft-top requires precision-cut fabric, a robust linkage and a solid locking mechanism against the windshield frame. It also needs storage space, when folded, and, in the case of power tops, space for the power cylinders, electric motors, or hydraulic accumulators.

The typical folding soft-top linkage consists of three sections of right-side and left-side roof rails, and a front roof rail. They are connected by bolts, hinges, and a series of connecting links and bows. The entire linkage system is hinged in the body at the C-post base. The front roof rail is locked at the windshield header by two hook-type locks that are an integral part of the two locking handles. Convertible top trim incorporates a spring-loaded, hold-down cable along the left and right roof rails. These cables are installed through retaining pockets in the top material and fastened at the front and rear side rails by attaching screws. The cables are designed to hold the top material tight against the side roof rail stay pads. This minimizes air leakage between the top material and the stay pads.

The part that holds the backlight is called a curtain. All curtain assemblies have an elastic webbing as an integral part of the curtain's upper valance. The elastic webbing is located in the upper corners of the curtain. This webbing reduces tension on the zipper assembly at the radius and offers improved zipper operation. The newst curtain assemblies include a hard-curved glass backlight. This backlight is dielectrically bonded to the vinyl back curtain. Other backlights are made of pliable plastic.

Particular safety considerations relating to convertibles center on rollover protection. The most-favored solution is to install a stylized B-post rollover bar (Fig. 7-8).

MATERIAL SUBSTITUTION

Iron and steel account for about three-fourth of an automobile's weight; it is easy to see, then, why it is important to investigate carefully the uses of automotive materials to find ways to reduce extra weight as much as possible. Careful selection of

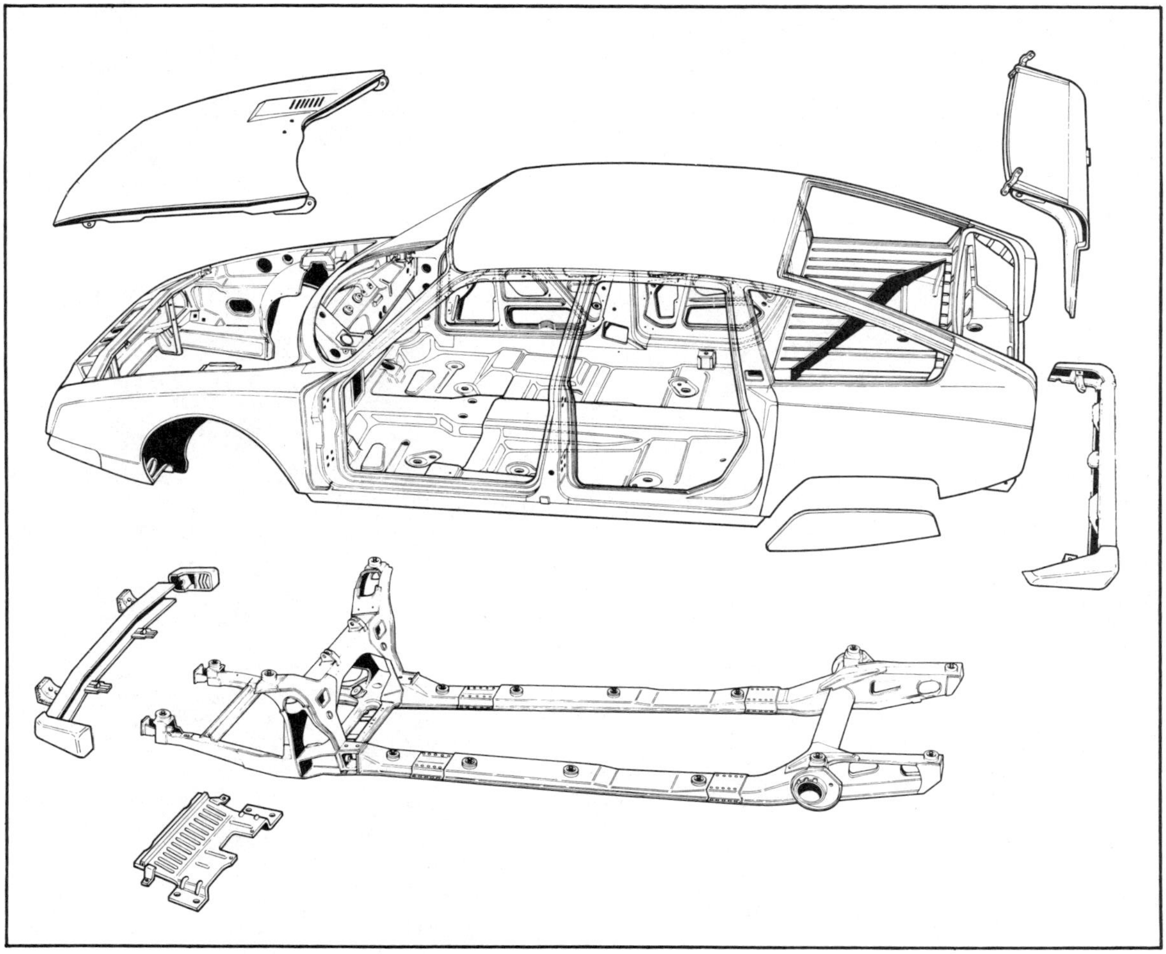

Fig. 7-5. The Citroen CX body shell and frame.

Fig. 7-6. A Fiat Ritmo convertible with the top up.

Fig. 7-7. A Fiat Ritmo convertible with top down.

Fig. 7-8. A Talbot Samba convertible with integral roll-bar.

materials can contribute to weight reduction. Two major trends are the adaptation of stronger materials (e.g., high-tensile sheet steel), and the use of lighter materials (e.g., plastics, aluminum).

The tensile strength of regular sheet steel is 20 tons per square inch. High-tensile sheet steel, however, is rated at 40 to 55 tons per square inch. That means thinner plates can be used, yet still maintain the needed strength. Thus the total weight of the sheet steel used throughout an automobile decreases. High-tensile sheet steel is now being used in doors, for bumper reinforcement, and elsewhere in the car body.

Plastics are very light materials (around one-fourth to one-eighth the weight of steel). Because plastics are very easy to mold, they can be made simply into even very complicated shapes, and they lend themselves to efficient production as well. Those qualities make the study of plastics for future weight-reducing applications very important. Plastics also offer advantages in terms of their noncorrosive, shock-absorbing and antivibration properties. Those are qualities obviously very useful in auto parts design.

Plastic is already being used in a wide range of automotive applications as well as for interior equipment and fittings. It is used for radiator grilles, in tail lamps, meter housings, clusters, fans, bumpers, door handles, and fluid reservoir tanks. Various types of gears, bushings, and so forth are also now made of plastic.

Complete plastic bodies are possible. The Chevrolet Corvette has been built with a plastic body on a conventional steel frame since 1953. Frames can be made of plastic (composites) but that means buying lightness for 8 to 10 times the price of a steel frame. Unit-construction with plastic bodies is also possible.

The 1958 Lotus Elite had a fiberglass-reinforced plastic body in which the mechanical components were mounted just like in an all-metal body shell. The Elite front end had a steel subframe in a sandwich structure to carry the engine, but the rear suspension system was anchored right in the body.

Experiments with wood auto bodies are rare in modern times, but the 1960 Marcos GT from Luton, England, was built with all-wood construction—including a plywood frame. It was designed by Frank Costin, an aircraft engineer, familiar with the de Havilland Mosquito.

Birch, fir, and spruce were used throughout. The frame was built as an assembly of torsion boxes. All joints were glued, and all wood was impregnated under pressure with protective and

Fig. 7-9. The Porsche 20-year car.

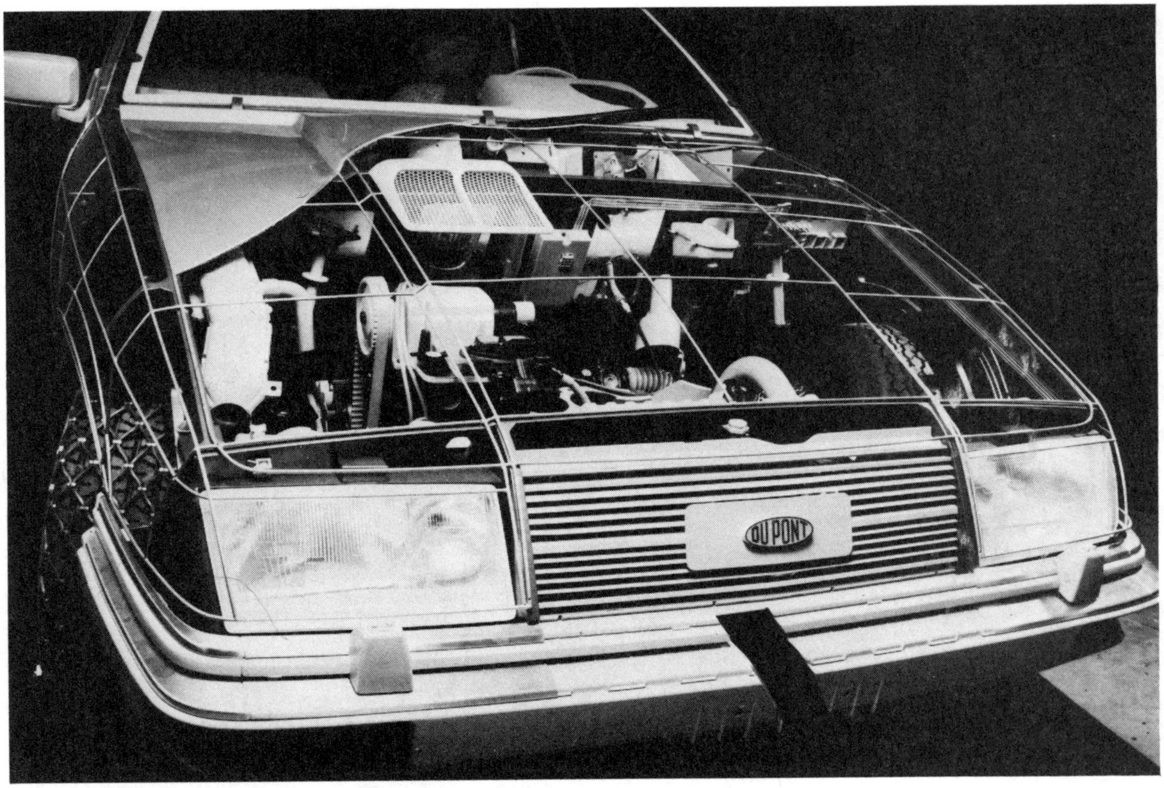

Fig. 7-10. DuPont's plastics are used for more than 150 applications in a car.

fire-resistant compounds. Body panels were mainly plastic. Several hundred such Marcos cars were built until 1969 (when a tubular steel frame was adopted).

Wood is light, but it has low strength and its cost is high. It would be unthinkable for a large-scale, industrial auto application.

THE LIGHT ALLOY

Aluminum, which for its size is about one-third the weight of steel, is another lightweight material that is being used widely to replace steel in automobiles. The Rover P3 of 1950 had an aluminum hood, trunk lid, and door skins. The European auto industry soon began using aluminum for radiators, hoods, and bumper bars on mass-production cars (Fig. 7-11).

Ford went to an aluminum hood on the Lincoln Versailles in mid-year 1977. Oldsmobile introduced aluminum hoods on its B-body and C-body cars for 1977. Made by Alcoa, the inner and outer aluminum panels were joined by a combination of hem flanging and adhesive bonding.

Buick had an aluminum deck lid on the 1978 Regal and an aluminum hood on the Century and Regal with certain power-train combinations. Bumper-face bars were made of aluminum for the 1978 Buick Century Sport Coupe and the Century Sport Wagon.

Innumerable other applications have followed

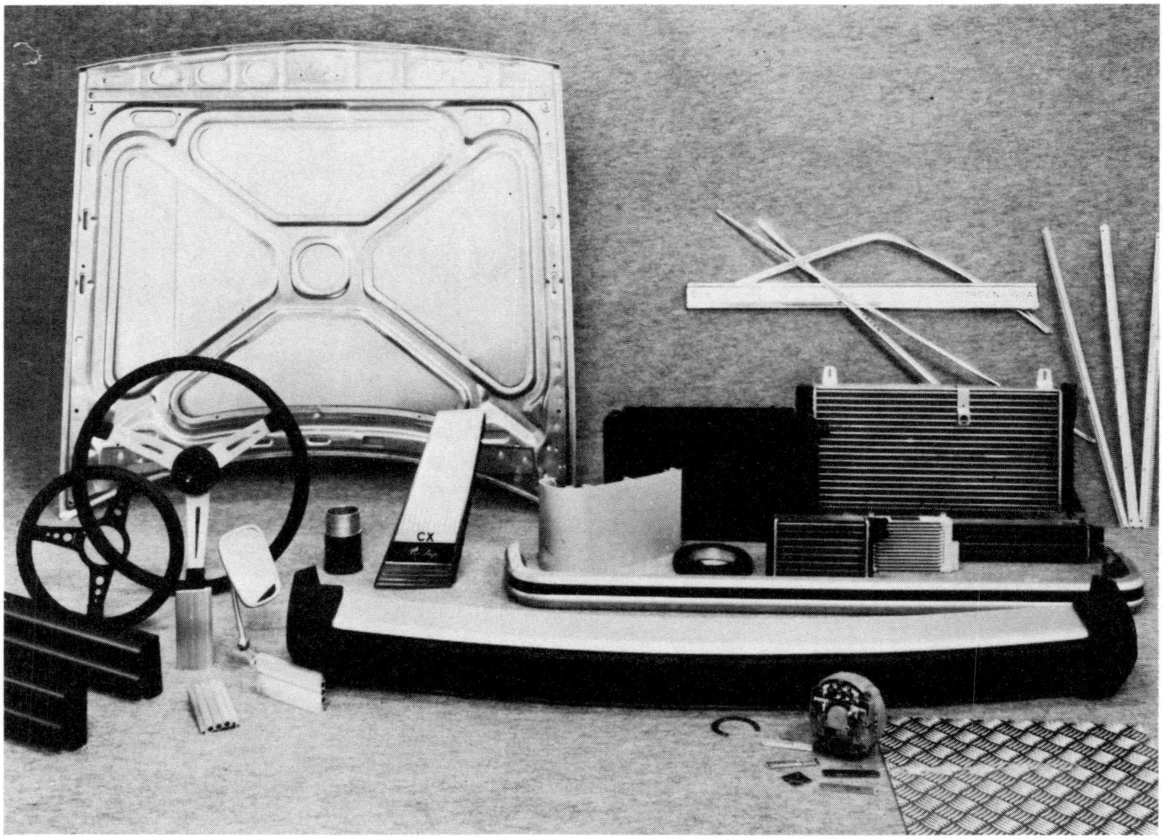

Fig. 7-11. Pechiney aluminum parts include radiators, hoods, and bumper bars.

in later years, but the total amount of aluminum used in cars remains low. Experts have predicted vast growth in the use of lightweight materials such as plastics and aluminum in cars for many years, but the revolution stubbornly resists taking place. Even in modern, small and lightweight cars it is steel and iron that dominate. Table 7-1, compiled by Teksid (Fiat's metallurgical subsidiary), shows where the industry stood in 1980.

SAVING WEIGHT IN SMALL DOSES

No auto maker yet builds a car that combines all the products making use of plastics, elastomers, fibers, films, pigments, fabricated products, finishes, and electronics. But in 1979, the marketing chiefs of Du

Table 7-1. Automotive Material Usage.

Make	Model	Weight (pounds)	Steel & cast iron	Aluminum alloy	Plastics
Alfa Romeo	2000	2,205	—	7.8%	—
Mercedes-Benz	250	3,088	70.3%	4.2%	—
Fiat	132	2,304	73%	4%	7%
Peugeot	104	1,587	67%	4%	7%
Fiat	128	1,764	80%	3.6%	4%
Fiat	126	1,280	72.2%	3.3%	6.7%
Fiat	131	2,293	72%	3%	7%
Volkswagen	Passat	1,951	82%	2.8%	7.2%
Austin	Allegro	1,905	81.2%	2.4%	2.4%
Isuzu	Gemini	2,040	85.6%	1.6%	1.4%
Volvo	244	2,977	71%	1.5%	4.2%
Ford	Escort	1,720	80%	0.1%	3.9%

Pont de Nemours International engaged Ogle Design to create an attractive shape for a "fictitious" car (Fig. 7-10) that started out as an empty cage. It was filled, part by part, with existing production units drawn from multiple sources.

The air-intake grille comes from a Fiat. It is made of Minlon, a mineral-reinforced nylon, that is also used for the heater-control flap and fuel-filler flap.

The gearshift casing is taken from an Opel and is made of Zytel, a glass-reinforced nylon. The radiator fan is a Jaguar part, the exterior door handles come from a Citroen, and the fuel-tank antisurge baffle is taken from a Fiat—all Zytel parts. The instrument panel is taken from a Citroen, having a base of Rynite thermoplastic polyester with an instrument cluster containing a flexible printed circuit on a base of Mylar polyester film.

The Volkswagen fuel pump body and Fiat seat adjusters are made of Delrin acetal resin. This is also used for the heater valve and fuel-filler cap.

The fuel-tank filler neck, steering wheel, and various minor parts are made of Zytel. An ionomer named Surlyn is used in the windshield trim. Brake hoses, V-belts, and steering-joint dust caps are made of neoprene, while VW, Audi and Opel shaft seals are made of Viton fluorelastomer.

Heat-resistant spark plug caps and throttle-cable outer casing are made of hypalon synthetic rubber. Teflon, the familiar fluorocarbon resin, is used for the carburetor spindles. Brake pads, clutch plates, and radiator hoses contain Kevlar aramid fibers.

Nylon BCF carpeting, taken from the Renault 18, covers a sheet of Keldax sound-deadening material. One front seat uses Dacron polyester and the other has Lycra elastane fibers. Door panels are also covered in Lycra.

An EPS tape is featured in the Toyota passive seat-belt system for transmitting the driver from the electric motor to the actuating mechanism when the driver is seated and the door is shut.

STEEL'S WORST ENEMY: RUST

Porsche made headlines in 1973 when it displayed its 20-year car (Fig. 7-9) at the Frankfurt auto show. Its raw bodies were zinc-coated in all exposed areas. This long life study was not a production prototype but rather a pathfinder for new development trends. Left behind was the "throwaway car" concept in favor of longer life span, rust-protected bodywork, lightweight construction, and simple exchange of body parts or components. Porsche then began selling its cars with a six-year guarantee against rust, followed by Ford of Europe and Talbot in 1980.

Zinc-coating and zincrometal inserts work well, but they are expensive. The industry has instead been moving towards electrophoretic treatment. This demands enormous plant investment, but gives low unit cost in mass production. Ford was the leader in electrophoresis applications, starting in 1963, with two luxury car bodies.

In Ford's electrophoretic process, the car body is dipped for 2.5 seconds by an overhead conveyor into a 90-foot-long tank. A feature of the system is that the car is given a positive electrical charge, while the water-based, thermosetting primer in the 50,000 gallon tank carries a negative charge.

The result is an instant mating of car body and antirust primer that leaves no body surface uncovered and even protects hard-to-reach structural members that an ordinary spray-painting system or dip would not touch.

SPACE-AGE SPINOFFS

In 1976, Chrysler engineers began working on exotic materials such as graphite fiber composites, fiberglass-reinforced thermoset plastics, and thermoplastic polymerics. These organic materials are attractive because they're light, tough, durable and don't corrode. After two years of preliminary research, Chrysler lab experts began evaluating these materials in test parts in a 1978 Chrysler LeBaron.

They dubbed the vehicle "Poly-car" or PXL for "Polymeric Extra Light." That's a takeoff on the XL car they did earlier, concentrating on light-weight metals. Twenty test parts in the Poly-car include graphite-composite drive shaft, rear-axle suspension, leaf springs, door hinges, engine accessory bracket, and valve pushrods. Among less radical applications were fiberglass-reinforced plastic wheels, door inner panels, grille opening panel, bumper energy absorber, transmission cross member, oil and transmission pans, engine fan, window-crank handles, and heater core top. Non-reinforced plastic headlamps, side windows, and sun visor bracket were also part of the package.

Concurrently, Ford was building an experimental car based on the 1979-model LTD. Major body parts were made with aerospace technology at the Milford Fabricating Company, a subsidiary of the Budd Company (which now belongs to West Germany's Thyssen Steel).

The Plastic Research and Development Center of the Budd Company made a joint study with Ford to survey and test the adhesives to be recommended for body assembly. Two methods were found applicable; they were an epoxy-type glue offered by Hysol and a urethane product made by Goodyear. Goodyear's Pliogrip was used for bonding the 20 plastic components of the body.

The Vought Corporation produced certain components using graphite-fiber reinforced plastic. Ford supplied its own prepregnated graphite material that was processed using state-of-the-art procedures. The semifinished parts were returned to Ford and cured by the Autoclave method.

The experimental Ford's frame was also plastic (with graphite-fiber reinforcements). The propeller shaft was a filament-wound, graphite-fiber reinforced plastic unit.

Instead of the standard V-8 engine, this car was powered by the four-cylinder (ex-Pinto), 2.3-liter engine matched with the C-3 automatic transmission.

SHHH!—PASSENGERS IN TRANSIT

Quietness in a car conveys an impression of quality. Low noise levels are important for the comfort of all occupants, for conversation, and for listening to radio or tape players. High noise levels contribute to driver fatigue and therefore have a negative bearing on safety. The quest for silence affects the work of designers and engineers from an early point in every new car project.

Automotive engineers have learned that all mechanical systems and structures are to some extent elastic in nature, and consequently not immune to vibration. The modern unit-construction body is a highly complex vibration system. Noise control is based on the concept that the vehicle structure must be prevented from acting as an acoustical transducer.

The car interior has to be shielded, above all, from noise emanating from the engine, transmission, and exhaust system. The designers and engineers are faced with the problem that the more successful they are in this task, other sources of noise take on greater importance. One source is wind noise, mainly an aerodynamic phenomenon, and the other is road noise.

In order to make a vehicle quiet, many steps from its conception to production are required. Very simply, there are four approaches to keeping noise away from the passenger. First, make the source of the noise quiet. Elimination of noise at the source is the most difficult. It means going inside engines, transmissions, and other working units—during the development stage—with acoustical experts to whom other design objectives are foreign.

Second, isolate the passenger from the source. Isolation means building barriers the noise cannot penetrate. One type of isolator is a laminated mat.

Fig. 7-12. Testing for noise in an anechoic chamber.

A typical example is made up of three distinct layers (each to serve a specific function). The vibration isolation layer of foam is attached to the structure itself and decouples structure-borne vibration from the lead septum that acts as a floating wall. The lead septum, in addition to acting as a floating wall, is an effective sound barrier; it reduces noise penetration through the enclosure. The outside layer of embossed foam provides absorption by soaking up maximum sound energy with minimum thickness within the enclosure.

Third, insulate the responding compartment. Acoustical insulation means absorbing the noise outside of the insulated cell. In practice, sound absorption consists in combining vibration-damping agents with materials that soak up noise.

The need to absorb noise inside a machine enclosure often accompanies the need to damp out vibrations on the same structure. A common example incorporates a thick layer of embossed foam bonded to a thin sheet of precured, high-efficiency damping compound. In this configuration, it provides a single solution to a twofold problem. The thickness of the foam layer is related to the frequency of the air noise (the higher the frequency, the thinner the material required). The damping sheet is effective for temperatures from 45 degrees F to 180 degrees F and over a wide frequency range. The material remains permanently bonded at temperatures below freezing to 400 degrees F.

The fourth method is called damping. Damping means dissipation of structure-borne noise (usually occurring in the form of traveling bending waves). This dissipation can be accomplished by converting the waves into some other form of energy—usually heat. That is usually accomplished by the application on the noisy area of a material with a high, internal energy-absorbing capacity.

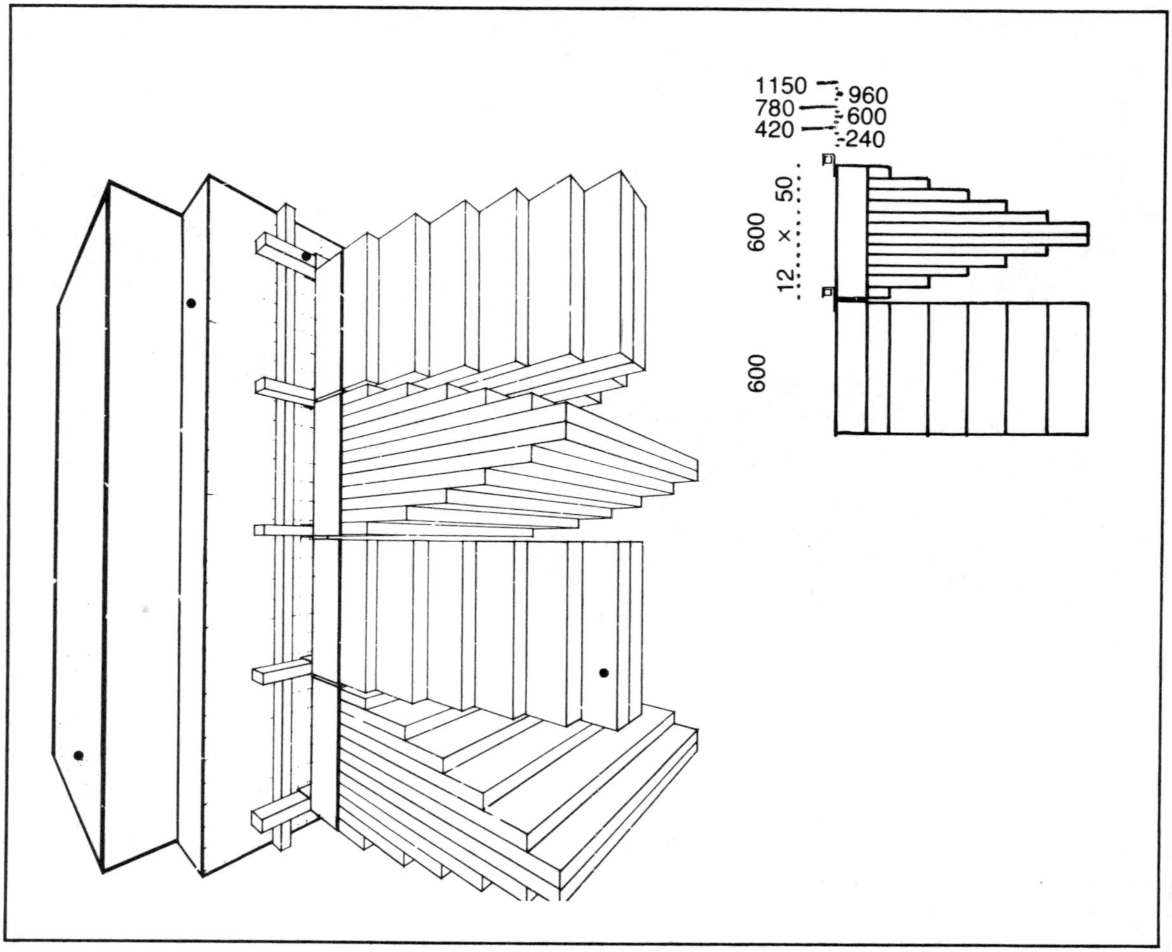

Fig. 7-13. The principle of the sound absorbing wall.

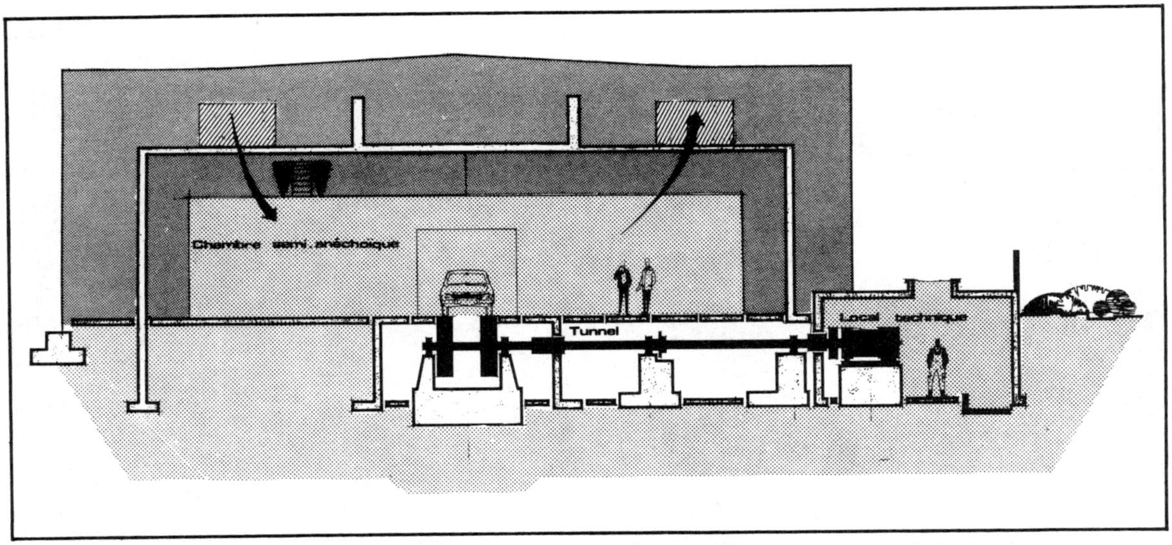

Fig. 7-14. The rollers and their drive system must also be acoustically insulated from the anechoic chamber.

Doing these things means starting with a vehicle concept that is compatible with the goals and being sure that each design decision, whether it be a styling feature or provision for a manufacturing gauge point, is also compatible. Body and chassis design is followed by development of quiet components, optimization of isolators, and the application of acoustical materials.

Where structural alterations are not easily made, resonances in external sheet metal are often most easily dampened by bonding on pads of acoustic material. Their application is only effective if every significant area is treated and even the smallest holes are closed. This attention to detail leads to the addition of a seal to the bulkhead disconnector (a plastic wiring connector on the dashpanel).

SLEUTHING FOR SOURCES OF NOISE

Cars under development are noise tested in anechoic chambers (Fig. 7-12). Such rooms must be spacious. The walls are generally covered with foam blocks projecting to varying distance into the room. The room is often described as without echo or acoustically "dead." It is designed to study interior noise for the comfort of passengers, and exterior noise for pedestrian acceptability and compliance with legislation.

The anechoic room provides a free-field environment for acoustic testing that gives a near-perfect simulation of outdoor conditions (on a lonely country road). All of the sound emanating from a source is absorbed in the foam-cover on the walls and ceiling (Fig. 7-13). Consequently, there are no reflections (echo) and the spatial sound radiation pattern of each source can be determined.

The car is positioned on rollers so that it runs in a virtual vacuum. The rollers and their drive system must also be acoustically insulated from the anechoic chamber (Fig. 7-14).

Peugeot is world famous for making quiet cars and, in its acoustical center at Belchamp, the roller bench is powered by a silent-shaft, 272-hp Ward-Leonard electric motor that is capable of giving car-wheel rotational speeds up to 125 mph (with acceleration and deceleration rates up to 0.25 g).

Fig. 7-15. Accelerated corrosion tests on the Peugeot proving grounds.

One of the limitations of the system is that it cannot give an accurate reading on road noise unless it has a set of interchangeable rollers that simulate a variety of road surfaces.

The level of road noise is dependent on many factors (above all the nature of the road surface). Car factors include the type and design of the tires, the characteristics of the suspension systems, the principles and execution of their anchorages to the frame or body shell, and the design and engineering of the body.

The British Motor Industry Research Association has defined road noise as the sound-pressure field maintained inside the passenger compartment by vibrational energy originating in the tire footprints and transmitted to the vehicle structure via the suspension systems.

Road noise manifests itself as excitation of the body structure by forces acting on the tires. It causes irritation, discomfort, and fatigue.

THE RISK OF ACOUSTICAL OVERKILL

J. D. Shedlovsky of Pontiac has pointed out that it is possible to spend much time and money developing and building cars with good isolation, excellent panel damping, good acoustical absorption, and still end up with an acoustical lemon.

On the other hand, a quiet car does not necessarily have efficient acoustical packaging. Some quiet cars carry far more deadeners, isolators, absorbers, and noise barriers than they would have needed if the basic acoustical map had been properly drawn. Acoustical overdesign, as it is called in the industry, costs extra money and adds to the car's weight.

Large luxury cars of the 1960s often carried 150 to 200 pounds of noise-control materials. Newer cars are smaller, and lighter-weight materials have been developed so that the average noise-control materials for a medium-size family car now need not exceed 50 pounds.

Part 2
Historical Review

Chapter 8

Changing Shape and Structure

MOST OF US RECALL TO SOME EXTENT, THE broad strokes in the evolution of the shape of automobiles. The structural evolution is less talked about, but inseparably intertwined with the history of car design. The first cars were primitive in layout, structure, and body design. The first self-propelled vehicle, Cugnot's three-wheeled steamer from 1769, was just a ladder-type frame with a seat on it. All of the machinery was concentrated around the single front wheel.

INFLUENCE FROM TWO SIDES

Gottlieb Daimler's first car, built in 1886, was simply an engine mounted in a horse-drawn carriage with the shackles removed and replaced by a steering mechanism. The frame and body were made of wood (Fig. 8-1). The first motor vehicle built by Carl Benz, in 1885, was a tricycle built for a one-cylinder engine. It was built up on a tubular steel frame (Fig. 8-2). A single, narrow-diameter tube, bent into a horseshoe shape (with the open end at the rear) had a bracket for the front-wheel fork at the other end, and engine-mounting brackets underneath. From these two pioneering examples can be seen the two major influences brought into play in the initial, formative stages of car design: the horse-drawn carriage and the bicycle.

After going to four-wheeled cars, Benz remained faithful to the steel tube frame (inherited from the world of bicycles) for many years. He had good reasons for this. A tube's stiffness is directly proportional to its weight, and increases as the square of its diameter.

Gottlieb Daimler was essentially an engine man. His vision was to put Daimler engines to work in all transport systems—from carriages and wagons to boats, locomotives, and streetcars. The car chassis did not interest him a great deal, but his assistant, Wilhelm Maybach, took an early orientation to lightweight car construction. In 1889, he

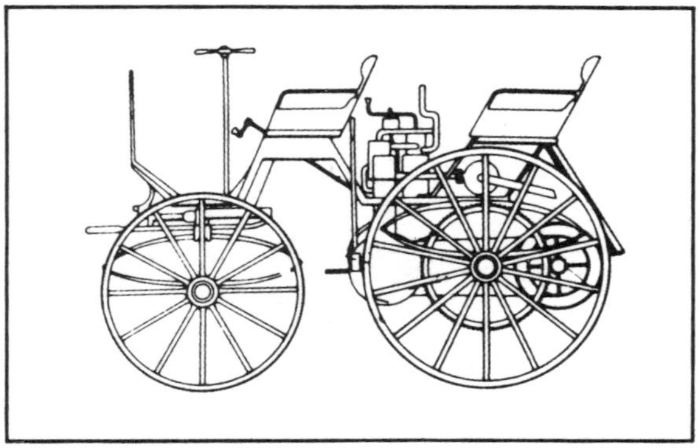

Fig. 8-1. The 1886 Daimler was a converted horse-drawn carriage.

built the so-called Stahlradwagen (wire-wheeled car) using axles with wheels of nearly equal size, and a steel tube frame that served not only to support the scanty bodywork, but also to carry the cooling water for the rear-mounted, V-twin engine (Fig. 8-3).

While the Stahlradwagen remained in the prototype stage, Peugeot adopted the tubular frame in 1891 (inclusive of the water-circulation principle). In America at that time, the bicycle was at the peak of its popularity, and Charles E. Duryea used a steel-tube frame, inspired by bicycle engineering, for his three-wheeled carriage of 1893. It was driven by its rear axle and it had a single front wheel to steer. The engine was mounted under the seat and ahead of the rear axle.

SUBFRAMES AND AXLE SUSPENSION

Later when Duryea built his Four-Wheeled Trap, he mounted the axles directly on springs attached to the body—without a subframe. A subframe was a support structure flexibly attached to the body. Its origins go back to carriage-making traditions. For centuries, coaches had been built flexibly suspended in the carriage frames. At first they had leather straps hanging down from posts rising at the four corners of the frame. That was the only springing because the axles were rigidly attached to the frame. Later, metal springs were adopted for coach suspension.

Car bodies based on carriage experience had rigid structures. The carriage-makers' methods of suspension could not be applied because they were too cumbersome in construction and had too much freedom of body movement for the higher speeds that cars were capable of reaching.

The subframe, interposed as a flexible medium between axles and body, was a primitive solution that won favor with several American pioneers. The Stanley Steamer had a tubular subframe that was rigidly attached to the axles. Front and rear axle shafts were inserted into straight cross tubes that were braced by arched tubes crossing above them. The arched tubes at front and rear ends were linked by two straight tubes.

The subframe had a certain freedom to twist. This accommodated axle movements, when the wheels hit roadway irregularities, without transferring the full torsional stress to the body structure. In the case of the Stanley Steamer, the body was mounted on springs resting on the subframe—a transverse leaf spring in front, and two longitudinal leaf springs in the rear—so as to assure "perfect" isolation of the body from road shocks. Subframes faded from popularity when leading car makers began putting springs between the axles and the frame.

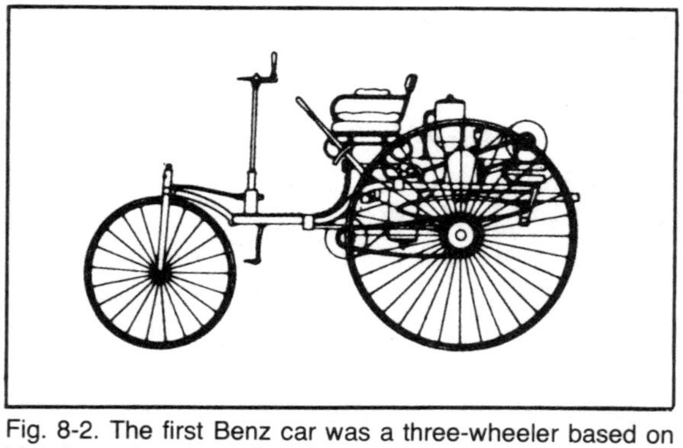

Fig. 8-2. The first Benz car was a three-wheeler based on bicycle-engineering principles.

Fig. 8-3. The Daimler Stahlradwagen of 1889.

Daimler adopted axle suspension (as inherited from the more advanced horse-drawn carriages) but remained faithful to the wooden frame until 1900. Panhard & Levassor, in Paris—who had a license to make the Daimler engine—at first used wooden subframes sheathed with angle-iron and bolted to the frame proper. In 1894, Emile Levassor moved the engine up front, placed under a hood, and adopted a ladder-type frame, made of wood, with spring suspension of the axles. This type of construction set the pattern for a whole new generation of cars (Fig. 8-4).

ENCLOSED PASSENGER COMPARTMENTS

Most automobiles of the pre-1900 era were open and without weather protection. At best they had a collapsible top above the seat(s). But the closed car was not long in coming. In 1894, Benz fitted a canopy top on a four-wheeled carriage, and, in 1896, Benz built a car with a fully enclosed body.

Of course, the closed body was not a new invention. It had been used for carriages since the days of antiquity. The Romans had four-wheeled carriages with canopy tops. Roof beams for permanent cover appeared in medieval times, and by the end of the 16th century, carriages with solid structures had come into use in Europe.

Richard Trevithick, Walter Hancock, James Fraser, Julius Griffiths, Goldsworthy Gurney, and Francis Macerone built steam carriages with closed bodywork between 1803 and 1840.

What was new in the Benz developments was

that the closed bodies were designed for an automobile with particular regard for the dimensional constraints and weight considerations affecting the car. The first examples were not successful because their height was excessive and the additional weight was preposterous. As a result, the standard car body was to remain open, for many years, while much ingenuity was spent in devising mechanisms and materials for collapsible tops.

In 1899, Charles Ferrand, designer for Rothschild & Fils, in Paris, introduced the rear-entrance tonneau body on the 6 CV Panhard & Levassor chassis. Tonneau is the French word for barrel; the word was used for an open rear body section, tapering to the front and rear, with bulging sides.

This body style immediately caught on with the motoring public because of its greater versatility. Seats could be mounted facing each other across a central corridor, facing forwards or facing backwards. For a forward-facing rear seat, this meant putting hinges on the center section of the rear seat, swinging away to open a path to the floor inside. Some coachbuilders made long tonneaus with vis-à-vis seating for four passengers, behind the driver's seat.

MOTORING FOR THE HARDY

All open bodies, even with the best protection of the collapsible top, had serious shortcomings with regard to comfort. Comfort and safety often go hand in hand. In the earliest days of the automobile, both were lacking.

For the most part, passengers rode blithe-

Fig. 8-4. The 1894 Panhard & Levassor with a front-mounted engine.

ly out in the wind, and sun, and dust. Goggles, gauntlets, caps and duster coats helped, but when the weather turned foul there wasn't much to do about it. Some cars had tops that were merely symbolic. Side curtains were developed later. They were, however, just a shade the lesser of two evils. Isinglass peep-hole windows were difficult to see through. And there were gaps galore through which drafts and rain or snow could enter the passenger compartment.

The driving task was so intricate that it could befuddle a present-day professional. Many cars had tiller steering (with many disadvantages). Its big swing prevented better use of the space. A sudden, single-wheel bump could kick it loose from the driver's hand. The Lanchester balanced tiller was safe, pointing forward from its pivot axis, but it was no lasting solution. The steering wheel was accepted everywhere as a superior principle during the span of three or four years.

The amount of controls calling for the driver's regular use was overwhelming. The driver had to adjust the spark advance manually. On many cars, he had to pump oil to the bearings by hand. His gearshift lever often moved in a quadrant, engaging each gear in succession. There was, of course, no synchromesh. Clutches were often tricky, and engines stalled easily.

The hand brake was often the main service (a long lever standing up beside the driver's seat). Brake pedals were secondary aids to slowing down. Some cars had no accelerator pedal, but merely a hand throttle.

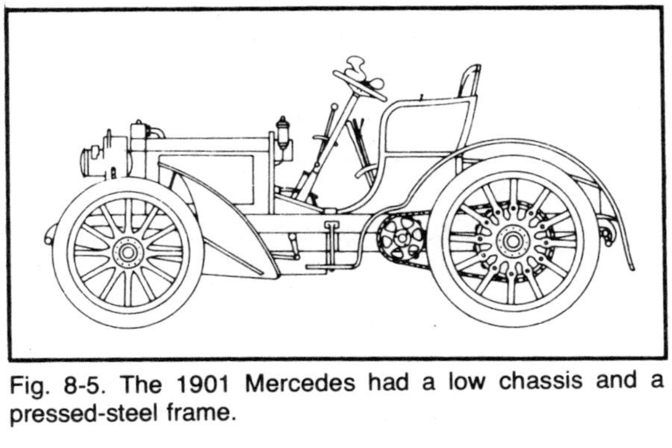

Fig. 8-5. The 1901 Mercedes had a low chassis and a pressed-steel frame.

On most cars, instrumentation was sparse. Most of the driving was done by ear and by feel. In the absence of warning lights, malfunctions would go undetected till they started to make expensive noises or quietly called a halt to the vehicle's terrestrial progress.

FROM WOOD TO STEEL FRAMES

On the structural front, the steel frame began to replace the wooden type about 1900. The original Mercedes of 1901 is usually cited as the first example (Fig. 8-5), but it is on record that Arbel made pressed-steel frames for Decauville and Delahaye prior to 1900.

A deeper look into the matter reveals that the pressed steel frame was invented in 1877 by Samson Fox, an employee of the Leeds Forge Company in Yorkshire, England. He intended it for use on railroad rolling stock as a means of combining maximum strength with minimum weight.

J. S. Critchley, engineer of Daimler in Coventry, devised a channel-section pressed steel frame for a prototype in 1900. But production Daimlers used wood frames until 1910. This case is fairly typical. The pressed-steel frame was slow to reach general adoption. In 1905, there were still more cars being built with reinforced wood frames.

A PLETHORA OF BODY STYLES

As coachwork became more elaborate, the bicycle-type, steel-tube frames were phased out. Ladder-type frames with straight side members, wood or steel, were favored by the vast majority. And body types proliferated.

Before the year 1903, there were cars with bodies called Landau and Landaulet, Victoria and Clarence, Berline and Brougham, Mylord and

Break, Diligence and Limousine, Phaeton and Caleche. Many of these were derived from horse-drawn carriage usage. The word limousine comes from the French town of Limoges (famous for its porcelain) where a roomy and luxurious type of closed-body carriage had come into use about 1800.

An Amedee Bollee limousine, built in Le Mans in 1901, (Fig. 8-6) has a fully enclosed rear seat above the rear axle. No side doors are fitted on the passenger compartment. The car has right-hand drive, and the front passenger's seat is tilted and folded away to give access to the passenger compartment via a subway-type sliding half-door.

The Berline came from Berlin. It was a coach with all passengers inside and the Driver's seat outside. The name is still in use in France and Italy to indicate a sedan. The term sedan, incidentally, is not used in Europe. To the British it's a saloon. The Germans use the word *innenlenker* (steered inside) as a generic designation. Four-door models are usually called limousine and two-door models coach or coupe. The first Berline was built about 1850 for the Elector of Brandenburg.

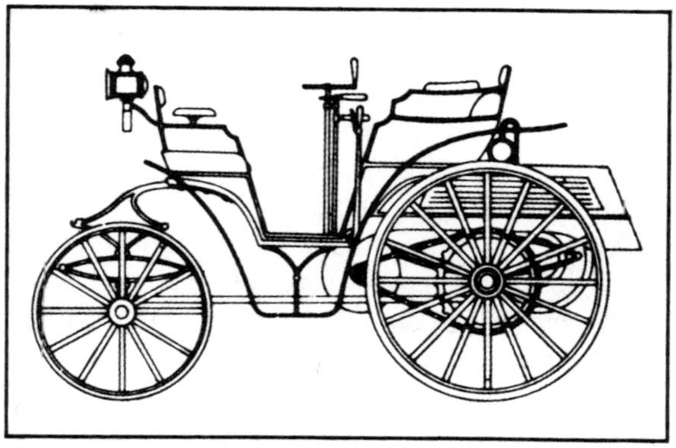

Fig. 8-7. The 1893 Benz Vis-ã-Vis.

Fig. 8-6. The limousine body on a 1901 Amedee Bollee.

The word sedan has its origin in the Sedan chair (from the French town of Sedan) that was an enclosed, one-person seat, not unlike a phone booth in shape, carried on long beams with handles for porters front and rear to lift. The idea originated in China, perhaps a thousand years earlier, but we know it as a Sedan chair.

Benz made cars for both vis-ã-vis and dos-ã-dos bodies as early as 1894. Both were inherited from contemporary carriage types. The terms refer to the seating arrangement: via-ã-vis means face-to-face and dos-ã-dos means back-to-back.

With the vis-ã-vis, the driver sat in the back seat, facing his passengers. See Fig. 8-7. With the dos-ã-dos, the driver occupied the front seat, and the passengers sat behind him—facing the road behind. In old carriage lingo, all seats were for passengers because the driver had his own seat up front.

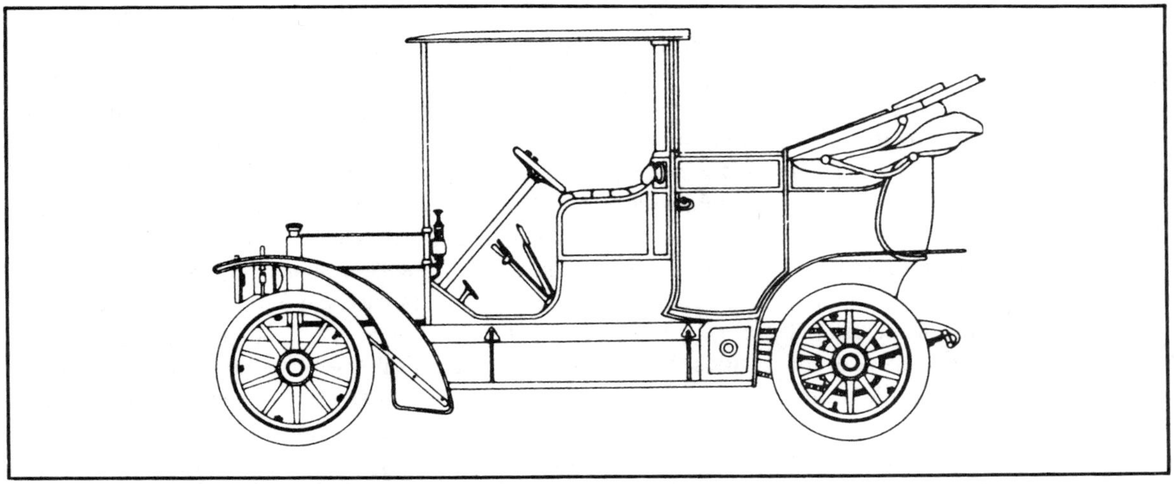

Fig. 8-8. The 1902 Mercedes Simplex with a landaulet body.

ORIGINS OF APPELLATIONS

The Landau was a long-wheelbase, open carriage (with collapsible top) with seating for four passengers, in vis-ã-vis formation, and a high seat for the driver above the front axle. It originated in the town of Landau in the Palatinate. The Landaulet was simply its derivative, originally carrying a smaller top, or a top only for the rear seat of a four-seater. One of the most advanced automotive versions for its time was the Mercedes Simplex Landaulet of 1902 (Fig. 8-8).

The Victoria is a carriage with a collapsible top. It was made popular by England's Queen Victoria. The original Victoria had seating for only two passengers. The driver's seat was above the front axle, outside the top. Some early Victorias had an

Fig. 8-9. The Brougham body on a 1911 Benz.

outrigger seat behind the rear axle. Later, some light carriage with vis-à-vis seating for four passengers were also called Victoria.

The Brougham is a formal town carriage with a closed body (Fig. 8-9). Invented in 1834 by Lord Brougham, England's Chancellor of the Exchequer, for his personal use, it was built by Robinson of London.

The Clarence was a closed coach, originating around 1830, and credited to the Duke of Clarence. It was a forerunner of the Brougham.

The Phaeton was named, for reasons that remain unclear, after the illegitimate son of the sun god, Melios, in Greek mythology. Phaeton's connection with wheeled traffic stems from the story of how he was allowed to drive the chariot of the sun across the sky for one day, but lost control of his horses and was felled by a thunderbolt thrown at him by Zeus, out of concern for preventing more widespread damage.

The typical horse-drawn Phaeton was a high-wheeler with a collapsible top over the front seat, located above the front axle, and a fully open back seat or load platform behind. The Stanhope and Tilbury were subtypes of Phaeton. The Mercedes Sport Phaeton of 1910 provides an early automotive example (Fig. 8-10).

The Mylord coupe was a Benz creation, dating from 1897 See (Fig. 8-11). It was a medium-weight carriage with a driver's seat above the front axle, out in the open, and a tall, closed (or fitted with a collapsible top) passenger compartment in the middle and a Benz flat-twin Kontramotor behind. The name was first used about 1860 for a light, four-seater vis-à-vis carriage with a collapsible top.

The Diligence was a stagecoach. Cars using that designation were usually roomy, closed coaches with no particular luxury or distinction.

Coupe means "cut off" in French. Many automotive coupe bodies were Berlines or Broughams—cut off behind the driver's seat—with luggage space, fuel tank, and spare wheel(s) behind. When it was cut off from the front, leaving the rear seat inside a partial Brougham or Berline body, but kept the driver's seat outside in the weather, it was called a Coupe de Ville.

The Torpedo (Torpille) body was invented in France and carried its name with it all over Europe. It was too new and too different to have been applied to horse-drawn carriages (the naval torpedo

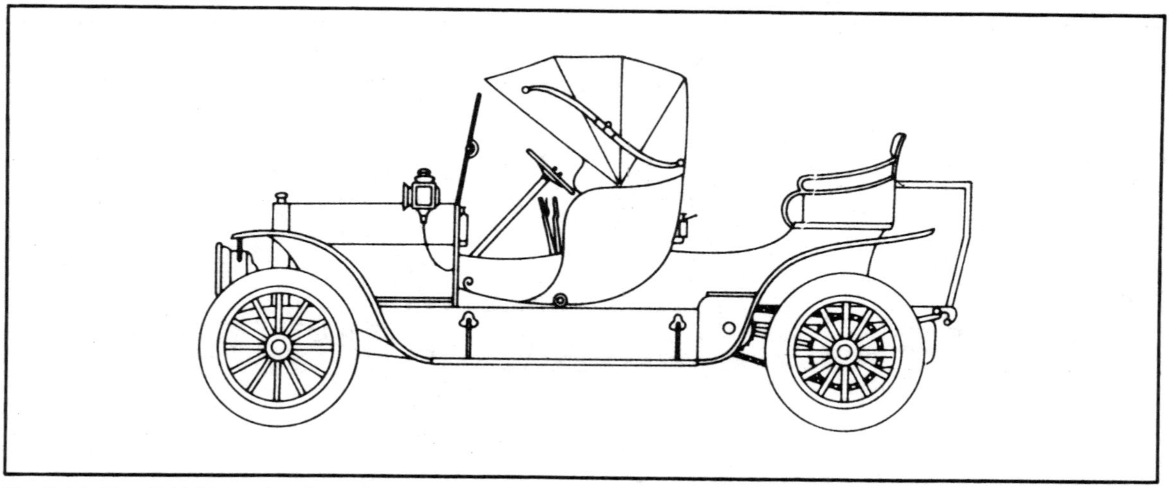

Fig. 8-10. The 1910 Mercedes Sport Phaeton.

Fig. 8-11. The 1897 Benz coupe with a rear-mounted engine.

was first tested in 1865). The idea was to imitate the torpedo shape for an open body, low and narrow, with a straight line from the hood into the belt line—which remained horizontal throughout. Benz, Audi, Austro-Daimler, Vauxhall, Mercedes, and others adopted it for the high-performance, four-seater touring cars they built for the Prinz Heinrich Trails of 1909-11. See Fig. 8-12.

FROM FOX HUNTING AND FOLK DANCING

Cabriolet is a term derived from the French word *capriole* that indicates a folk-dancing maneuver: jumping and turning around, partly or completely. The term was used to describe the excessive ride motion of the light, two-wheeled cart that came into use in France about 1840. It had seating for two under a folding top. In later years, it's the folding top that has preserved the use of the name Cabriolet (still in use in Europe to denote a convertible).

The Caleche was a folding-top carriage. The name was taken from a collapsible bonnet worn by fashionably dressed ladies in the early 1800s. The name was absorbed in the English language (*calash*) in the restricted meaning of a canvas folding top, and not the carriage.

The Dogcart came from the world of game hunting. On big hunts, as practiced in Britain, the

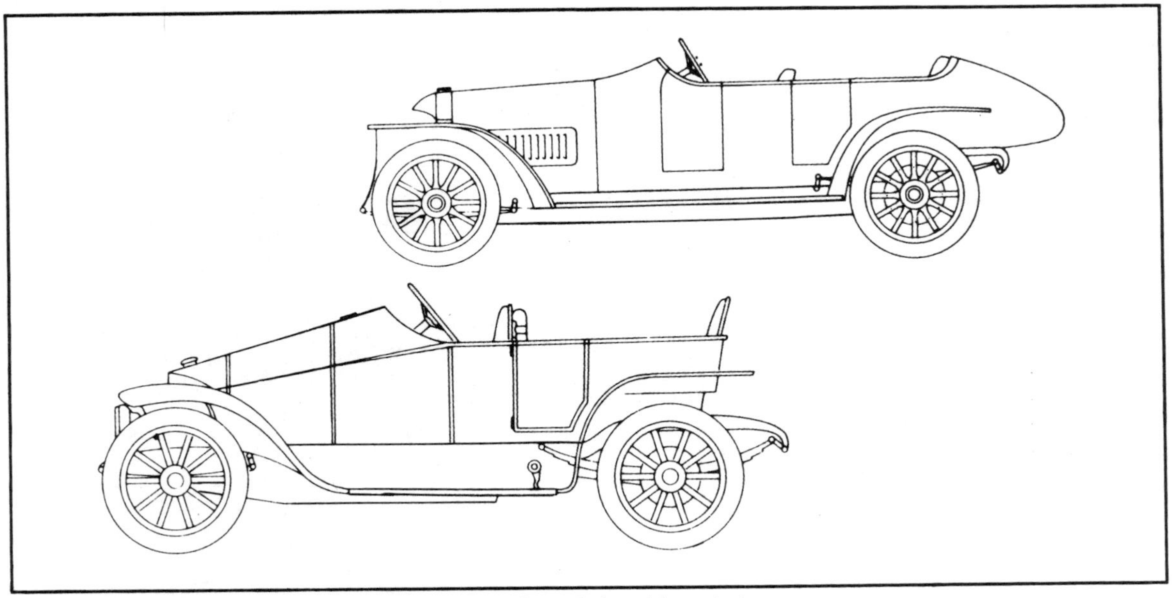

Fig. 8-12. Early examples of the Torpedo body (top: Benz, below: Mercedes).

dogs would be brought to the site of action in a roomy cart, with a separate compartment behind the seats. The Break was from the same scene. A hunting party had a shooting brake or break to carry the guns, ammunition, and other equipment. To the French, a station wagon is a break to this day. The original horse-drawn brake was open, but motor-driven ones soon acquired complete coachwork.

SPACE ALLOCATION

Vehicle architecture progressed along with manufacturing techniques. As the chassis was lowered, the floor came down and the belt line and roof followed. In the pattern laid down by the 1901 Mercedes, the engine stood vertically, just behind the front axle, and the front seat was placed midway inside the wheelbase. The back seat was mounted above the rear axle, where it pushed the roofline to a higher level, if normal headroom was to be retained.

On luxury cars, wheelbases were stretched and back seats moved forward of the rear axle. That could not be done within the normal dimensions of a medium-priced family car.

Larger cars with better seating space and more elaborate coachwork needed more powerful engines to provide the kind of performance that their clients demanded. More power meant, basically, a larger engine. Prior to 1915, that usually meant a longer engine, with bigger cylinders and more of them, all arranged in-line. That pushed the cowl further back and necessitated further lengthening of the frame, wheelbase, and body to provide proper passenger accommodation. Chassis engineers were most reluctant to move the engine forward in the chassis. This was mainly due to concern over weight distribution (more weight on the front axle meant harder steering).

How far the low-priced American car had come by 1908 is shown by the Model T Ford (Figs. 8-13 and 8-14). The frame, although it was kicked up over the rear axle, had its upper face well above wheel-hub level. The floor level was high, the doors were small, and even the running boards

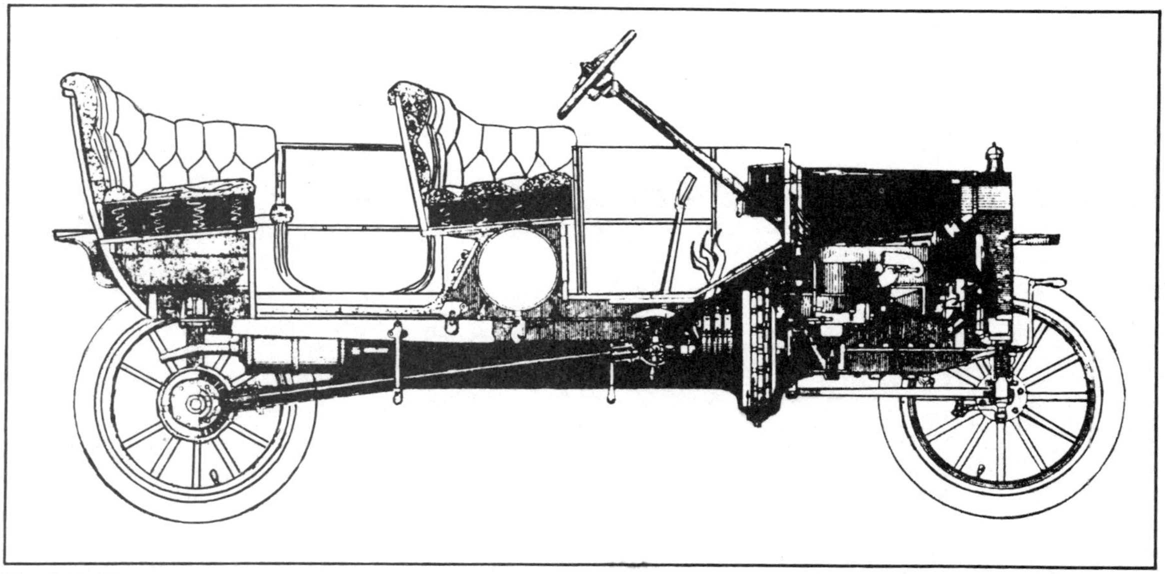

Fig. 8-13. The architecture of Ford's Model T.

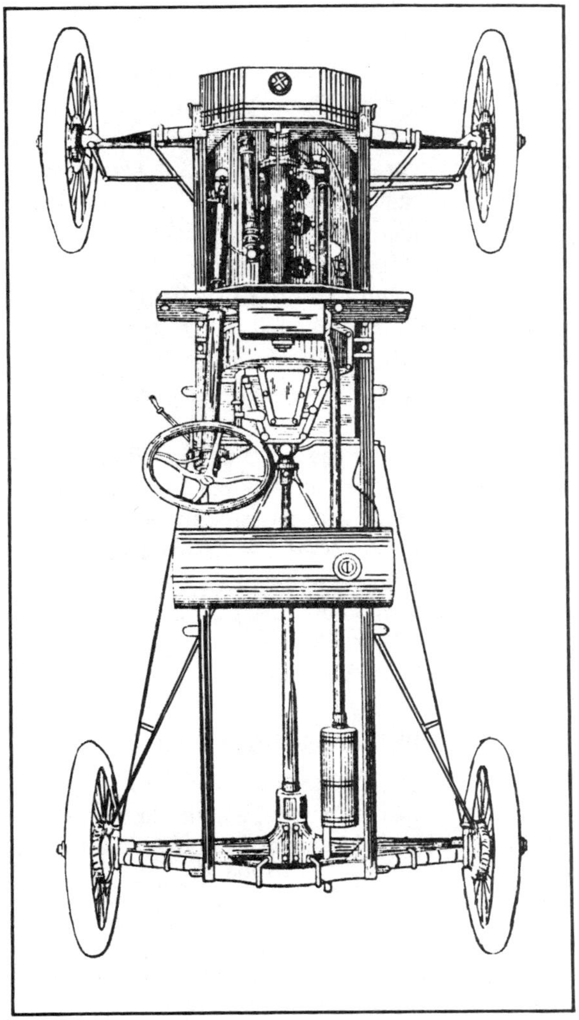

Fig. 8-14. The Ford Model T chassis.

were high up. The fuel tank was positioned under the front seat, and the back seat was pushed back to a location squarely above the rear axle. The engine was mounted so far back that its center of gravity fell outside the front wheel periphery (more than a foot behind the axle). The frame was short; it ended at the axles.

THE SWING TO STEEL BODIES

Ford used a pressed steel frame and steel body panel for the Model T Ford. Cadillac went to angle-steel frames in 1903, and Packard and Peerless followed in 1904. Steel parts came into extensive use for body components, but the all-steel body was still in the future. A forerunner of the all-steel body was built by Wilson Haynes on an Eastman Electric car in 1897. But the mass-produced steel body was created by the Budd Company.

Edward G. Budd was born in Delaware, in 1870, and in 1889 found a job with the American Pulley Company in Philadelphia. One of its most successful products was a metal pulley assembled from two pressings. It was lighter and less expensive than the usual castings. This pulley was invented by a mechanic named Corscaden. Budd never forgot the lesson.

In 1902, Budd joined Hale & Kilburn. There he got involved with the problem of making a one-piece pressed-steel bowl for the de Laval milk separator. The company had a large production of stamped hubs for wood-spoke wheels and made railroad-carriage seat frames from assembled steel pressings.

The Pullman Car Company came to Hale & Kilburn, in 1904, with a project for an all-steel railroad carriage. All problems were ironed out, one by one, and Hale & Kilburn built several thousand such carriages until 1911. Pullman then began manufacturing its own rolling stock.

The first contacts with the automobile industry had been made, in 1908, when Emil Nelson, chief engineer of Hupmobile, placed an order for pressed steel panels to cover the wooden framework that constituted the stress-carrying body structure.

While wood might seem strong, its strength is deceptive. It will carry a normal load, but then under an overload it will suddenly fail completely by snapping or splintering. On the other hand, steel will "roll with the punches" of heavy overstress. It yields with stubborn resistance. This helps absorb overloads without suddenly collapsing in complete failure. A wooden component usually can be made stronger only by making it larger. On the other hand, a steel pillar, a roof rail, or a brace can be shaped like a girder for maximum strength. With steel, greater strength can be gained through engineering design rather than sheer size.

STEEL-STAMPING TOOLS AND TECHNIQUES

Budd recognized the advantages of steel and so did one of his colleagues at Hale & Kilburn, an engineer named Joseph Ledwinka. They talked about making the complete body structure from steel pressings and convinced themselves it was feasible. When Budd resigned to form his own company, Ledwinka went with him.

Budd got his start, in 1912, with an order for body panels from Charles W. Nash, then president of General Motors, for a tryout on a small number of Oakland cars.

When Budd found out about the Dodge Brothers's plan to build cars in their own name, he talked to the Dodge engineers about making the structural members as well as the body panels from steel.

Joseph Ledwinka kept up with the metalworking news from all over. For instance, he found out that one-piece, pressed-steel bathtubs had been made since 1907. The techniques used by the bathtub makers opened the way for compound-curve, pressed-steel fenders and other stampings of complicated shape.

Next, Thompson & Lachman invented a system of spotwelding, in 1910, that proved to make considerable improvements in the speed of mass-production body assembly.

Deep-draw, double-acting presses had come into use about 1908, but Budd found no presses large enough to handle the main structural parts for the Dodge body. Consequently, the first Dodge bodies, in 1914 (Fig. 8-15), were welded up from more than 1200 separate stampings. But the breakthrough had been made and the individual pieces were, little by little, combined into big ones.

The Keller machine was developed about 1921. It was a die-making machine, and the first means for accurately reproducing a complicated surface from a hand-made die model. In 1923, Budd began supplying the first all-steel bodies to Dodge.

In 1927, Budd developed the body for the 1927-model Dodge Victory Six by using new spot-welding techniques and all-steel construction (Fig. 8-16). When the steel roof arrived in 1931, wood framing was finally eliminated from mass-produced American cars.

Perhaps one of the greatest advantages of an all-steel body is the way it is put together. Wood pieces can be joined together only with screws, bolts, nails, or glue. Steel can be welded. At each weld, extreme heat actually fuses the metal of two different parts together. See Fig. 8-17. The roof, however, was still a fabric or oil-clothlike material laid over a chicken-wire support.

PIONEERING IN LIGHT ALLOYS

Aluminum was also thought of—and used—not only for unstressed exterior panels, but also for structural parts. The first known use of aluminum on a motor vehicle occurred in 1894 when Montupet & Primet cast the oil pan for the ¾-hp engine for the new de Dion Bouton tricycle. By 1900, aluminum gearbox casings, engine parts, and even rear axle casings were quite common in France. They were used by Buchet, Aster, and Aries.

In 1901, Fonderie de Crans supplied aluminum rear axle casings to Brasier, Mathis, Panhard & Levassor, and Rochet-Schneider used aluminum running boards for a new chassis. In 1898, C. S. Rolls had specified an aluminum body when ordering a new racing car from Panhard & Levassor.

The first use of aluminum body panels in America was probably by Smith & Mabley, in 1902, fitted on a 18-hp Panhard & Levassor. J. M. Quimby

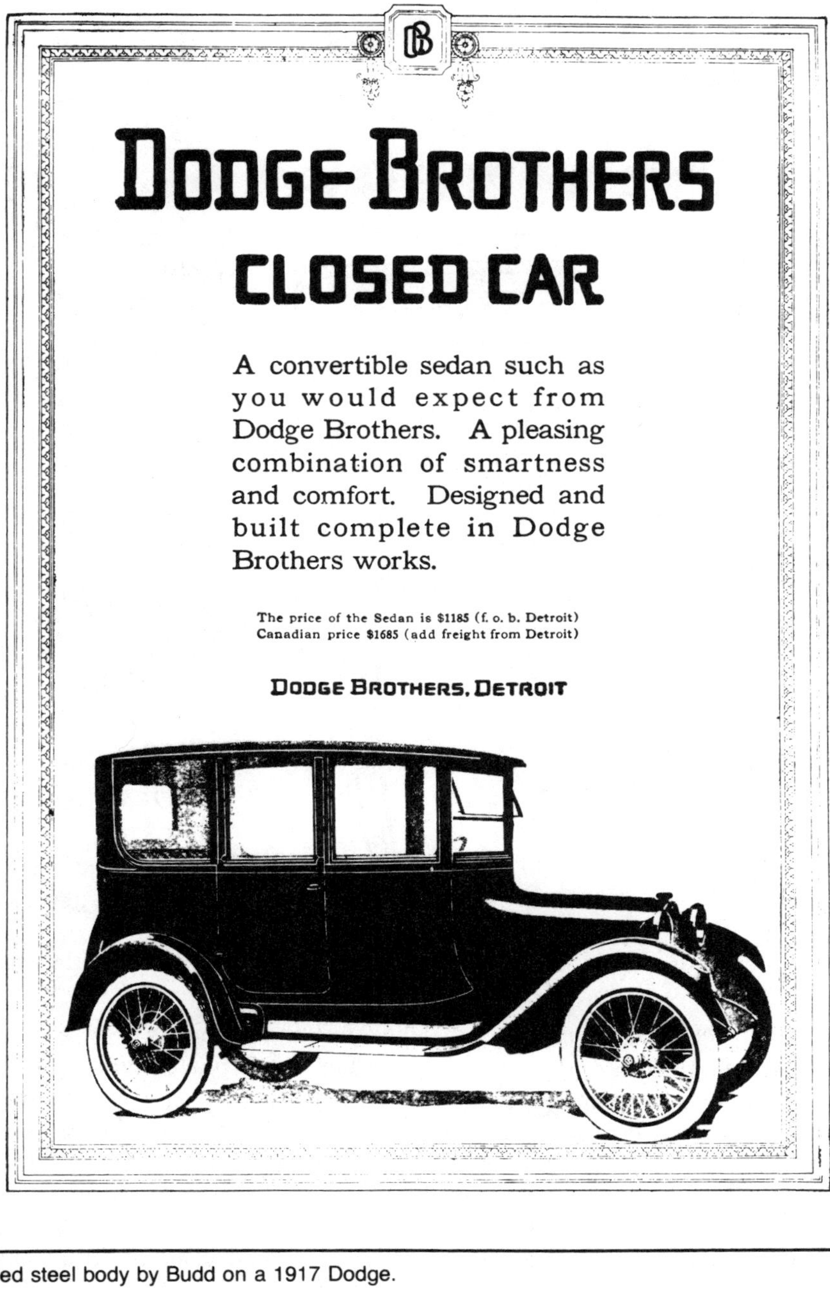

Fig. 8-15. A closed steel body by Budd on a 1917 Dodge.

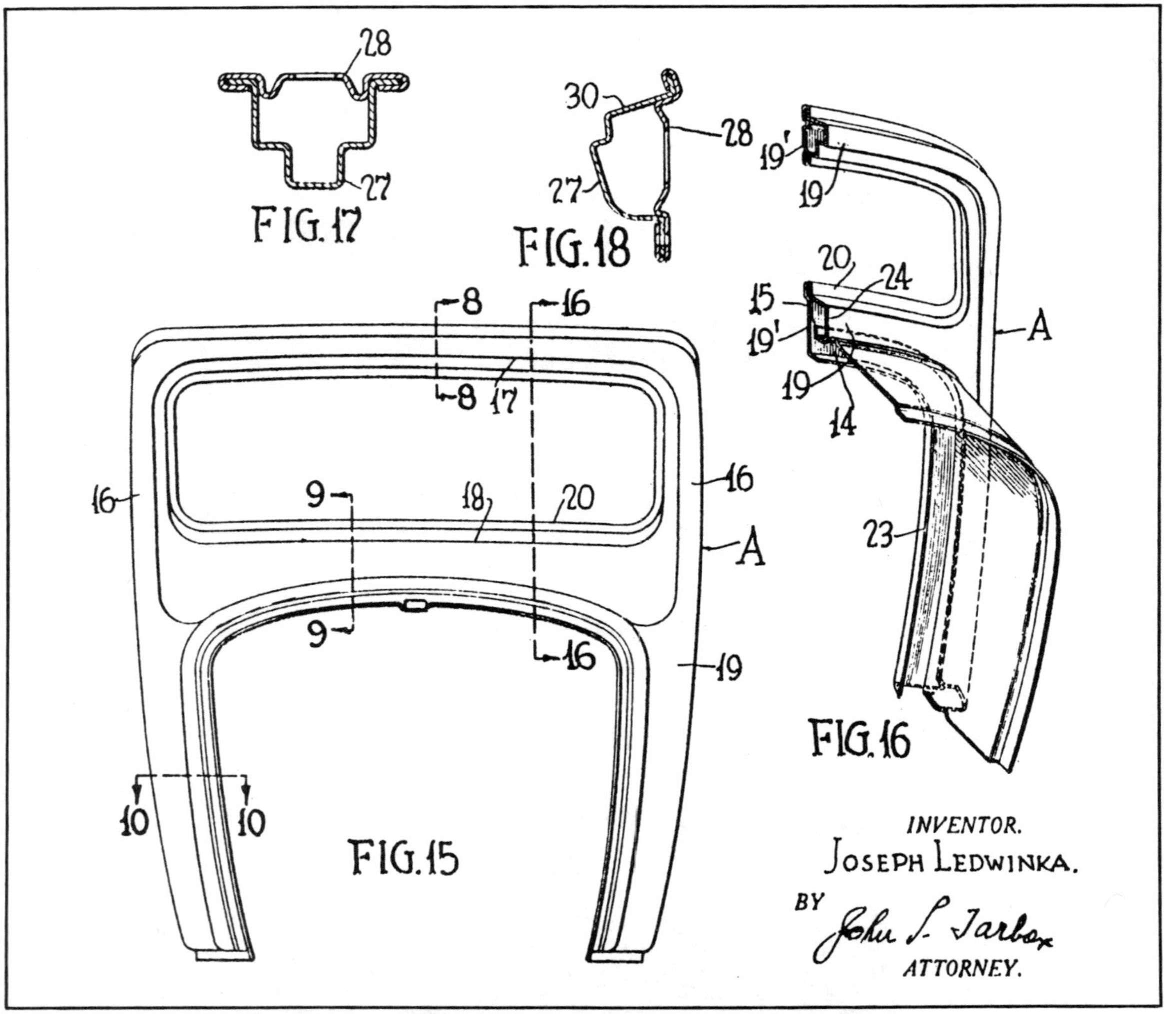

Fig. 8-16. The Budd patent drawing for all-steel body construction (showing lightweight reinforcements).

of Newark, New Jersey, began building auto bodies with aluminum panels in 1903.

Aluminum body panels came into popular use in Europe and America about 1910. The first closed bodies with aluminum panels were built by Winton in 1913. The 1912-14 Pierce-Arrow model 48 had an aluminum-panel body as a standard. Marmon was a leader in the use of aluminum, and the Model 34, of 1917, had aluminum body panels to the extent of including the radiator shell.

In 1919, French coachbuilders, such as Henri Labourdette and Lagache & Glaszmann, made some bodies where all panels were aluminum.

Charlesworth of London produced an all-aluminum torpedo body on a 25-hp Talbot in 1920. And that same year, the Aluminum Company of

America (Alcoa) sponsored the construction of several all-aluminum cars, designed by Laurence Pomeroy and built in the Pierce-Arrow works in Buffalo, New York. The complete chassis frames, engines, and axles were aluminum. The body structures were built up from a combination of aluminum castings and stampings, fitted with aluminum panels.

In 1921, Fiat and Rolland-Pilain replaced their wooden cowl structures with cast-aluminum cowls. Shortly afterwards, Renault made the main chassis cross member of aluminum for one model, and, in 1923, Montupet cast a complete body structure of aluminum for Panhard & Levassor. On Sensaud de Lavaud's 1927 prototype, the complete chassis frame had been replaced by a giant, one-piece cast-aluminum platform.

The 1939 Amilcar Compound, built by Hotchkiss to designs by J.A. Gregoire, had a cast aluminum frame with integral cowl structure. Gregoire followed the same formula for a small car he created for Jean Dupin of aluminum Francasis, in 1943, and on the Hotchkiss-Gregoire production car of 1951.

THE WEYMANN SCHOOL

Aluminum has never stopped being news, but the price is still a problem. There was an interlude of almost 20 years when metal was seriously challenged as a body panel material (though only in

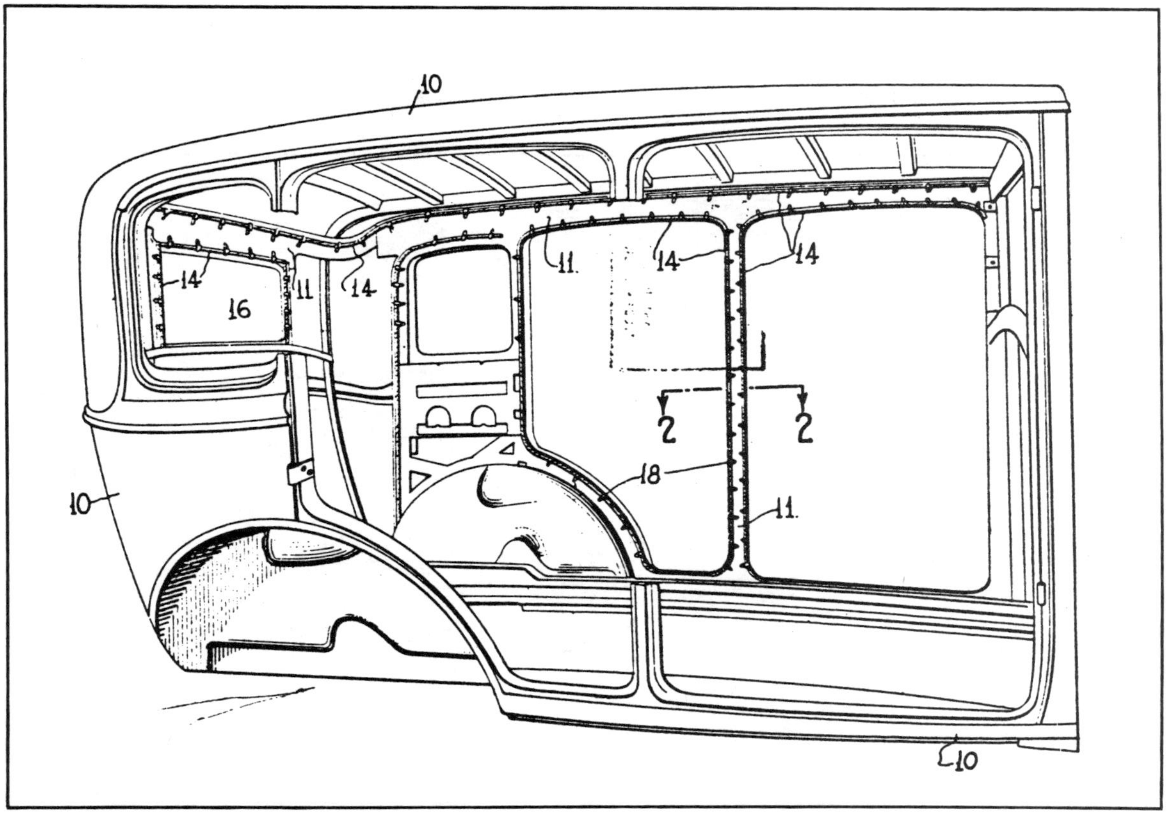

Fig. 8-17. Frame side members are shown welded into the body structure in this patent from 1930.

limited production volume). What became known as Weymann body construction was patented in 1922. It consisted of a conventional wood framework. For body panels, however, Weymann specified an impregnated fabric backed by wire-mesh reinforcement. It was light, tight, and generally free of rattles.

The fabric skin was not the central idea in Weymann's mind. Instead, it was the air space he used to separate the wooden pieces of the body framework from each other. This air space allowed them to yield to chassis flexing without fighting each other and making noise.

Charles T. Weymann was an American engineer who attended school in France (where his mother's family lived). He was a prolific inventor who made a fortune on a number of automobile accessories. He opened a coachbuilding factory in Paris to supply special bodies to industry and individual clients, while also selling licenses all over the world.

In 1923, Rover was one of the first to buy a license for Weymann body construction. Stutz began to use Weymann body construction in 1926. Weymann set up a subsidiary in Indianapolis that closed in 1931. Among Weymann license holders in 1925 were no less than 15 French car manufacturers, 25 French coachbuilding firms, six Italian ones, four in Germany, and two in Belgium.

Another Paris coachbuilder, Daste, grasped Weymann's basic idea and, in 1930, devised a system to use Silentbloc bushings to insulate the pieces of the wood framework for a body, regardless of panel material (Fig. 8-18). The Silentbloc is a small, hollow rubber cylinder encased in a metal cartridge. Daste fixed the Silentbloc to one piece and a steel core to another. The core filled the central hole in the rubber bushing, and a flexible,

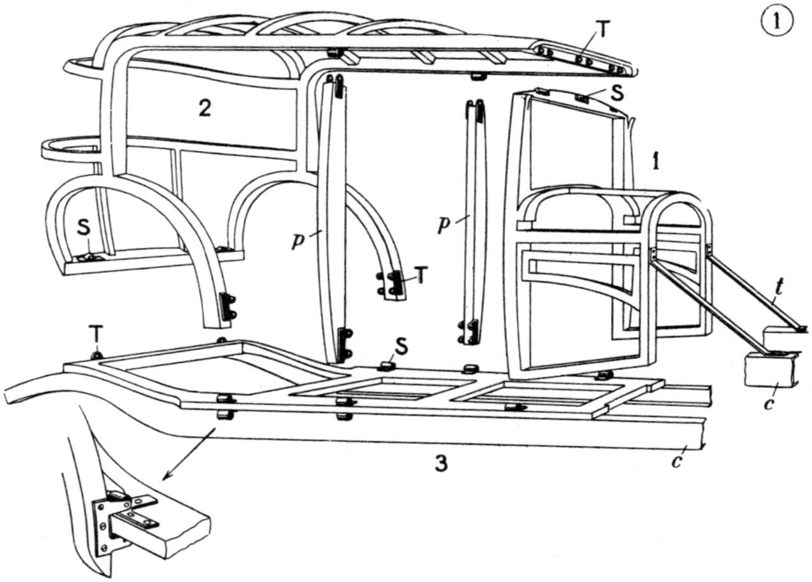

Fig. 8-18. Daste body structure with Silentbloc joints.

yet sturdy, connection had been made (Fig. 8-19). Daste proposed Silentbloc mountings for all major junctions. But Daste was too late, he was overtaken by the all-steel body.

BACKBONE CHASSIS CONSTRUCTION

Frame design evolved slowly in the years 1900-1940. The ladder-type frame dominated, and pressed steel had replaced wood entirely by 1919. A few car manufacturers took up the central tube "backbone" frame and used it for cars of all sizes. It saved weight in the frame, but necessitated body reinforcements across the floor and in the sill areas.

From 1921 to 1933, Austro-Daimler used tubular backbone frames for cars of all sizes. There were four- six-, and eight-cylinder engines mounted in front and driving the rear wheels. These cars were designed by Karl Rabe. Hans Ledwinka, chief engineer of Tatra, used tubular backbone frames for cars with front or rear-mounted engines from 1921 to 1940.

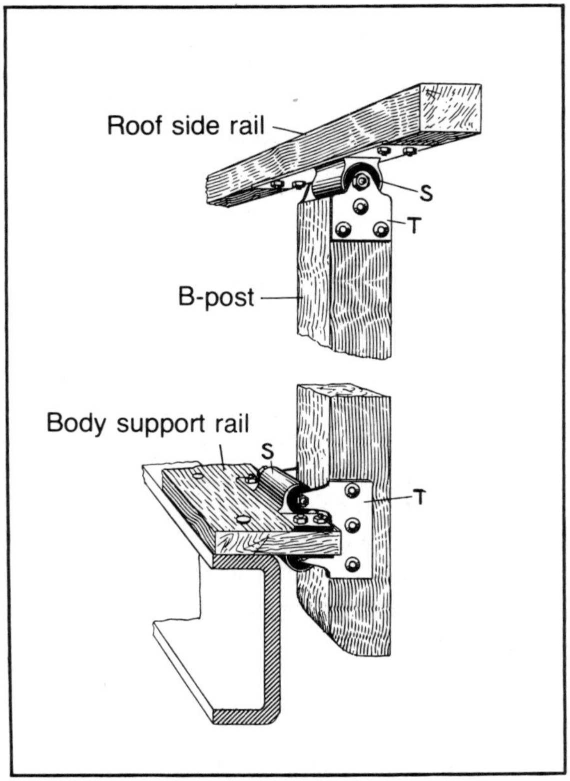

Fig. 8-19. Silentblock mountings.

The 1922 Julian, designed by Julian Brown and built in Syracuse, New York, had a tubular "backbone" frame, placed at hub level. This gave a chassis of very low build for its day. The engine was a rear-mounted, 60-hp, six-cylinder radial unit. The car, complete with two-door coupe body, weighed only 2400 pounds.

The first example of a backbone frame was actually the 1904 Rover designed by Edmund W. Lewis. Rover did not persist long with the design even though it was technically full of merit. The frame was a tubular aluminum casting to which were bolted the gearbox at the front end and the final drive unit at the rear. The suspension was carried on a subframe. The main bracket serving as a base for the front part of the body was also a large aluminum casting.

CRUCIFORMS AND DOOR OPENINGS

The backbone principle may have given rise to the creation of the cruciform frame. The Fiat 1500, of 1935, and the Mercedes-Benz 170 V, of 1937, had true cruciform frames without side members. Two channel-section beams ran together down the center line of the car under the passenger-compartment. There were forked ends for the front and rear suspension systems. Both had front-mounted engines and rear-wheel drive. Their standard bodies were four-door sedans.

On the Fiat, they were hinged at opposite ends and, in the absence of a B-post, locked into both doorsill and roof beam. With both doors open on one side, the opening was complete. Yet it did not greatly facilitate entry or exit, and the loss of body rigidity was considerable.

The Mercedes-Benz door arrangement was just the opposite: both doors were hinged on the

B-post. The front door opened toward the cowl (which many find preferable for getting in and out), but has the drawback that accidental opening of the door while traveling will result in the door flying wide open and slamming against the rear one under the force of the air flow. This risk existed also for the rear door of the Fiat 1500.

From 1920 to 1940, American cars had front doors hinged on the A-post, but rear doors hinged on the C-post (and therefore open to the same criticism). Beginning in 1941, Fisher Body (General Motors) put hinges on the leading edge of all doors. Ford and Chrysler followed after 1945.

French designers—and probably the public—always preferred to have front doors hinged on the B-post. For safety reasons, Renault began in 1951 to switch with the Fregate, Peugeot switched in 1954 with the 403, and Citroen switched in 1958 with the DS-19.

Cruciform frames have reappeared on the 1951-50 Mercedes-Benz 300, on Chevrolet and Cadillac during the late 1950s, and on the 1963-65 Buick Riviera (Fig. 8-20).

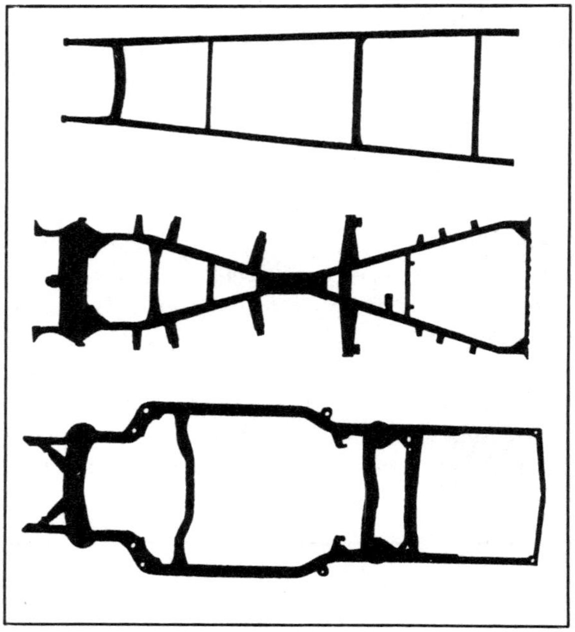

Fig. 8-20. Frame evolution (Top: ladder-type frame with straight side members, Center: cruciform frame without side members, Bottom: perimeter frame).

FROM STRAIGHT LADDER TO PERIMETER FRAME

The typical ladder frame began to change after 1920 (Fig. 8-20). Side members that had been straight and level along most of their length were reshaped to allow a lower build for the car. They were pinched in at the front end and widened inside the wheelbase and in the rear overhang. From the side, the profile showed large kickup over the rear axle (some even bent the frame to pass below the rear axle), and a curve into a gentle arc above the front axle. The adoption of independent front suspension also brought frame modifications.

As the frames lost their simplicity of design, they needed reinforcement to avoid loss of rigidity; this added weight. The adoption of heavier, stiffer frames in the 1920s stemmed mainly from a desire to improve the ride comfort. The chassis engineers found that they could use softer springs if they could take some of the flexibility out of the frame. Excellent results were obtained not only for the ride but also for the body. Twisting bodies cannot be quiet, and some of their durability is lost.

The Hotchkiss AK, of 1921, had a full ladder-type frame with additional X-bracing in the middle. That made for excellent torsional stiffness, but the X-bracing approach had the drawback of forbidding any substantial lowering of the floor.

X-braced frames began to appear on American cars about 1927. Auburn was among the first users to take advantage of the central X to eliminate two or three cross members, while obtaining a definite weight saving.

Most of the U.S. industry—Ford, General Motors, Chrysler, Hudson, Packard, Studebaker, and Nash—followed during the 1930s. In Europe, frames were beginning to disappear as the unit-construction body gained ground. The next step in frame evolution would not have been possible without unit-construction technology.

Oldsmobile and Pontiac went to perimeter frames in 1961. Ford followed in 1965, and then Chevrolet, Buick and Cadillac joined them (Fig. 8-20). With transverse torque boxes linking the widest sections to the narrow ones between the wheels, this frame was notable for its deliberate sacrificing of torsional strength.

It became an extra springing element between the road and the car's interior, and its use on modern sedans was made possible only by strengthening the body structures using the lessons learned in the production of unit-construction body shells. Body shells were made very stiff, and the number of body mounts reduced to a minimum (6 or 8 load-carrying mounts plus 4 or 6 locating mounts).

The adoption of the perimeter frame was proclaimed as the last word from Detroit (except at Chrysler, with unit-construction bodies on all models except the Imperial). The method of using it was new, but the perimeter frame design was almost 100 years old. It was used on an 1872 steam car built by Amedee Bollee.

Chapter 9

Budd and the Unit-Construction Body

TERMS LIKE *MONOCOQUE, INTEGRAL BODY,* and *frameless construction* are just different ways of saying the same thing. Most of the people who build them call them unit-construction bodies. How to define the unit-construction body? The alternative terms do a good job of describing it. It is a one-piece shell with no separate frame. The unit-construction body is the all-supporting structural unit for the vehicle.

The first serious attempt at unit construction for an automobile was made in 1903 by the Vauxhall Iron Works in London, England. The car was an open two-seater powered by a single-cylinder engine that was not unlike the Ford, Cadillac and Oldsmobile of the same year in its overall concept. But the Vauxhall did not have a separate frame. It was the sheet-metal body panels that supported the whole car (Fig. 9-1).

Though the upper body structure was rudimentary, this Vauxhall holds its place in history as the first unit-body structure. It was developed by a former marine engineer, F. W. Hodges, who had come to Vauxhall in 1897. Why did he choose this type of construction? Most likely as a weight-saving exercise. This 1903 prototype was the first Vauxhall ever. He wanted to keep it simple even to the extent of dispensing with the reverse gear.

British law stated that cars weighing less than 5 cwt (550 pounds were not required to have a reverse gear. Consequently, Hodges took a leaf from the construction of iron ship hulls and applied it to this car. As a matter of fact, the Vauxhall prototype weighed almost 50 pounds over the legal limit, and the 1904 model was duly equipped with a reverse gear. The 1904 Vauxhall also had a conventional ladder-type frame.

The first to use a closed body with true shell-type unit construction was the French Deperdussin company—but not for a car. It was the fuselage for a 1912 monoplane. Deperdussin built up a cagelike, wooden framework and covered it with stressed-skin plywood sheets.

Fig. 9-1. Shipbuilding techniques are evident in the unit-construction body of the 1903 Vauxhall.

Another curious forerunner of the unit body was produced in 1913 by Lagonda of Staines, Middlesex, England. It is indeed a curious example. It came into being for cost-cutting purposes and not because the unit body offered structural advantages. The body structure was built up from two thin side rails; sheet-steel panels extended upward from the rails. The rails were held together by a central cross member. This solid casting also provided a base for the rear engine mounts. Only the sheet metal braced the rails at the rear end.

The only body style was a two-seater roadster. The 11.1-hp Lagonda was a small and light car. It had 90-inch wheelbase and an overall length of 120 inches. The bending loads were minimal.

Lagonda offered larger and higher-powered cars at the same time (a 20-hp, 4-cylinder model and a 30-hp, 6-cylinder model) using conventional frames. The unit-body experiment led to nothing because civilian production came to a standstill in 1914. A few 11.1-hp cars were assembled in 1919, but the following year Wilbur Gunn, the Ohio-born founder of Lagonda, died, and the new management decided to concentrate on more expensive models. The next step in unit bodies was taken in Italy.

While Vauxhall and Lagonda had actually worked with stressed-skin structure, Lancia developed an inner shell for non-structural exterior sheet-metal panels. Vincenzo Lancia, former racing driver and founder of the Lancia auto works in Torino, Italy, began toying with the idea of a space frame during World War I. His thinking was directed towards lightness, and his basic approach was to space out the frame elements so as to radically increase the frame's torsional stiffness while retaining lightness and simplicity.

His solution appeared, in 1923, on the Lancia Lambda, (Fig. 9-2); it enjoyed a long and successful production life (until 1928). Lancia created an internal body structure based on a framework of wood and sheet metal. The frame structure did not reach above the body belt line, but it was the first educated, scientific attempt at this type of unit construction. With the next-generation model, Lancia reverted temporarily to frame-type chassis.

The next chapter in the history of unit bodies was written mostly in Philadelphia, Pennsylvania. The instigator was an Austrian-born engineer who served as engineering director of the Budd Company. His name was Joseph Ledwinka, and he had been with Budd since Edward G. Budd started his business in 1912. Ledwinka had met Budd while both men were working for the Hale & Kilburn Company in Philadelphia.

As more and more cars adopted the all-steel body, Ledwinka's thinking about automobile structures matured, and he began to prepare a series of inventions connected with ways to combine the body with the frame.

Joseph Ledwinka obtained a basic patent for combined body and chassis construction on September 29, 1931 (Fig. 9-3). The application was filed on June 2, 1928. That shows how far in advance of his time Ledwinka's thinking really was, and it helps to explain how Budd gained its predominance in this field.

The 1931 patent specifies a frame and a steel body to be constructed separately, and then assem-

Fig. 9-2. The closed body on a Lancia Lambda. The body shell included a platform, but did not rise above the belt line.

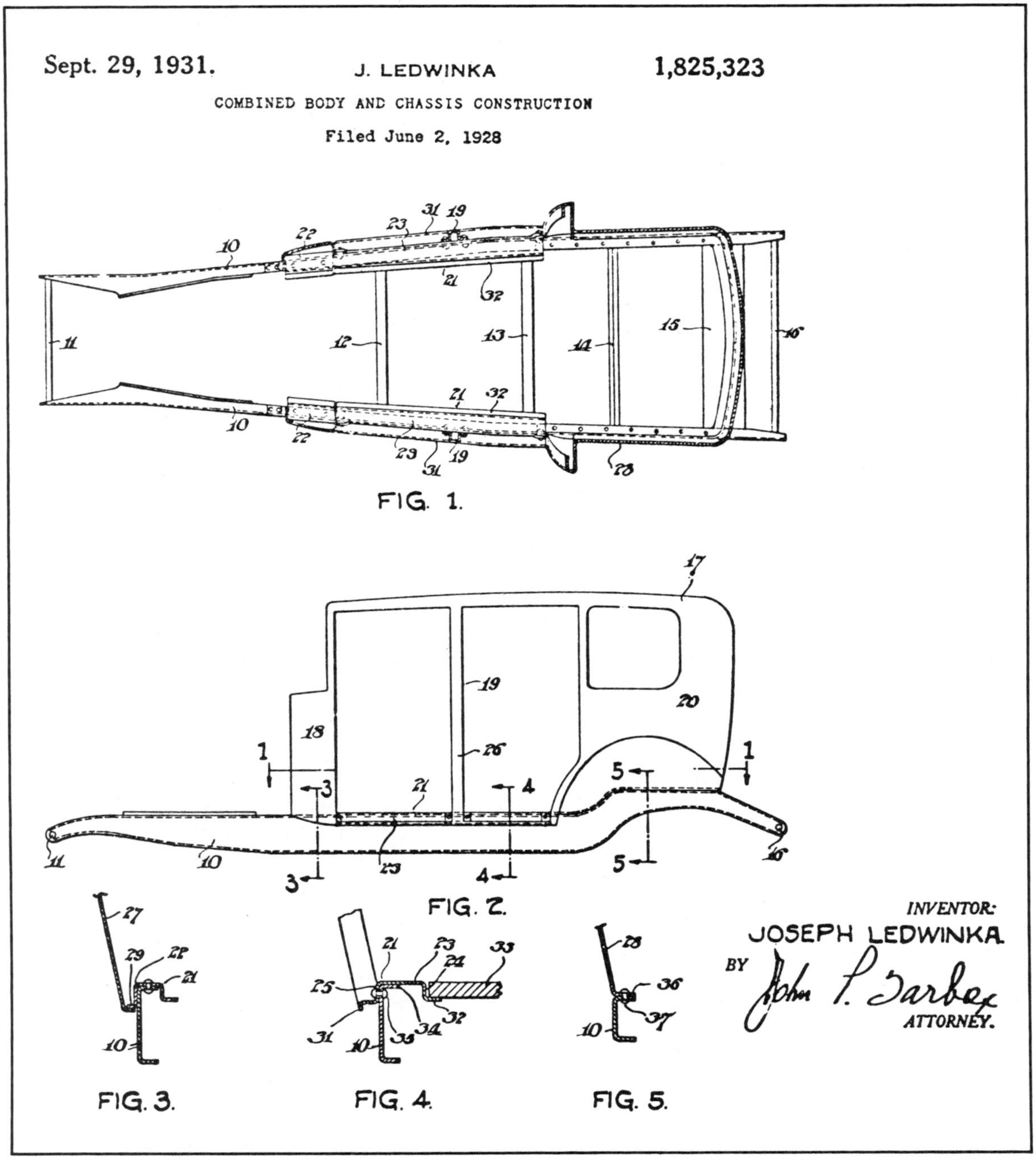

Fig. 9-3. A Budd patent for joining the body structure to the frame.

bled into a complete, single unit. This opened the way for gradual modification of frame members to better fill the needs of the main body structure. A series of subsequent patents demonstrates how Ledwink's theories evolved, and how the frame per se ceased to exist and was split up into various reinforcement sections for the body shell.

Another Ledwinka patent, issued on May 16, 1933, covers "automobile bodies constructed for the most part of substantially entirely, of sheet metal, and more particularly to bodies of this class in which the walls are constructed of unit subassemblies consisting in the main, of large unitary die stampings." The application for this patent was dated August 16, 1927.

Described in a 1936 article in *FORTUNE* magazine as "shy and softspoken," Joseph Ledwinka had by then formed a strong team of younger engineers around him at Budd. The most prominent—who contributed to his unit-body patents and sometimes applied for patents in their own names—were George Trautvetter, John W. Kelly, A. R. Lindsay, and Theodore Ulrich.

Budd had some hopes for a contract with Ford. It was no secret to Budd in 1931 that Ford was preparing a V-8 engine. The engineers of Budd were also well aware of the fact that Ford lacked a modern chassis to go with it. In an attempt to sway Ford, Budd engaged a pioneer of front-wheel-drive engineering, William J. Muller—the technical brains behind the Ruxton—to create a front-wheel drive prototype with a V-8 engine, and, incidentally, a unit-construction body according to Budd patents. The car was built, demonstrated to Ford, and Budd engineers made the most out of all the finer points of the new architectural design freedom offered by their unit-construction body.

But Ford was not swayed. The V-8 went into a conventional frame-type chassis with rear-axle drive. The prototype was not wasted, however, because Andre Citroen saw a demonstration of it while on a visit to Detroit and Philadelphia at the end of 1931. Citroen saw the logic of going to front-wheel drive in order to maximize the benefits from unit-construction. These advantages would be the same with four- and six-cylinder engines, Andre Citroen reasoned, as with the V-8. He brought the drawings with him back to Paris.

To get his new pet project under way, Andre Citroen completely bypassed his established engineering staff and hired an outsider.

This outsider was an exceptional young engineer who had worked for years with Gabriel Voisin, but had moved to Renault when Voisin's fortunes declined during the economic depression that spread from Wall Street around the world in 1930-31. His name was Andre Lefebvre.

A graduate of the Ecole Superieure d'Aeronautique, Lefebvre was a specialist on aircraft construction rather than an auto engineer. Lefebvre's logical and analytical mind had been convinced of the attractions of front-wheel drive for some time, and he had privately prepared designs for a lightweight, low front-wheel drive car before joining Renault (who preferred to ignore them).

Early in 1933, Lefebvre moved into a small office at the Citroen plant with two assistants. By August of that year, the first prototype was running.

Body design and engineering followed closely along the lines of the Budd prototype. Andre Citroen engaged a young Italian sculptor, Flaminio Bertoni, to create a new look for the front-wheel drive car. Bertoni had never worked on a car design before, and he had to rely heavily on the knowledge and advice of the chief body engineer, Raoul Cuinet. It was Cuinet's responsibility to study the producibility of every component as Bertoni shaped the car and made production drawings. He had several Budd experts assisting in questions of methods and tooling.

The car was designated 7 CV and was introduced on March 7, 1934. Bertoni's body was a magnificent work of art. The accent was on fashion rather than streamlining.

The body (Fig. 9-4) was built up from a pressed-steel platform frame to which were welded sheet-steel sections forming the cowl structure, body sides, rear panel, and roof. The engine was carried in a pressed-steel cradle, taking the form of a stub frame, that was bolted to the cowl. The front fenders, radiator grille, hood, and bumper were easily detachable.

The design elements of the 7 CV were taken advantage of to make the entire car as low as possible. Because there was no propeller shaft to the rear axle, the floor could be lowered. It was so low that passengers stepped down from the doorsill to the floor inside. Only considerations of ground clearance kept the car from being built even lower. The body had no running boards; this emphasized the low look.

The 7 CV, four-door sedan weighed 1980 pounds. For an overall length of 175 inches, it had a wheelbase of 102.5 inches—leaving minimal overhang at either end. For an overall width of 63.75 inches, it had a front track of 52.75 inches and a rear track of 52.0 inches. This gave the car a wide stance and a definite look of stability. And it was low at less than 60 inches from the ground to the top of the roof.

Rather underpowered—with its 32-hp, 1.3-liter, four-cylinder engine—the 7 CV was short on performance but long on fuel economy. It could reach 62 mph and it returned 26 mpg in all-around driving.

The same basic design remained in production for 21 years. The small engine grew to 1.5, 1.7, and 1.8 liters by 1937, and its name changed to 11 CV. For 1939, it was given a six-cylinder, 2.8-liter companion, and an enlarged front end on the same body.

Other European car companies were quick in

Fig. 9-4. The 1934 Citroen 7 CV body shell.

Fig. 9-5. The 1933 Lancia Augusta with a unit-construction body shell.

their reaction to Citroen's use of unit body construction. Lancia of Torino had actually preceded Citroen, but the Italian firm was not a mass producer like Citroen. At the industrial level, Citroen's accomplishments stand undiminished. At the level of technical innovation, Lancia stands unchallenged.

The techniques used on the old Lambda had not been forgotten. They had been polished up for the new, small four-seater Augusta of 1933 (Fig. 9-5). For the first time, here was a closed body structure incorporating the necessary reinforcements needed due to the elimination of the frame. In addition to the standard sedan, Lancia made the Augusta with a platform frame for special coachwork. See Fig. 9-6.

By 1937, Lancia was in production with its Aprilia. It had a more intelligent layout, but also more expensive components, in a unit-construction body of unusual shape and structure (Fig. 9-7). As with the Augusta, Lancia produced a platform-chassis Aprilia for special coachwork. All subsequent Lancia production cars have had unit-construction bodies.

The first company in Detroit to take an interest in Budd's unit body construction was an old customer—Hupmobile. In 1931/32, this producer was tempted to go the full route—unit body *with* front-wheel drive—but could not afford the requisite tooling. Hupp Motor Car Company ended up merely using the Budd prototype styling concept (dressed up by Raymond Loewy).

In mid-1932, when this was going on, a number of Hupp engineers decided it was time to leave. One of them was the assistant chief engineer, Earle S. MacPherson, who was to create the front suspension system that bears his name for Ford many years later. His first step after leaving Hupp was to join the Chevrolet experimental staff. He was accompanied by Ted Ulrich, Al Lindsay, and Carl Cenzer.

Their first assignment at Chevrolet was a

Fig. 9-6. The Lancia Augusta pressed-steel platform with cowl (as supplied to custom coachbuilders).

Fig. 9-7. The 1937 Lancia Aprilia.

small car project with a unit body. Nevertheless, it did not get into production at Chevrolet. It was taken over by General Motors Overseas Operations, and handed to Charles S. Crawford, chief engineer of Opel in Russelsheim, Germany, and C. E. King, chief engineer of Vauxhall Motors in Luton, Bedfordshire, England.

They were told to produce "sister" cars while both relying heavily on Budd patents. Opel worked closely with Ambi-Budd of Berlin that had been established, in 1926, as a German-based advisory and manufacturing subsidiary of The Budd Company. Vauxhall was linked with the Pressed Steel Company Ltd. of Oxford that had been formed in 1927 by William R. Morris as a result of his acquiring a Budd steel-body license and setting about its implementation. The first result was the Opel Olympia of 1936. It was so named in honor of Hitler's political upstaging of the athletic contests that summer. Vauxhall followed with its new Ten (10 hp) in 1937. See Fig. 9-8.

In the United Kingdom, Hillman and Morris were quick to follow Vauxhall's example. Ford of Germany went to Ambi-Budd for its first Taunus unit body in 1939. Ford of England dropped the separate frame and adopted unit construction for the 1950 Consul and Zephyr, conceived and designed by Earle S. MacPherson. In 1944, Volvo of

Fig. 9-8. The 1937 Vauxhall Ten.

Sweden started production of its unit-body PV 444.

Way ahead of Ford and Chrysler, GM was building itself a bank of technology and experience with unit bodies. Not that the other car companies were idle or content to continue forever with body structures resting on separate frames. One of the first to act was Stuart G. Baits, chief engineer of Hudson, together with G. Betancourt, the body designer. The 1932 Essex-Terraplane had a body type that employed certain unit-body principles. This gave an unusually rigid body structure that permitted the use of a very light and relatively flexible frame. They were not welded together; they were bolted together using several times the usual number of bolts. This permitted using the body structure's rigidity to stiffen the frame.

Chronologically, the next was Chrysler, driven by its energetic executive engineer, Carl Breer. The Chrysler Airflow was mainly his brainchild. And he was anxious to develop a body structure somewhat in keeping with its advanced aerodynamics. He engaged aeronautical engineer Alexander P. de Seversky as a consultant. Seversky later designed the P-38 Lockheed Lightning.

The Chrysler Airflow and De Soto Airstream, of 1934, used "interlocking" body and frame construction (Fig. 9-9). The body structure incorporated structural framework for high rigidity, and the frames were light and flexible. The frame existed mainly for ease of assembly and not for any subsequent function. The original plan had been to weld the body to the frame. At the last minute, the tooling was not ready, and Chrysler decided to bolt the elements together.

The body contributed about 90 percent of the torsional stiffness; the frame contributed only 10 percent. Budd might have played a part in the development of the Chrysler Airflow body structure, but the production contract went to Briggs Manufacturing Co. in Detroit.

When Ford wanted a low-priced Lincoln added to the line, it became Edsel Ford's responsibility.

Fig. 9-9. The body structure of the Chrysler Airflow.

He engaged the talent of John Tjaarda, who had been building aerodynamic prototypes named Sterkenburg since 1930, for the overall concept. Briggs Manufacturing Co. was selected to produce the bodies.

The 1936 Lincoln-Zephyr used a bridgelike body structure without a separate frame. The body shell was heavily reinforced and provided maximum torsional rigidity. The frame was a simple, lightweight structure, welded to the body, giving a form of integral construction. Regrettably, the Zephyr body shell experience was not to lead to any change for the high-volume models at Ford Motor Co.

Concurrently with Ford's experience, one of the smallest (and moribund) car companies in the midwest was tackling its own version of the unit body. When Errett L. Cord wanted to resurrect the car bearing his name, he gave the engineering assignment to George Kublin, who had been chief engineer of Moon during the 1920s. The "coffin-noise" Cord body was designed by Gordon Buehrig. It used an all-steel unit structure with a front stub frame.

The Cord's main body was a masterpiece of stress distribution, with a heavily reinforced floor pan, strong vertical bracing, and a firm roof piece. Its main shortcoming was in insufficiently rigid attachment of the stub frame to the cowl section. The car suffered from bad shake, and front end vibrations were transmitted to the main body structure.

When Auburn-Cord-Duesenberg folded, the Cord 810/812 body dies were sold to Hupp. Hupp used the dies for a small number of Hollywood models in 1938-39. Later that year, Graham-Paige Motors bought the dies from Hupp and announced the Graham Hollywood for 1940/41.

The first American car company to go into production with a Budd unit body stopped short of also switching to front-wheel drive. It was Nash. This pioneering effort was triggered by Budd's Ted Ulrich, who sold the concept to Meade F. Moore, chief engineer of Nash since 1933. In turn, Moore convinced Engineering Vice President Erick Wahlberg to give him the go-ahead on the Budd proposal. At that point, Ulrich resigned his post at Budd and joined Nash as chief body designer.

Out of this project came the 1941 Nash 600 (Fig. 9-10). It was a true unit body with stressed skin (except for door openings, hood and trunk lid). There were several new elements in detail design. Joseph Ledwinka and George Trautvetter had kept right on refining their ideas and applying for patents after the Citroen went into production. The Nash design was truly the most advanced unit body design in existence at the time of its introduction.

The cowl structure extended forward to the radiator support brackets and provided spring towers for the front suspension on both sides. The front coil springs were carried on top of the upper control arm, and the suspension stress loads were

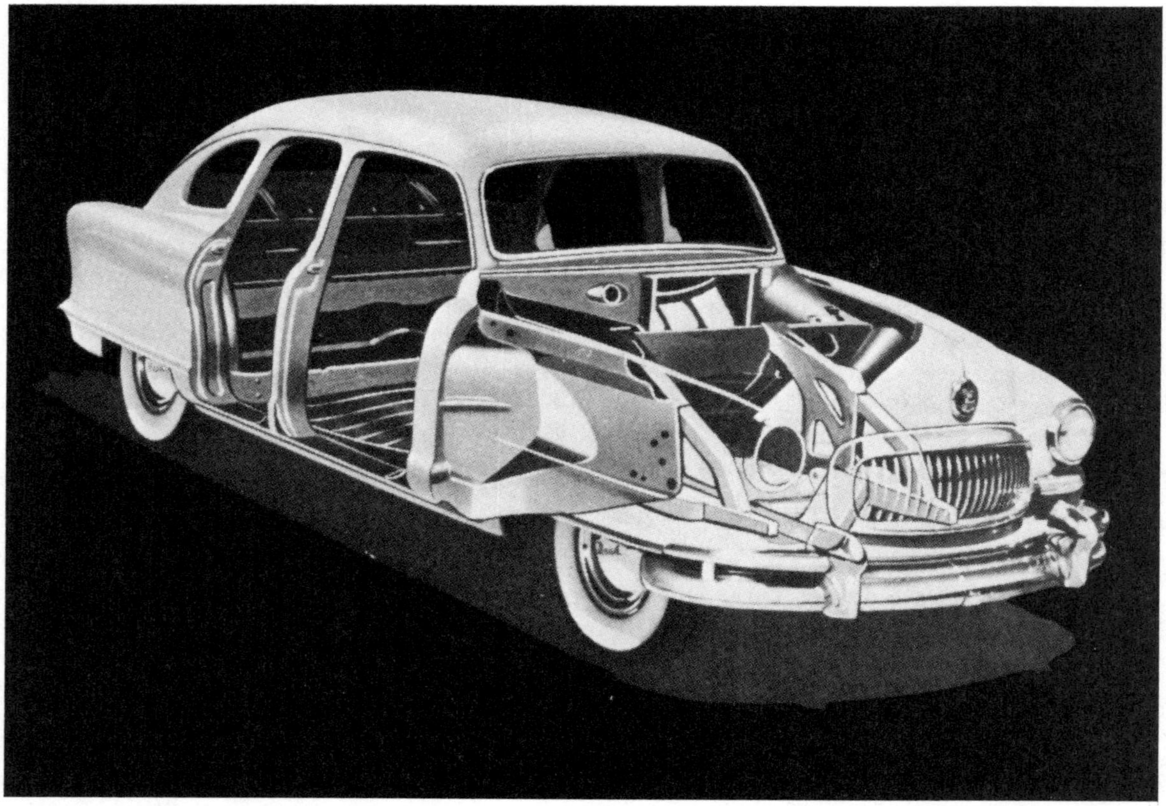

Fig. 9-10. The body structure of the Nash 600.

distributed over the full height of the body. The Nash 600 was no flash in the pan. It was suspended in 1942 as Nash converted to wartime production, but it was reinstated in 1946 and lived on with minor changes until 1950. Then it was replaced by the Rambler. The body structure of the Rambler was basically the same as that used on the Nash 600.

American Motors came up with an entirely new unit body for the 1963 models. The final shell was not greatly different from the one it replaced, but the production method was drastically new. The '63 models introduced the uniside principle. The entire side panel, minus door openings and front fenders, was made as one single stamping. Two or three styling changes have been made at American Motors since then, but the basic body of the '63 Rambler has survived the Rebel and was used on the Matador.

Hudson's early experiments had been abandoned, and World War II gave the engineers respite for tackling unit construction with a new approach. In 1939, the experienced body engineer (ex-Budd, ex-GM) Carl W. Cenzer was placed in charge of follow-up work on experimental body design at Hudson. During the war years, Hudson's chief body designer, Frank S. Spring, and one of the top engineers, Millard H. Tonoray (who became chief engineer of Hudson in 1950), talked a lot about the advantages of unit construction. They started a development program that became the responsibility of Cenzer's group. Out of this program came the 1948 Hudson with its "Monobilt" body (famous for its step-down design).

The Hudson body used a sturdy stub-frame, fore-structure welded to a cage that constituted the main body structure. Why the front stub frame? It was mainly due to space considerations on the assembly line. The plant could turn out a higher number of shorter bodies. The Hudson frame side rails ran outside the floor area and formed a kind of door sill. It gave the impression of stepping *down* into the car, and Hudson publicity made the most of that feature. There were entirely valid reasons for doing it this way. Spacing out the side members gave more leverage to resist torsional stress on the body structure.

The 1953 Hudson Jet used the same type of body construction—with far less spectacular styling. Then Hudson merged with Nash in 1954, and the Jet disappeared. The 1955 Hudsons were built with Nash unit bodies. The 1952 Aero Willys used unit bodies of a type Willys called "aero-frame." Actually it closely resembled the body structure used in the contemporary 100-inch wheelbase Rambler.

Now we are approaching the epoch when Chrysler revived its interest in unit bodies. Following the Airflow experience (1934-37), Chrysler was approached by Budd regarding the full-unit body.

The engineers were convinced. Production boss K. T. Keller gave his okay on one condition: that the assembly could be done in the existing Jefferson Avenue plant. A it turned out, the body length was a problem. It meant fewer bodies on the line or factory remodeling. Keller wasn't prepared to remodel the works so the unit body was shelved.

Nevertheless, Chrysler never stopped experimenting with unit bodies after the Airflow was dropped. One example that never made it into production was the Dodge A-227, an experimental front wheel drive car with unit construction. In 1952, Chrysler built its A-303 (a car of Rambler size and European styling influence). It was technically advanced with a 156.6-cubic-inch V-6 engine and torsion-bar independent front suspension. The only reason it did not get into production were negative conclusions in a market research study.

The next step of progress was the creation of the A-600 in April, 1955. It was a unit body for a full-size Chrysler, but 4 inches lower than the '55 production models. It was designed and developed by a team led by James W. Shank and Robert H. Kushler.

Fig. 9-11. The 1961 Lincoln body-in-white, nearing completion.

The A-600 was chosen as a basis for the next-generation production cars due in the fall of 1959 (as 1960 models). But production considerations were to play a major part in introducing modifications. The overall length of the body threatened to slow down the assembly line because fewer bodies could be processed on a given section. Shank and Kushler were forced to redesign the car with a separate front stub frame. The resulting prototype was called A-599 because, from a technical viewpoint, it was a retrograde step from the A-600.

The stub frame was simply bolted to the main body shell at the cowl. Top management was still not satisfied that the A-599 was the answer. They ordered the body engineers to try out other configurations such as a platform frame and a ladder-type unit body. Prototypes of each were built and tested. The platform-frame vehicle was called A-628 and the ladder-type body car was called A-685. Minor modifications and minor innovations were made all the time. When the most desirable features had all been combined in one single body, it was given code number A-706 and cleared for further development.

By mid-1958, Chrysler had designed, built, and tested more than 30 entirely different bodies. The A-706 vehicle was tooled for production and was ready in time for the 1960-model deadline. The Chrysler "unibody" was adopted for Plymouth,

Fig. 9-12. The 1956 Citroen DS-19.

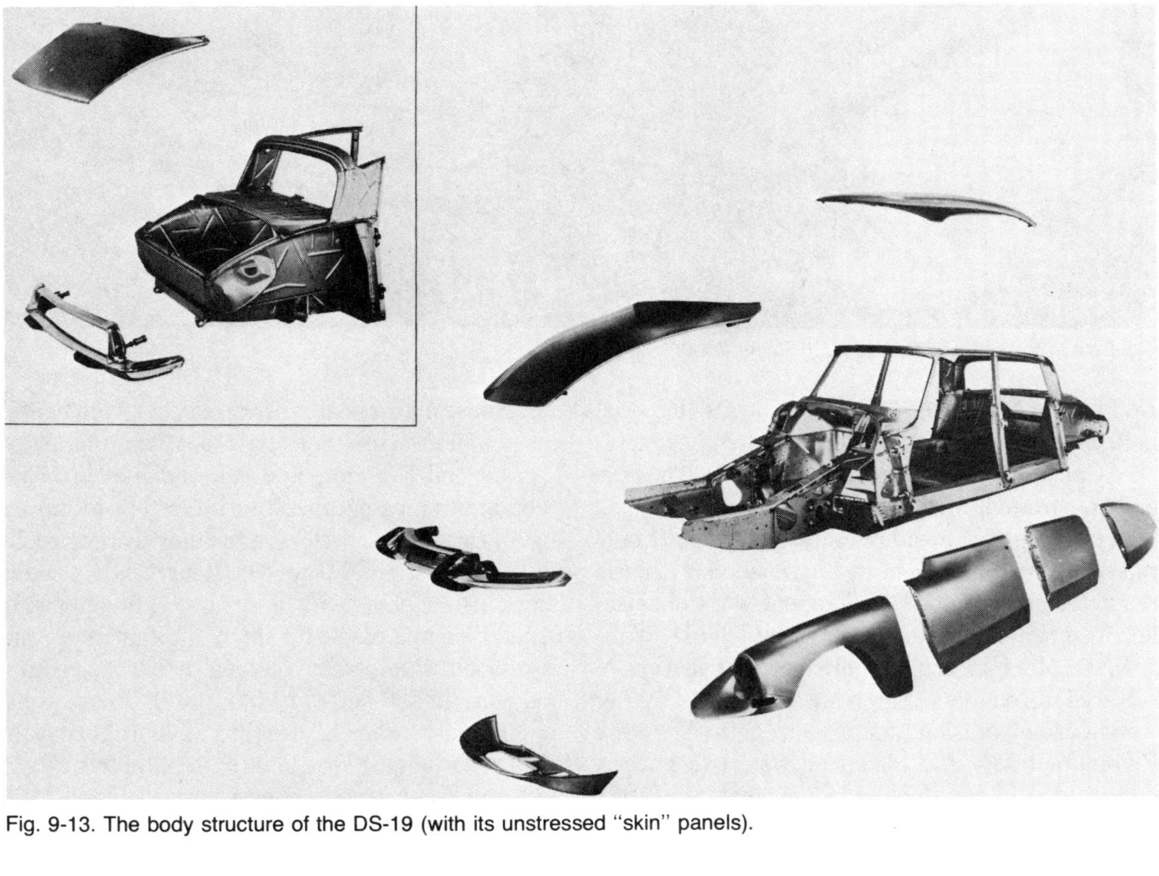

Fig. 9-13. The body structure of the DS-19 (with its unstressed "skin" panels).

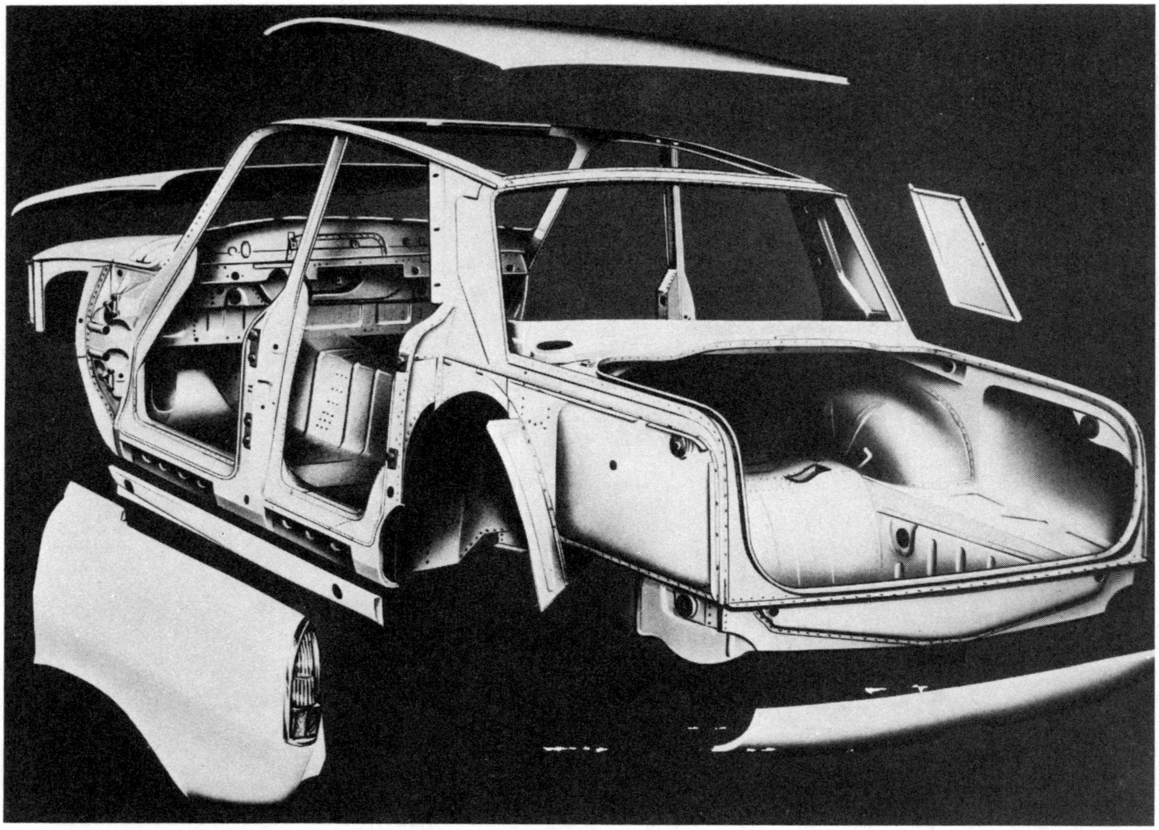

Fig. 9-14. The body shell of the 1963 Rover 2000.

Dodge, De Soto, and Chrysler (but the Imperial held on to its separate frame for another 10 years).

One might think that after such thorough study and preparation, the same basic body could enjoy a long production run without major changes. It only ran two years before a new, improved version was ready for production. Development work undertaken by an engineering group, led by John D. Withrow, resulted in a more efficient distribution of material throughout the body structure, saving weight and increasing torsional strength. The 1962 Plymouth body, for instance, was 185 pounds lighter and 37 percent stronger than the 1961 Plymouth body.

Several European mass producers went to unit construction with their new postwar models—Renault and Peugeot, in 1948, and Fiat in 1950. Separate frames became a rarity except for heavy luxury cars. Nevertheless, even they were to adopt unit-construction. Mercedes/Benz made a unit-construction body with a separate, flexibly suspended, engine cradle for its new 180 in 1953, and developed a more conventional unit-construction body for the 220-series in 1959. Rolls-Royce went to unit-construction bodies for its Silver Shadow in 1965. The biggest cars built in the Western world with a unit-construction body were the 1958-1960 and 1961-1965 Lincolns. (See Fig. 9-11.)

Two notable variants on unit-construction concepts appeared in 1955 and 1963. The first was the Citroen DS-19; it was another fabulous Lefebvre-Bertoni coproduction (Figs. 9-12 and 9-13). Here the body was not a complete shell, but merely a platform frame with a steel cage to which the sheet-metal panels were bolted. The roof was a plastic insert and, for some years, even the hood was made of plastic.

The Rover 2000 (Fig. 9-14) borrowed heavily from the DS-19. This type of construction has its importance in offering more flexibility in production, easier and less expensive collision repairs, greater opportunities for material substitution without major retooling, and greater ease of making detail styling changes. Some of their principles are planned for cars now in predevelopment stages.

In America, the breakthrough into the most popular makes was made in 1959 when the Big Three brought out compact cars with unit-construction bodies: Chevrolet Corvair, Ford Falcon, and Plymouth Valiant. A year later, the same basic bodies were extended to the Pontiac Tempest, Oldsmobile F-85, Buick Special, Mercury Comet, and Dodge Lancer. All subsequent compacts and subcompacts made in the United States have had unit-construction bodies. Today, the separate frame has practically disappeared from use in American cars. It took over 50 years for the ideas of Joseph Ledwinka to find universal acceptance.

Chapter 10

Coachbuilders and Industrial Designers

TODAY ALL THE MAJOR CAR COMPANIES HAVE their own styling departments or design centers. At the time of the birth of the auto industry, things were different. The profession of industrial designer had not yet come into being, and the car manufacturers felt their job was finished when the chassis was completed. The body type, design, and construction was the customer's affair.

The customer went—well, where else could he go?—to one of the established carriage-makers. Gradually the most successful car companies began to make arrangements direct with their preferred carriage makers for certain standard body types. If this general course of events seems somewhat oversimplified, it was certainly international in scope, applying all over Europe as well as in America. Some of the first of the famous builders of custom coachwork made their names in the world of automobiles.

THE BRITISH HERITAGE

Among Britain's first were two London-based carriage makers of great tradition. The oldest, Barker, was founded in 1710. Hopper dates from 1805. The origins of Arthur Milliner, in Northampton, can be traced back to 1760. The Mulliner family had pack-horse operators and wagoners who branched out into building some of their own equipment.

In 1863, James Young acquired the coachbuilding works of J. K. Hunter in Bromley, Kent. He produced a variety of fine styles in carriages. One of the most popular was the Bromley Brougham.

It was not until 1908 that James Young ventured into building automobile bodies. The first was a one-off on a Wolseley chassis for a private customer. After that, James Young began to seek orders for special car bodies. He did a number of unique creations on expensive chassis.

GALLIC ART AND ARTISANS

One of France's oldest carriage-making firms, Rothschild & Fils of Paris, was among the first to convert its factory from horse-drawn to horseless carriages. The factory was at Levallois-Perret. In 1905, Rothschild et Fils produced one out of every six bodies put on French cars. The Rothschild firm did not belong to the famous bankers. It was owned by a French industrial group, Rheims & Auscher.

Another example is Kellner. Georges Kellner was born in Alsace in 1831. He left home, at the age of 20, after completing his apprenticeship as saddlemaker. He was determined to get experience in foreign countries. He worked in the carriage-building establishments of Proux & Cammaerts in Brussels, and for Barker in London. When he came to Paris, in 1855, he was a skilled craftsman.

He joined Maison Aigner as a foreman and within five years was a director of the firm. He left in 1861 to start his own shop (making wagons and carriages). Notably, Kellner built an important number of ambulances during the Franco-Prussian war of 1870-71.

Customers for car bodies began knocking on Kellner's doors in 1895. By 1900, coachwork for car chassis was his main business. It grew fast in the years that followed, and in 1960 Kellner moved into a large, new factory at Billancourt that was close to the Renault plant.

Another pioneer of note was Belvallette. Alfred Belvallette took over his father's coachbuilding enterprise in Paris about 1873. He became involved with horseless carriages right at the start, and built all standard bodies for Panhard & Levassor from 1891 to 1896.

CONTINENTAL CLASSICS

North of the border with Belgium, d'Ieteren Freres was an old, established carriage maker that had originally been founded in 1805. Jean-Joseph d'Leteren left his native Holland and came to Brussels under Napoleon's rule. He started a shop to make parts for carriage makers. His first complete carriage was built in 1830. His sons Adolphe and Alexandre took over the business in 1832. They created the Roi des Belges style for a succession of royal carriages. Jean-Joseph d'Ieteran began building car bodies about 1902 (mostly on Belgian chassis, such as FN, Pipe, Minerva, and Metallurgique).

Guillaume van den Plas opened a workshop in Brussels, in 1870, to make—not carriages—but wheels and axles for carriage makers. Fourteen years later he moved to Antwerp and branched out into building complete carriages. Van den Plas became an important supplier of high-class bodies for automobiles at the turn of the century.

To the East, there were parallel developments in Germany. Heinrich Glaser of Dresden was a firm with long traditions in carriage building, the business got started in 1864. The first Glaser car body was a limousine on a 1902 Mercedes chassis.

The coachbuilding house of Wendler in Reutlingen was established in 1840 by Erhard Wendler, and the factory that Wilhelm Karmann took over in Osnabruck, Westfalia, in 1901, had been building carriages since 1874. Karmann's first contract for car bodies came from Durkoppwerke of Bielefield in 1902.

Before the turn of the century, companies were formed specifically for the purposes of building bodies for cars. Among the first were Ludwig Weinberger of Munich and Willi Erdmann of Berlin, both dating from 1898. Erdmann's business was refinanced by the wealthy merchant Eduard Rossi, in 1906, and thus was created the house of Erdmann & Rossi (which rose to great fame in the Nazi era).

WHEELS ON ROMAN ROADS

In medieval times and as long as royalty and nobility

traveled in horse-drawn vehicles, the Italians were the world's leaders in carriage design and construction. But the transition to auto body production did not occur smoothly. In most cases, the old coachbuilding houses went under and new ones were formed to cater for the needs of the car industry. The first generation of car coachwork builders, however, had been trained in the old carriage-making industry. That was the case with both Alessio and Castagna.

Marcello Alessio learned the carriage-making trade during his years of working for Locati & Torretta, in Torino, and started his own shop in 1896. Alessio first attracted attention for a mailcoach displayed at the Torino fair in 1898.

Alessio car bodies began appearing in 1899, notably for the Welleyes (forerunner of Fiat) and Lanza. Several car manufacturers placed orders with Alessio, such as Storero, Ceirano, and Fiat. Alessio secured the franchise for the Rochet-Schneider, and began importing chassis from France. He equipped them with bodies of his own design.

Carlo Castagna was only nine years old when he began his apprenticeship with the carriage building works of Francesco Mainetti in Milano. He later bought control of Mainetti and other carriage-making companies to start Carrozzeria Carlo Castagna in Milano in 1894. There was no important car industry in Italy until 1900, but in the years 1900-1903 Castagna converted his shops to coachwork for cars as its main product.

THE NEW WORLD ON WHEELS

On the American side, Studebaker might be the first name that comes to mind. But the Studebakers of South Bend were wagon builders, not carriage makers, and when the Studebaker car was born, bodies were bought from an outside supplier. On the other hand, there was an important carriage-making industry centered in the New England states. In the forefront were Brewster, Healy & Co., and Demarest.

James Brewster built up a famous carriage-making business in New Haven, Connecticut in the 1850s. On his retirement, his son, William J. Brewster, took over the firm. He set up a branch factory on Queens Plaze in Long Island city to serve the automobile trade in New York.

A. T. Demarest had its works in New Haven, Connecticut, and Healy & Co. was located in New York City.

Among America's first companies formed expressly for the purpose of building car bodies was Locke & Co. in New York City. Its founder, J. Vinton Locke, came from Watertown, Massachusetts, where he was born in 1863. He learned the coachbuilding trade while working for Healy & Co. Later he worked for Demarest. There he rose to the rank of superintendent before leaving to set up his own shop in New York City in 1902.

The distinction of being the first to set up a coachbuilding plant in Detroit goes to the C. R. Wilson Body Company. Charles R. Wilson was born in Coburg, Ontario, in 1844, and began his career in the grocery business. After he moved to Detroit, he got interested in road transport, and in 1875 started a shop for making wood parts for horse-drawn buggies. That enterprise grew into a major carriage-building works. About 1901, he turned his attention to the automobile, and became a supplier of bodies for Oldsmobile, Ford Cadillac, Peerless, and Elmore.

DESIGN SCHOOL

The first generation of men who shaped the American car were steeped in the traditions of carriage-making, and the ties extended well into the second generation. This second generation is epitomized by the young men who graduated from the Andrew F. Johnson school in New York City. Among them we find three of the Fisher brothers, Everett Lindberg, Otto Graebner, and Ray Dietrich. The

Fisher brothers started the enterprise that survives to this day as Fisher Body Division of General Motors.

Everett Lindberg became president of Briggs Manufacturing Company, Otto Graebner rose to the position of chief engineer of the Murray Body Corporation, and Ray Dietrich was a partner in the LeBaron enterprise before he went to Chrysler as chief of styling.

Just what was this precocious institute for training car designers? Andrew F. Johnson came from Westfield, in Nova Scotia, where he was born in 1854. He went to New York and studied carriage-design and construction under John D. Gribbon at the Young Men's Institute. He graduated with top honors, and the Carriage Builders Association presented him with a purse for a year's study at the Chamber of Coachbuilders in Paris.

On his return from France, he found a job as a designer for a carriage maker in Massachusetts and stayed there till he learned of Gribbon's death. He applied for the post and was accepted. Shortly afterwards, however, the Carriage Builders Association withdrew their financial support for the school, but Johnson, carried on under his own steam. His classes were essential in forming the next group of professional automobile designers in America.

Let us jump now to 1921. America at that time had three major auto body producers working on an industrial scale unheard of in Europe. They were Budd, Briggs, and Murray. Fisher was tied to General Motors and would soon join the big league. Hayes, Seaman, and Ames were in the second rank.

In contrast to these enterprises, there were a number of coachbuilders who specialized in one-off creations or very small series of identical bodies mounted on expensive luxury-car chassis. Most prominent among them were Brewster (Fig. 10-1), Locke, Holbrook, Demarest, Wood, Clayton, Bridgeport, Brunn, and Smith-Springfield.

MASS-PRODUCTION SUPPLIERS

Budd's rise to prominence was mainly due to its leadership in the steel-body technology. In the cases of Briggs and Murray, styling played a greater part.

Walter O. Briggs began his career as a switchman with the Michigan Central Railroad. A few years later he became a yard foreman of the C. H. Little Cement Company. The big change in his life came through a chance meeting with Barney Everitt (the E of EMF). Everitt had a lot of irons in the fire and told Briggs about a little wagon works he had been running as a sideline, but no longer had the time to give proper attention to. Would Briggs like to take it over? He would and he did.

Briggs Manufacturing Company grew, at the pace of the auto industry's expansion, from contracts with Ford, Chalmers-Detroit, Paige, and Abbott-Detroit. Briggs also grew by taking over other firms as well as by extending its base in Detroit.

In 1922, Briggs hired Herman C. Maise, who had presented an excellent technical paper on the design of closed bodies. Maise had been an engineer with C. P. Kimball & Co., carriage builders in Chicago, and later worked as head of the body design department at the Chicago Carriage and Trimming Company. Next, he worked for a year with the Springfield Body Company, and then joined the Hayes-Ionia Company. Maise was instrumental in the modernization of Briggs' products until his retirement in 1929.

Briggs took over the Michigan Stamping Company in 1923 and two years later was producing half a million car bodies a year. Willys ordered bodies for the Whippet from Briggs, who supplied 500 units a day, until Willys cut back production and discontinued the Whippet.

Briggs received large orders for bodies for Ford's Model A, in 1928, and continued to collaborate with Ford until 1942. Ford styling was mainly

Fig. 10-1. Brewster body on 1922 American Rolls-Royce. The style was called "oxford" phaeton.

done by Briggs designers such as Ralph Roberts, Philip Wright, and Holden "Bob" Koto.

In 1930, Briggs got the contract for supplying all Plymouth bodies. By 1934, Briggs was making the bodies also for Dodge, De Soto, and Chrysler cars. Ties with Chrysler grew even stronger when Ford began manufacturing its own bodies and, in 1953, Chrysler bought control of Briggs.

The Murray example is broadly parallel to that of Briggs. It began, in 1903, when John W. Murray, with his son James, started the J. W. Murray Manufacturing Co. in Detroit to produce bodies for the car industry. John W. Murray had been a bicycle manufacturer in the 1890s and, in 1903, he founded the Know Independent Wheel Company.

The J. W. Murray Manufacturing Co. enjoyed rapid growth on the strength of contracts with car makers like Royal, Hudson, Hupmobile, King, and various minor car companies in the Detroit area. In 1924, the company merged with the C. R. Wilson Body Company, the Towson Body Company, and the J. C. Widman Body Company, emerging under the new title of Murray Body Corporation of America.

One of the things that helped Murray win new clients was its strong styling ability. Much of that was due to Amos Northup, who joined Murray in 1925. Prior to that, he had done the masterly styling of the Wills Sainte Claire. He was to draw the 1932-34 Graham-Paige and the 1933 Willys 77.

In 1927, Murray built the bodies for Cleveland, Hupmobile, Jordan, Marmon, and Studebaker. John W. Murray died that year, and James W. Murray left the firm to go to California where he started the Murray Steel Products Company in Hollywood. He died, in 1933, at the age of 51.

Ford came through with a big contract for Model A bodies in 1938. By 1930, Ford orders had pushed out all other customers except Hudson and Reo. From 1940 to 1950 Murray worked exclusively for Ford. When that contract was canceled, the number of car makers had dwindled to a very few who mostly built their own bodies. Murray's last business were bodies-in-white for the Hudson Jet and Aero Willys in 1952-54.

UNDER DETROIT'S DOMINANCE

Long before Ford and Chrysler were taking steps to build their own bodies, General Motors acquired a majority interest in the Fisher Body Corporation. That was in 1919. It had been established in Detroit, in 1908, by Frederic J. Fisher who was the oldest son of a Norwalk, Ohio carriage builder. Fred Fisher also started the Fisher Closed Body Company, in 1910 in order to keep the two separate to deal with separate body types for separate customers. In 1916, the two were merged into Fisher Body Corporation.

Ford had been the first important client of the Fisher Body Company. Cadillac placed a large order for closed bodies in 1910. In 1911, Fisher received an order for 10,000 touring-car bodies from Studebaker, and began a vast factory expansion.

It was after its full absorption in the GM organization in 1926 that Fisher rose to front rank in the industry. The following year, GM established its Art & Color Section under the direction of Harley J. Earl. That was the start of GM Styling. Up to that time, the appearance of the cars had usually been determined by each division's body engineer.

Harley Earl came from a background of custom-built coachwork. Born in Los Angeles, in 1893, he grew up as the son of a carriage builder. After graduating from Stanford University, he joined the parental Earl Carriage Works.

After he was named general manager, in 1918, he showed drawings of the cars he designed to clients of a local Cadillac dealer, Don Lee, and took some orders. As a result of his Cadillac bodies, he met Lawrence Fisher and Alfred P. Sloan, the GM president. He was engaged to make proposals for the La Salle that Cadillac was then planning. When they were accepted, Sloan made up his mind to have Earl supervise the design of all future GM cars. Thus General Motors played its part in ushering in the era of the industrial designer. It signaled the end of the custom-built body companies in America.

Locke closed its doors in 1932. Before that, Herman A. Brunn had shut down his plant in Buffalo, New York. John Graham's Holbrook works at Hudson, New York were no longer in business. Brewster stayed alive until 1936.

Fleetwood was taken over by Fisher Body in 1927. It had been established in Reading, Pennsylvania, in 1909, by Harry C. Uhrich in order to build metal bodies. The name came from the Borough of Fleetwood, where the plant was located. Luxury cars with Fleetwood bodies won a great reputation for both styling and quality. Fisher Body still uses the Fleetwood name mainly for Cadillac models.

LeBaron's independent days ended in 1929 when Edsel Ford prevailed on Walter O. Briggs to take it over. The LeBaron name lives on as a Chrysler model.

THE END OF CUSTOM BUILDING

The LeBaron example emphasizes the problems of building custom bodies in the United States. The first LeBaron bodies were built on Packard chassis for a New York City dealer in 1921. They were designed by Ray Dietrich and Thomas L. Ribbard.

Ray Dietrich had gone to work for Brewster as an apprentice, about 1910, and Brewster sent him to study under Andrew Johnson at the Mechanics Institute. He stayed with Brewster until 1919 when he signed up with the Chevrolet design office in Manhattan (then preparing the 1922 models). He was back with Brewster in 1920, but later that year he and a colleague at Brewster, Tom Hibbard, left to start their own business they called LeBaron Carrossiers.

Hibbard had started his career in 1916 as an apprentice designer with Leon Rubay in Cleveland (who produced the bodies for all White cars). Two years later he left to accept the post of chief designer with the C. P. Kimball Company in Chicago. He came to Brewster in 1920.

The breakthrough came, in 1922, when Edsel Ford commissioned a sports phaeton on a Lincoln chassis from LeBaron. It was built at Smith-Springfield, and LeBaron looked forward to doing a steady business with Ford. But Tom Hibbard left after meeting Howard A. "Dutch" Darrin in Paris, where they set up a new partnership.

Ray Dietrich decided to carry on with LeBaron. Assisted by Ralph Roberts of Cleveland—who had come to visit on a day off from Dartmouth College and joined LeBaron on a full-time basis in 1921—he built the LeBaron name up to national fame. Until 1924, LeBaron was simply a styling studio with no facilities even for prototype construction. That changed when Dietrich bought the Bridgeport Body Company.

From that day on, LeBaron was the most prestigious of America's coachbuilders. They repeatedly astonished the industry with high-style bodies not only on American chassis, such as Pierce-Arrow, Cadillac, Stutz, Lincoln, but also imported makes such as Rolls-Royce, Mercedes, Delage, and Hispano-Suiza. Edsel Ford signed a contract for a series of special bodies on selected Lincoln models.

Murray scored a coup in 1925 by securing the exclusive services of Ray Dietrich. That left Ralph Roberts to carry on alone at LeBaron. After its purchase by Briggs, Roberts played a leading role in its styling effort. Ray Dietrich left Murray in 1931 and signed with Chrysler. When Chrysler bought Briggs, it also acquired the ownership of the LeBaron name.

As for Hibbard & Darrin, they built some spectacular cars in Paris, but the partnership broke up in 1928. Hibbard first joined GM Styling, but he left in 1934 to set up his own design studio in New York. He did a lot of work for Ford after World War II, and even put in a year as chief of Ford Styling 1947-48.

Darrin stayed in Paris and associated himself with a local banker named Fernandez. He built prototypes and one-off models for wealthy clients as well as doing some consulting work for Stutz, General Motors, Studebaker, and Dodge. In 1937, he returned to California to work as an independent consultant. His special designs for Packard became famous. After the war he was engaged by Kaiser-Frazer.

DAWN OF THE INDUSTRIAL-DESIGN ERA

No name did more to establish the industrial designer's role or the need for one than that of Raymond Loewy. In 1893, Raymond Fernand Loewy was born in Paris. He was educated at the Chaptal College and he holds an engineering degree from the University of Paris. His first job had nothing to do with cars. He was hired as a fashion illustrator with *Vogue* in 1919. Ten years later he was art director for Westinghouse Electric Company. Then he left Westinghouse to start his private industrial-design firm.

In 1930, he designed a duplicator for Gestetner that made him famous in the world of design. His first big job came, in 1934, when he was hired to redesign the Coldspot refrigerator for Sears, Roebuck & Co.

Loewy's first car design was a partly streamlined Hupmobile in 1934. In the years that followed he contributed designs to Reo Trucks, Frigidaire, Goodyear, International Harvester Co., Greyhound, and the Santa Fe Railroad. Then Studebaker engaged him to style the 1939 Champion—a totally successful package.

Norman Bel Geddes published his design philosophy in a book called *Horizons*, published in 1932, filled with pictures of airship-inspired cars. His first automotive work was done for the Graham brothers in 1928. It was intended for a succession of future models, but all of that went down the drain during the depression. Instead he was engaged to design railroad locomotives for Burlington. He did scales for Dayton and radios for Philco. Autocar hired him, in 1935, to style a new truck, but the car companies never wanted his help.

The truck connection is not unusual. Federal hired Henry Dreyfuss to design its 1938-39 trucks, and White Motor Co. engaged Alexis de Sakhnoffsky to design their all-new 1936 trucks.

In Kiev, Alexis de Sakhnoffsky was born the son of a prominent Russian nobleman. He fled his homeland after the Bolshevik Revolution and came to America. He worked first as an illustrator and then was named chief designer of Hayes Body Corporation, an office which he held until 1930.

The 1931 Marmon V-16 exterior was done entirely by Walter Dorwin Teague, Jr. who was a young student at the Massachusetts Institute of Technology. That was his first and last complete car design. Later he worked with his father's consulting firm, Walter Dorwin Teague Associates, for about 10 years, and then left in 1942 to join Bendix.

In 1907, Brooks Stevens was born, in Milwaukee, the son of an automobile engineer. He studied architecture at Cornell University and decided on a career as an industrial designer. He proposed new car designs and built some prototypes. He soon had clients for whom he evolved new designs for washing machines, bedside lamps, bicycles, and outboard motors.

But not cars. So—he decided to build his own. He displayed his Valkyrie, in 1954, with a body built by Hermann Spohn. The Gaylord followed in 1955. He was hired as a consultant for Studebaker, in 1960, and designed the Jeep Wagoneer (1963) for Kaiser.

After the war, there were fewer places for a would-be car designer to work in America, but there were more places where he could receive formal education as an industrial designer.

Edward A. Adams had established the Art Center College of Design in Los Angeles in 1930. It was one of the first institutes to have a department of industrial design, and it is still considered a leader in this field.

Car designers can also study their chosen subject at Pratt Institute in New York City, the Cleveland School of Design, and the Art School of the Society of Arts and Crafts.

EVOLUTION OF NATIONAL STYLES

In Europe, the evolution of car styles took a different course. During the 1920s, most car manufacturers set up their own body manufacturing plants or took over a coach-building firm. Thus there were no direct counterparts to Briggs and Murray, but Budd did have its own European subsidiaries.

With regard to design, national trends became discernible. From about 1927 onwards, one can see evidence of a French style, British style, German style, and so on. The "French" line was due to Figoni. He first drew what became known later as the typical Figoni coupe as early as 1928, but the first one was not built until 1935. His 1938 Delahaye 135M (Fig. 10-2) still looked modern in 1948. By that time, all his rivals had copied it.

Joseph Figoni was born in the Paris suburb of Boulogne in 1898 as the son of an Italian immigrant worker. With little or no schooling, because he went to work as a shepherd when he was eight years old, he got his start in coachbuilding by becoming an apprentice sheet-metal worker in the Mulbacher shops in Paris. Later he also learned wood sculpting and cabinetmaking in other jobs. Prior to World War I he had some experience in airframe construction. It was in 1921 that he started his own coachbuilding shop, with one apprentice as his only helper.

Figoni bodies soon appeared on Ballot, Bugatti, Delahaye, Talbot, Hispano-Suiza, Packard, Chrysler, and Alfa Romeo chassis. Joseph Figoni revealed himself as an extremely talented designer. His method was to build a scale model before he started to prepare full-scale drawings. He

Fig. 10-2. The Figoni convertible on a 1938 Delahaye 135 M.

hand-carved the models himself and hired designers to do the drawings.

For over 10 years, the firm was known as Figoni & Falaschi. Figoni had gone into partnership with Ovidio Falaschi, a successful businessman from Florence, who financed Figoni's operations in the difficult years before and after World War II. Figoni's biggest rival was Chapron, who had a bigger factory and consistently outproduced him. They were contemporaries.

FRENCH FASHIONS

In 1886, Henri Chapron was born in the Sologne area of France famed for its game-rich forests and village inns. At the age of 14, he left home to start his apprenticeship with a coachbuilder in Paris. He "lost" (as he once told me) seven years to military duty, and finally established his own shop at Neuilly in 1919.

The first Chapron bodies were utility vans and ambulances on Ford Model T chassis. In 1921, he broke into the luxury-car field with a sumptious body on a Cadillac chassis.

Chapron moved to larger premises at Levallois in 1923 and began series-production as well as special one-off bodies. In his peak years, 1928-31, he produced about 500 bodies a year. His body on a Delahaye 135 from 1938 created a sensation in the fashion world (Fig. 10-3).

After World War II, Chapron built bodies for the Talbot Lago Grand Sport (Fig. 10-4), Salmson 2300-S, Hotchkiss 2050, Delage, and Delahaye. Independently of Citroen, he created the DS-19 convertible in 1958. It was so successful that it was still being built in 1970. The most spectacular of the Paris coachbuilders was probably Saoutchik.

Jacques Saoutchik was originally a cabinetmaker, of Russian origin, assumed to have fled the Russian Revolution and come to Paris about 1920. No one remembers exactly when he first became interested in cars, but the first Saoutchik bodies date from the early 1920s. Right from the start, the Saoutchik bodies were remarkable for their ornate finish and opulent look. His early styling themes were borrowed from the railroad carriages (Wagons-Lits) of long-distance trains such as the Orient Express.

In later years, Saoutchik drew great inspiration from Figoni's work and copied some of his designs, merely adding frills, louvres, and chrome. His company went under about 1953.

Many notable creations were signed Pourtout. Marcel Pourtout formed a coachbuilding company at Rueil-Malmaison, in 1925, and built bodies on Peugeot, Bentley, Hispano-Suiza, Talbot, Delage, Delahaye, and Renault chassis. He invented what might have been the original hatchback. He patented and sold it as the Eclipse roof. It tipped to give access to the trunk.

Other coachbuilders of the French school in the 1930s were Letourneur & Marchand, Guillore, Labourdette (Fig. 10-5), Van Vooren, Henri Binder, and Georges Gangloff. Among the old guard, it was Kellner that held out the longest.

During World War I, Georges Kellner kept the plant busy with production of airframes for the fam-

Fig. 10-3. The Chapron convertible on a 1947 Delahaye 135 M.

Fig. 10-4. The Chapron coupe on a 1950 Talbot Lago Grand Sport.

Fig. 10-5. The Labourdette convertible on a Delage D8-120 from 1939.

ous SPAD fighter/reconnaissance plane. Georges Kellner died in 1931. His son, Jacques, directed the business until 1941. He was arrested by the Gestapo and then executed in 1942.

The d'leteren company lost its clientele, in the depression of the 1930s, and began assembling Studebakers. The company still survives as a Volkswagen dealer.

TEUTONIC TRENDS

In 1904, Hermann Graber was born the son of a wagonmaker, in Wichtrach, near Berne, Switzerland. He took over the company in 1925 after years of experience with coachbuilding in Paris and Geneva. In 1927, he built his first car; it was a cabriolet on Fiat's 509 chassis. Graber had a contract with Alvis for coupes and convertibles during the 1950s and 1960s.

In Germany, the style-leader was Erdmann & Rossi (Fig. 10-6). Karosseriewerk Sindelfingen had been taken over by Mercedes-Benz and they built a variety of styles for cars of all sizes (Fig. 10-7).

Karmann took the industrial route in 1921 by starting to produce a large series of bodies for Aga, NAG, and FN. After 1927, Karmann specialized in convertible bodies. Adler and Hanomag were large customers.

Gustav Drauz set up a shop in Heilbronn to build car bodies, in 1900, and soon received orders from NSU. Drauz became a supplier to Ford, in 1931, and expanded its facilities two years later to be able to make all-steel bodies. In 1935, Drauz took over the firm of Alexis Kellner in Berlin. They had been building car bodies since 1910. Drauz was eventually taken over by NSU.

Ambi-Budd was established in Berlin in 1926,

Fig. 10-6. The Erdmann & Rossi two-seater convertible body on a 1939 Mercedes-Benz 540-K.

Fig. 10-7. A Sindelfingen four-seater convertible body on a Mercedes-Benz 540-K.

and soon grew to take the place of the largest supplier of bodies to Germany's car industry. The main customers in the early years were Adler (Fig. 10-8), Hanomag, Ford, BMW, and NSU.

Hermann Spohn of Ravensburg built most of the bodies for Maybach, beginning in 1921. In the 1930s, Gläser became famous for its high-quality convertibles supplied to most of Germany's car makers of that time, including Auto-Union's Horch (Fig. 10-9). Weinberger became a BMW dealer; from 1933 to 1939, he specialized in BMW convertibles.

The origins of Karosseriewerk Joseph Neuss go back to 1857 when Joseph Neuss began building carriages in a shop in Aachen. His son became chief of the firm, in 1899, and Neuss bodies became famous after the move to Berlin in 1910. He sold out, in 1925, to Karl Trutz who restricted Neuss bodies to luxury car chassis. Trutz retired in 1933, however, and ownership of the Neuss company fell into the hands of Erdmann & Rossi.

BRITISH COACHCRAFT

The British style was exemplified by the knife-edge treatment given by Barker, Hooper, Park Ward, and others to Rolls-Royce in the 1930s. Some did very well during the depression years. Swallow Coachbuilding Co. blossomed out into Jaguar Cars, Ltd. The brothers Richard and Alan Jensen made their name a make of car instead of merely a body. Just as the coachbuilders of France and Germany failed to survive the immediate postwar period, the British ones went under—but at a later date.

When the Bentley Continental (Fig. 10-10) appeared in 1952, anyone would have predicted a great future for its creator, H. J. Mulliner. Instead it was taken over by Rolls-Royce and merged with Park Ward in 1961. Park Ward had been organized in London in 1919; it became Park Ward & Co. in 1924. Its bodies of the 1930s were conservative, but low in roofline (Fig. 10-11).

Hooper (Fig. 10-12) and Barker were taken over by Daimler, which is now part of British Leyland. Thrupp & Maberly fell into the hands of the Rootes Group. Tickford's was taken over by Aston Martin.

Vaden Plas, originally a London branch factory of the famed Belgian house, worked mainly for Bentley during the 1920s. It was taken over by Austin in 1946. Mulliners of Birmingham was taken over by Standard-Triumph International, in 1958, and Lancefield closed its doors in 1960. Others who fell by the wayside include Freestone & Webb, Windovers, and J. Gurney Nutting.

Only in Italy has the coachbuilding industry known how to survive and prosper. It's not just because the Italians had the artistic flair that enabled them to take the lead in world styling trends after 1945. It is also because they have understood how to adapt their functions and services to the ever-changing realities of the car market and the needs of the car manufacturers.

Fig. 10-8. A 1929 Adler with Ambi-Budd, all-steel body.

Fig. 10-9. A Gläser convertible on a 1937 Horch straight-eight.

Fig. 10-10. A 1952 Bentley Continental, two-door sports saloon by H. J. Mulliner.

Fig. 10-11. A Park Ward body on a 1937 Bentley 4¼-Litre.

DOOM AND RENAISSANCE IN ITALY

The ingredients in this success story do not have ancient roots. Throughout the 1920s, you will look in vain for anything that set the Italian coachbuilders apart from those of Paris, London, or Berlin. Italy's skilled artisans were second to none, but they lacked creative leadership. That was to change dramatically over the space of five years at the outside. But no analysis of this period of change can replace the concrete examples provided by history.

In the background were the old-established houses of Sala, Castagna, Garavini, Alessio, and others that were headed for extinction. Garavini was one of the pioneers of the Italian luxury-car body. Eusebio Garavini was born, in 1881, in Carpinello di Forli, between Bologna and the Adriatic. At the age of 18 he went to look for work, in Torino, and was hired by Diatto. After he had learned all about railroad-carriage construction, he moved to Locati & Torretta, and from there to Carrozzeria Taurus, where he held the title of technical director. In 1908, he started his own coachbuilding business, Carrozzeria Piemonte, in Torino.

Diatto became interested in Garavini's creations, but found his factory too small. That led to the formation of a joint enterprise, Diatto-Garavini, with a work force of 50 men in 1911. The Diatto affiliation ended during World War I.

During the 1920s and 1930s, Garavini built fine bodies on Fiat, Isotta-Fraschini, Lancia, Itala, Alfa Romeo, Delahaye, Citroen, and Bugatti chassis. After World War II, the Garavini company branched out into making bus bodies and drifted away from cars. Eusebio Garavini died in 1947. The company was taken over by Viberti.

Carlo Castagna died in 1914 and his son, Ercole, took over the business. Ercole Castagna learned his coachbuilding skills in his father's shops. He also studied body engineering at the school of car mechanics and design in Milano.

The designer of the Castagna bodies (Fig. 10-13) from this period was Emilio Castagna, younger brother of Ercole, who had studied art and design at the Academy of Belle Arti at Brera. He retired in 1934, but a younger man named De Lorenzi had been acting as chief designer since about 1924 (Fig. 10-14). Castagna was then famous for luxurious bodies on Alfa Romeo, Fiat, Isotta Fraschini, OM, and Bianchi chassis. The activity declined in the 1930s and Castagna declared bankruptcy in 1953.

Cesare Sala began building car bodies in Mil-

Fig. 10-12. A four-door Hooper body on a 1934 Rolls-Royce 20/25 HP model.

Fig. 10-13. The Torpedo body by Castagna on an Alfa Romeo RLSS chassis, 1925.

Fig. 10-14. A Sports tourer by Castagna on Alfa Romeo 6C 1750 GTC chassis, 1931.

ano, in 1901, and became famous for fine bodies on chassis such as Isotta-Fraschini (Fig. 10-15), Ansaldo, Lancia, Fiat, Alfa Romeo, and Bianchi during the 1920s. This was mainly due to Oliviero Bergomi who was chief designer for Sala from 1920 to 1927. Sala built its last car bodies in 1933.

As for Alessio, the end came in 1929. After the founder's death in 1912, the business was led by his son, Giovanni Alessio. He landed a large contract with Itala, in 1922, and produced smaller series for Isotta Fraschini and Lancia in the years that followed. By that time, new names were coming into the limelight—Zagato, Ghia, Farnia, Bertone, Touring, and others.

FRESH STARTERS IN MILANO AND TORINO

Ugo Zagato was only 15 years old when in 1905 he was apprenticed to a general engineering and coachbuilding factory in Cologne, Germany. He stayed there for more than 10 years, and returned to Italy after having absorbed the typically German insistence on robustness and strength as well as their passion for quality and precision.

On his return to Milano, he worked for a short period with Carrozzeria Varesina. He spent some time at the Stabilimenti Pomilio to learn the principles of airframe construction.

After setting up his own shop, in 1919, Ugo Zagato built his first bodies on Fiat 501 chassis. He was a leader in lightweight construction and among the first to use aluminum panels exclusively. His roadster body on the Alfa Romeo 6C 1750 of 1929 became a trend setter (Fig. 10-16).

Ugo Zagato was also a director of the Pomilio

Fig. 10-15. A Sedan de Ville by Cesare Sala on a 1924 Isotta-Fraschini 8A chassis.

Fig. 10-16. An Alfa Romeo 6C 1750 S with "spider" body by Zagato.

aircraft works in Torino. That enabled him to follow developments not only in lightweight materials and construction techniques, but also in aerodynamics.

In 1936, Ugo Zagato signed a 10-year contract which bound him to work exclusively for Alfa Romeo. Zagato bodies made contributions to many racing victories. Other makers, notably Lancia, were eager to obtain the same advantages. Lancia became the principle client after 1948; others were Siata and Fiat. Ugo Zagato died in 1968, but his sons Elio and Gianni had been partners in the business for some years, and they are keeping Carrozzeria Zagato alive to this day. Its production program centers on the Zele battery-electric commuter-car.

In the 1950s, the Zagato-bodied Fiat-Abarth coupes were frequent race winners (followed by Porsche Abarths in 1961). When the Maserati brothers began building OSCA sports cars, they came to Zagato for aluminum bodies. Zagato even produced a special coupe on a shortened Aston Martin DB-4 chassis in 1961. Its Lancia and Alfa Romeo sports coupes were seen in races and rallies all over Europe. Then the company fell back to prototype construction and made an interesting Volvo 244 coupe in 1975 and a Briston 406 convertible in 1976.

THE BERTONE FORMULA

One day a youngster named Giovanni Bertone left the family farm near Mondovi in Piedmont and went to work for Verrua, a local cartwright. After learning the trade thoroughly, he went to Torino in 1907 and found a job with Diatto. The Diatto brothers were cartwrights, too, but on an industrial scale. In addition to wagons and carriages, they built bodies for streetcars and railroad rolling stock.

Giovanni Bertone left Diatto in 1912 and started his own shop in Torino. At first he had no thought of automobile bodies, but began making and repairing horse-drawn wagons and carriages.

About 1912, he got an order from SPA for wooden platforms with seats. When SPA gave him an order for 9000 torpedo bodies, Bertone discontinued the construction of bodies for horse-drawn vehicles. After the SPA contract, he received orders for bodies from Fiat, Itala, Chiribiri, Aurea, SCAT, and Ceirano.

Carrozziera Bertone assumed industrial dimensions when Lancia gave him a contract, in 1928, for bodies for the Lambda, Dilambda, and followed up with body orders for the later Artena, Augusta, and Astura.

Bertone's son, Nuccio, joined the business when he completed his formal schooling, and tried his hand at car design. His first complete creation was a sports roadster for a 1933 Fiat Balilla chassis. But his father decided to rely mainly on outside consultants. Mario Revelli drew several special bodies for Fiat chassis for Bertone, beginning in 1934.

After the war, Nuccio Bertone did his own

Fig. 10-17A. The Siata Amica convertible by Bertone, 1950.

Fig. 10-17B. The Scaglione-designed body for the 1952 Fiat-Abarth 1500.

styling for some years (Fig. 10-17) before he began consulting with Franco Scaglione (an eminent stylist and engineer). Franco Scaglione was born in Florence in 1916 and began his career as an industrial designer in 1946 in Bologna. Moving to Torino in 1952, he began to collaborate with Carlo Abarth. His Fiat-Abarth 1500 received worldwide acclaim (Fig. 10-17A). That car also led to a long association with Bertone, where he did a series of streamliners on Alfa Romeo chassis. See Fig. 10-17B.

Giovanni Bertone retired from active participation in the business in 1945, and he died in 1972. Nuccio Bertone has been in sole charge throughout the postwar era. The transformation of Bertone into a modern industrial concern came in 1954 at the instigation of Dr. Rudolf Hruska, who was then director of production for Alfa Romeo.

Scaglione had created a lovely coupe on the Giuletta platform, and Hruska wanted about 10 times more of them than Bertone had capacity to build. Hruska offered certain guarantees, and Bertone built a new plant, for the Giuletta Sprint. The collaboration with Scaglione ended in 1957, and the following year Bertone lured a young designer away from Fiat. His name was Giugiaro.

Giorgetto Giugiaro was born in 1938 at Garessio in Cuneo province. At the age of 19, he broke off his art studies at the Albertina Academy in Torino and joined Fiat. After moving to Bertone, Giugiaro designed the Alfa Romeo 2600 coupe in 1961 and the Giulia GT coupe in 1962. They were followed by the Lamborghini Miura, in 1965, and the Fiat 850 Spider later that year. His last design for Bertone, the Fiat Dino coupe, did not appear until 1966. Giugaro left Bertone in 1965 and worked for a couple of years with Ghia before he established his own firm, Ital Design, in 1968.

Nuccio Bertone did not have to search long for

his successor. One day a young man came to show him some drawings he had made; he was hired on the spot. Marcello Gandini started off by designing the Lamborghini Marzal and Bravo. His list of credits include the Alfa Romeo Montreal coupe, the Alfa Romeo Carabo, the Maserati Quattroporte, the Alfa Romeo Navajo, the Ferrari Rainbow, the Jaguar Ascot, and the 1978 Sibilo study on a Stratos chassis. Among mass-production cars, count the Audi 50 and VW Polo of 1975.

The advent of the X-1/9 turned Bertone into a full-scale car assembler (receiving the mechanical elements from Fiat and building the complete car). While Bertone has designed some scooters for Innocenti-Lambretta, boats like the Clipper-37, and a line of accessories for motorcycles such as the Stratos crash-helmet, the firm remains firmly anchored in the automobile industry.

THE FARINA FAMILY

In 1910, a young craftsman in Torino, Giovanni Farina, decided to set himself up as a car-body manufacturer. He was then 27 years old, and had been working for Alessio since he was 12. A younger brother, Carlo, became his partner in the formation of Stabilimenti Farina. There was a third brother, Battista, then barely 18, whom they also invited to join the firm. Apparently the younger brother's talent for design had already impressed them. When the Farina brothers were negotiating for a batch of bodies on Fiat chassis, Battista "Pinin" Farina was introduced to the Fiat representatives as "our designer," much to their amazement.

The Fiat contract led to others; and the business prospered. During World War I, Stabilimenti Farina built 100 army truck bodies a day. In the

Fig. 10-18. The Lancia Astura four-seater convertible by Pinin Farina, 1936.

Fig. 10-19. The front end of Pinin Farina's new Lancia Astura sports convertible, 1937.

early 1920s, Farina became the exclusive supplier of bodies for the Temperino car, at the rate of 1000 to 1300 units a year. After Lancia introduced the Lambda, Farina established a regular business with this firm, and young Pinin formed a fast friendship with Vincenzo Lancia.

Stabilimenti Farina also built bodies on the prestigious Isotta Fraschini and soon ranked with Castagna, Sala, Garavini, and Alessio as one of the leading coachbuilding firms of this era.

Always the most forward-looking, Pinin Farina made a study tour of American industry, but turned down an offer from Ford so he could go back to Torino and marry his sweetheart, Rosa, and continue his work at the Stabilimenti. About the time he was 33, however, he was no longer satisfied with his position there, and Vincenzo Lancia encouraged him to start his own business.

The break with his brothers came in 1928. At the end of that year, Pinin was the owner of a small shop, with 18 men, capable of building 50 bodies a year. New financing was secured in 1930, and his firm was incorporated as Carrozzeria Pinin Farina. They also moved into new premises around the corner from the Lancia works.

In his first decade as head of his own enterprise, Pinin Farnia built special models not only on Fiat and Lancia chassis (Figs. 10-18 and 10-19), but also Alfa Romeo, Hispano-Suiza, Rolls-Royce, Cadillac, and Mercedes-Benz.

His brother, Giovanni, continued as head of the Stabilimenti Farnia. The company began to diversify outside the coachbuilding industry after the loss of Pinin's designing talent. A separate department for aircraft engines was started in 1929. Giovanni also distinguished himself as an inventor of some note. He patented a hydraulic power top for convertibles in 1933. That year he also began production of a dual-circuit hydraulic brake system that was marketed on the Sabif trademark. In 1939, he added a line of Sabig adjustable hydraulic shock absorbers (anticipating the Koni).

Pinin Farina's fame and fortune grew fast as the result of a firm basis in the excellence and innovation of the products. His streamlined Lancia coupe for the 1936 Mille Miglia was spectacular. Even if its lines were drawn purely by intuition, it turned out to be aerodynamically correct as measured against today's knowledge. By 1939, he had 500 workers and the factory was turning out two cars a day. During the war, the activity was practically at standstill, but a fresh start was made in 1946. The following year, Pinin Farina built 186 auto bodies (with a work force of 425 men).

Stabilimenti Farina, in contrast, dropped out of coachbuilding to concentrate on component manufacture and engineering subcontract work. Giovanni retired with the company in the hands of Attilio and Giuseppe (the latter securing the title of world champion driver in 1950 as a member of the Alfa Romeo team). But by 1953, the Stabilimenti was closed down. Giovanni died in 1957.

INNOVATION TEMPERED BY GOOD TASTE

Pinin Farina made big waves when his lovely coupe on the Cisitalia 202-C chassis was shown in 1947 (Fig. 10-20). Here was something new, a synthesis of what was good in the various trends that had been pulling Italy's car designers this way and that since the war ended. It uniquely interpreted the public's expectations of a true postwar car. The flowing lines blended with a certain ovality in which all surfaces played a part.

After the Cisitalia, Pinin Farina refined his new style for bodies on Maserati and Alfa Romeo, and even tried to adapt it to a Bentley. In 1951 came the Lancia Aurelia GT B-20; it went into production as a catalogued model (Fig. 10-21).

In the meantime, Nash had signed up Pinin Farin as a design consultant (with a 5-year contract). He was given a lot of credit for the "upside-down bathtub" Nash of 1951; actually the Nash was based on a scale model made in 1943 by Ted Pietsch and Bob Koto. Nor did he have much to do with the Nash-Healey roadster even though Nash liked to attach the Farina label to it. The contract was not renewed.

Pinin Farnia did develop other contracts that were longer lasting. Peugot became a client in 1952, and today is Pininfarina's biggest customer. Traditions that started with the Model 403, in 1954, live on in the whole range of 1982 Peugeot cars, from the 104 to the 604.

About the same time, Pinin Farina formed a relationship with the newly formed British Motor Corporation, and contracted for a series of new bodies for Austin, Morris, Riley, and Wolseley cars. Inevitably, some turned out to be direct derivatives from Peugeot designs. This agreement was in force until BMC came under Leyland rule at the end of 1968.

In one 1968 prototype, using the Morris 1100 as a base, Pininfarina created an aerodynamic sedan that now must be seen as positively prophetic. It set the pattern for Opron's Citroen GS and CS, and influenced Giugiaro in his design of the Alfasud and David Bache in his shaping of the Austin Allegro.

It was in the mid-1950s that Pinin Farina branched out from design and prototype construction to add actual manufacturing. The first model to be produced on an industrial scale was the Alfa Romeo Giulietta spider of 1954. Nearly 28,000 were built over a 11-year period, and then its place was taken by the Giulia. In 1966, Pininfarnia also began producing the Fiat 125 Spider (100,000 had been built by May, 1975).

Fig. 10-20. The Pinin Farina coupe on Cisitalia 202 C, 1947.

Fig. 10-21. The Lancia Aurelia B-20 coupe by Pinin Farina, 1951.

The company's growth was staggering. In 1957, its 672 workers produced 3148 auto bodies. Three years later, Pininfarina had a work force of 1600 and produced 11,130 cars. By 1963, output nearly doubled (to 20,816 cars) with the same number of workers. And in 1967, the daily production was about 80 cars.

This was not accomplished in the old factory at Corso Trapani. In 1955, land was purchased for a new plant at Grugliasco, on the northwestern outskirts of Torino, and a new manufacturing complex and design center was completed in 1958.

In 1960, the trade name was contracted into Carrozzeria Pininfarina. At the same time, the family name was legally changed from Farina to Pininfarina. In 1979, the corporate title became Industrie Pininfarnia, S.p.A. in order to reflect the broadening of its activities.

PARTNERSHIPS WITHOUT FRONTIERS

Since 1958, Pininfarina-designed bodies for certain Ferrari models have been produced by Scaglietti of Modena. Scaglietti now exists only as the coachbuilding division of Ferrari (an inevitable example of industrial consolidation). Sergio Scaglietti was 32 years old when he opened his body repair shop in 1951. Four years later he built some special bodies that attracted the attention of Enzo Ferrari, and orders for Scaglietti bodies of new Ferrari chassis began to flow in from Maranello.

Pininfarina's masterpieces continued to flow, unabated, from the Ferrari 250 GT Berlinetta Lusso of 1963 to the Ferrari 512 Berlinetta Boxer in 1971.

After Battista Farina's death in 1966, the reins of the company were taken over by his son, Sergio, and the husband of his daughter Gianna, Renzo Carli, both of whom had long held positions of responsibility in the organization.

Sergio Pininfarina was born in 1927 and was graduated from the Politecnico di Torino, in 1950, with a degree in mechanical engineering. He is less the artist, and more the modern business executive, then his father was, and he has taken the initiative to keep Industrie Pininfarnia in the forefront of its section of industry. He had a full-scale wind tunnel completed by 1972, for instance, and sponsored the research and patent applications of Professor Alberto Morelli for his mathematical formula for low-drag, zero-lift bodies that makes it possible to have a computer design an ideal shape for any type of car.

Sergio Pininfarina has the final say over the design work, too, but that is really the province of his chief designer and engineer, Leonardo Fioravanti, who is the creator of the latest generation of trend-setting styles.

Pininfarina has design experience with aircraft, motorboats and skis, and manufacturing experience with sheet metal for a variety of products from bicycle fenders to refrigerator panels. On the car-production side, Pininfarina builds the complete Fiat 124 Spider, Lancia Beta Montecarlo, Alfa Romeo 2000 Duetto Spider, the Peugeot 504 coupe and convertible, and the Talbot Samba convertible.

The company now has 2500 workers. In 1979, production came to 27,000 bodies. The land at Grugliasco is now filled, and further expansions must take place elsewhere. Pininfarina now has a new research center at Cambiano. Its staff makes studies going far beyond car design. They look into new types of energy and their effects on the use of cars, manufacturing technology and humanization of the work place, and the education of new talent. The Cambiano establishment will also house the future Pininfarina museum.

FLUCTUATIONS OF GHIA

Giacinto Ghia worked as an apprentice for several coachbuilders in Torino, including Rapid and Diatto. He set up his own shop in 1915. During the 1920s and 1930s, he became known better for his styling innovation than for his craftsmanship. Many Ghia bodies revealed a search for genuine beauty on the part of the designer, but they were not notable for lightness or aerodynamic efficiency.

Giancinto Ghia died in 1944, and two years later the Ghia family sold the firm to Mario Boano. Mario Boano was born in Torino in 1903, was graduated from the San Carlo technical college, and went to work for the Stabilimenti Farina. In 1930, Boano chose to go with Pinin rather than stay with Giovanni. During the wartime years, he had his own business that made wood patterns and parts for the industry.

Boano did a lot of design work himself, but he also brought in an independent design consultant, Mario Revelli de Beaumont, who was a Roman of considerable experience, despite not being quite 40 years old. Among his qualifications was a period he had spent with Garavini before becoming an independent consultant.

Ghia built a Saoutchik-inspired body for a Fiat 1500, in 1947, and an even wilder body on an Alfa Romeo 6C2500 in 1948. Calmer lines followed with a beautiful Alfa Romeo 1900-C in 1952.

Where Boano really needed help was in money matters. For that, Ghia hired Luigi Segre as business manager, in 1950, and Segre also bought a piece of the company. It was Luigi Segre who sold Virgil Exner, Chrysler's styling director, a plan for the construction of prototypes and show cars, much faster and at lower cost that it could be done in Detroit. The collaboration lasted for 15 years.

Among the most memorable results were the 1950 Plymouth X-500, 1951 Chrysler K-310, 1953 De Soto Adventurer, 1953 Chrysler d'Elegance, 1954 Plymouth Explorer, 1954 Dodge Firearrow, 1955 Falcon and Gilda, 1956 Norseman (which went down on the *Andrea Doria*), 1956 Dart, 1958 Plymouth Cabana, 1960 Plymouth XNR, and 1963 Chrysler gas turbine car.

Boano concentrated on design for European clients. The Volkswagen Karmann-Ghia coupe was mainly his work.

As a result of a quarrel between Boano and Segre, the partnership broke up in 1953, and Segre became majority owner of Ghia. Boano set up his own shop in Torino. There he built prototypes to designs by both Virgil Exner and Raymond Loewy.

FROM BOANO TO FORD

As a consultant, Boano helped Renault do the Dauphine body with its heavily domed steel panels. That was done to gain strength, but it turned out to have benefits also in preventing noise and vibration. His lack of business acumen proved his undoing, however, and Carrozzeria Boano went under in 1957.

Ghia designs were mainly due to Mario Savonuzzi, from 1953 to 1958, and the work of Sergio Sartorelli from 1958 to 1963. Suddenly, Luigi Segre died in 1963, and Gino Rovere ran Ghia as a caretaker until Luigi Gaspardo Moro took the reins of Ghia. In 1967, Rowan Controllers bought Ghia and Alejandro De Tomaso became president. Ghia acquired control of Vignale in 1969. The Vignale story is intertwined not only with Ghia, but also with Michelotti and Pinin Farina.

Alfredo Vignale's father had been chief of the paint shop at Stabilimenti Farina. Young Alfredo worked as a panel beater for Pinin Farina when the first all-steel bodies were built there about 1933. During the war, he had a small garage, and he made a living on auto repairs of all kinds.

In 1945, Vignale was using an old run-down Fiat 500 for business. The following year he built a new body for it, and some of his friends and clients talked him into renting space for displaying it at the Turin Auto Show. Orders came flowing in, and Vignale got his start building new bodies on old chassis. When Fiat and Lancia was able to supply new chassis, he began competing with other small coachbuilders. Not entertaining any exalted illusions about the limits of his own designing talents, Vignale began an association with Michelotti.

Born in Torino in 1921, Giovanni Michelotti joined Stabilimenti Farina in 1937 and was still there in 1945-46. He had also worked with Vignale, Balbo, Bertone, Allemano, and Ghia before setting up his own styling studio in 1949. He entered an exclusive contract with Vignale for Italy, while remaining free to work for other clients in other countries.

Michelotti had facilities for prototype construction, but never got involved with production. He designed the 700 for BMW and then the 1500. The Hino Contessa came from his studio. Triumph in Coventry became an important client, first for the Herald in 1958, then the TR4 in 1961, the Spitfire in 1962, the 2000 in 1963, and the 1500 in 1965. Michelotti also did the DAF 55 body. Tragically, he died at the age of 58 in 1980. The Michelotti studio continues with Tateo Uchida, who had joined Michelotti in 1966, as chief designer and Edgar Michelotti as president.

But back to Ghia. After the two years when Giugiaro was chief designer, an American, Tom Tjaarda, was brought in. He was the son of John Tjaarda, head of advanced design for Briggs in the 1933-40 period, and he had some experience from working in the Pinin Farina studio. It was Tjaarda who designed the Pantera that Ford and De Tomaso would build together. De Tomaso and Ford turned Vignale into a production plant for the Pantera bodies, but it was closed down in 1974.

In 1972, Carrozzeria Ghia had become Ghia Operations, a branch of Ford Motor Company, working exclusively for the manufacturing divisions of Ford around the world. Tom Tjaarda left in 1977 to go to Fiat. He worked in association with Gian Paolo Boano (son of Mario) and Sergio Sartorelli. Filippo Sapino, whom Ford had made general manager of Ghia Operations, also took on the responsibilities of styling director. Sapino worked as a stylist with Ghia from 1960 to 1967, but left to join Pininfarina, and returned via Ford in 1973.

SUPERLEGGERA: SHADES OF WEYMANN

All the coachbuilders of Torino produced steel bodies. Perhaps that was natural because Fiat is a large steel maker. In Milano, however, the coachbuilders worked mainly with aluminim. Perhaps that was due to the proximity of the Montecatini works. We have seen how Zagato became a champion of lightweight construction. Now we shall consider the case of Carrozzeria Touring.

It had its start in 1926 when Felice Bianchi Anderloni and Gaetano Ponzoni purchased Carrozzeria Falco in Milano, with its plant and equipment, from Vittorio Ascari. Falco had been airframe constructors, and they had developed new techniques in lightweight construction.

Felice Bianchi Anderloni came from Rome where he had studied law. But instead of practicing at the bar, he joined Isotta Fraschini, in 1904, where he took part in widespread activities from test-driving to sales promotion.

Touring engaged the talent of Giuseppe Seregni to create new styles. As early as 1927, Touring was noted for its beautiful shapes covering Lancia, Alfa Romeo, Ansaldo, OM, Fiat and Isotta-Fraschini chassis. The Touring bodies of 1927 were built under Weymann license.

By 1936, Bianchi Anderloni had perfected his own construction principles that he summed up in one word: Superleggera (super light). A cagelike frame work of small-gauge steel tubes enclosed the chassis and passenger compartment. The aluminum body panels were not welded to it; they simply rested on graphite-insulated cushions attached at a multitude of points on the cage. This gave tough, rattle-free, and very lightweight construction, but one that did not lend itself to mass production.

Apart from building advanced, one-off bodies on Alfa Romeo chassis and setting the postwar style in sports cars for the BMW 328 streamliners built for the 1940 Mille Miglia, the company was an airframe subcontractor from 1935 to 1943.

After the war, Bianchi Anderloni's son, Carlo, became Touring's chief designer. He designed the Ferrari 166 "Barchetta," Alfa Romeo 6C 2500 SS Villa d'Este, Pegaso Z-102 Thrill, and the series-production Alfa Romeo 1900 C (Fig. 10-22) and 1900 S.

Touring sold a license for Superleggera body construction to Tickford (who built the bodies for Alvis and Aston Martin). Another license was sold to Bristol; they used Superleggera construction for the 400, 401, and 403.

But Touring failed to modernize its methods, and Superleggera construction lost out to modern monocoques when computers came in to help designers lighten their unit-construction steel structures. Among Touring's last masterpieces were coupes and convertibles on Lancia and Maserati chassis.

A large contract with Rootes, in 1961, for the assembly of the special bodied Sunbeam sports coupe led to a major plant expansion that never paid back the investment. Matters came to a head when Chrysler bought control of Rootes and canceled the contract. Carrozzeria Touring declared bankruptcy in 1967 and was closed down. Carlo F. Bianchi Anderloni was signed up as design director of Alfa Romeo; there he created the Giulia sedan, 1750 sedan, 2000 sedan, and Alfetta sedan.

FAILURES FOR WANT OF IDEAS

Smaller coachbuilders who had prospered in the

Fig. 10-22. The Touring coupe body on the 1955 Alfa Romeo 1900 SS.

immediate postwar era had proved ill adapted for finding a complementary role at the industrial level. They were always minor factors in terms of production volume, but influential nevertheless. They had served for a considerable time as a necessary part of the coachbuilding industry's fabric and structure. Among them were Balbo, Savio, Allemano, Monviso, Viotti, Canta, and Boneschi.

Alfonso Balbo opened a coachbuilding shop in Torino, in 1911, for working mainly on Fiat and Lancia chassis. He died in 1926 and Carlo Follis took over the leadership. Balbo was building car bodies until 1954; some of the bodies were to Scaglione designs.

The brothers Antonio and Giuseppe Savio began building bodies on Lancia, OM, Itala, Isotta Fraschini, and Fiat in 1928 in Torino. Savio also served as a subcontractor for Alessio and Baravini in the early days, and for Boneschi and Ghia in later years.

Stabilimenti Monviso was active in Torino from 1944 to 1955 (when it was absorbed by Ghia). Monviso bodies were smart and attractive, usually on Fiat and Lancia basis.

In 1921, Vittorio Viotti established his coachbuilding shop in Torino. Viotti bodies on small cars like the Fiat Balilla, Lancia Artena, and Augusta attracted favorable attention during the 1930s. His small convertibles and sports coupes maintained his reputation after 1945. Vittorio Viotti died in 1956, and Viotti designs from 1956 to 1960 were by Pietro Frua. Frua had learned the trade as an employee of the Stabilimenti Farina from 1928 to 1937.

Pietro Frua operated his own works from 1944 to 1958 and then sold out to Ghia. He continued as a private design consultant and prototype builder. He produced many excellent Maserati bodies, and among his latest creation is a four-door sedan on a Lamborghini chassis.

Sebastiano Canta began building elegant bodies on small cars, mainly Fiat, in 1949. Later, Canta also produced some bodies on Lancia chassis. The firm discontinued car bodies in 1963 to concentrate on truck and bus bodies.

Serafino Allemano started an auto repair shop in Torino in 1928. His work on damaged bodies was so good it also attracted customers for modifications, and his shop was gradually transformed into a small coachbuilding works. By 1935, he was making bodies to his own or client's designs. Carrozzeria Allemano built car bodies until 1965.

Giovanni Boneschi learned his trade by working for Castagna. In 1921, he established a coachbuilding shop in Milano. The name of Boneschi first became known for a magnificient limousine on a Rolls-Royce Silver Ghost chassis.

For many years, Boneschi built bodies under Kelsch patents (a lightweight construction somewhat similar to Weymann's). The founder died in 1946 and, in 1960, the firm became associated with OSCA. The company also occasionally built bodies for Alfa Romeo, Fiat, and Lancia. About 1970 Boneschi diversified into vans and ambulances.

SERVICE TO THE INDUSTRY

New automotive-industry companies have come into existence, but they do not fit the old definition of coachbuilding. Instead they are organized to provide specialized services to the auto industry. It is as such that the Michelotti studios survive.

Another is Coggiola of Torino. Sergio Coggiola was born in Torino in 1928 and worked for Ghia from 1951 to 1966. He established his styling studio in 1966. He has facilities for prototype construction but no production. He is known to have done consulting work for Saab and Volvo, among others.

By far the most prominent, however, is Ital Design. It was established, in 1968, by Giorgetto Giugiaro after he had finished the Iso S.4 sedan for Count Rivolta. The first Ital Design car was the Bizzarrini Manta prototype of 1968. Then Giugiaro redesigned the Skoda sedan for the Czechoslovakian auto industry and sold two designs (a city car and a small bus) to Suzuki in Japan. The Ital Design Iguana on Alfa Romeo chassis appeared in 1969 and the Maserati Bora appeared in 1971.

His first mass-production car was the Alfasud; it went into production in 1972. That year he also produced the designs for the Maserati Merak, Maserati Boomerang, and Lotus Esprit. He clothed the VW Passat (Dasher) in 1973 and the VW Golf (Rabbit) and Scirocco in 1974.

In association with Karmann, Giugiaro designed the Ace of Spaces in 1973 and the Ace of Diamonds in 1976. John DeLorean chose Ital Design for his DMC-12, and Alfa Romeo Ital Design for the Alfetta GT and Alfasud Sprint. The Ace of Clubs was designed for Isuzu in 1979 and went into production in 1982.

In the case of the South Korean Hyundai Pony, Ital Design supplied the body design, body engineering plans, a study of the required tooling and series-manufacturing systems, an analysis of production rates and the cost of the finished product.

Ital Design works primarily in the automotive field, but is also ready to carry out planning and design for boats, furniture, and electrical appliances. It also has the capability for creating the entire manufacturing system needed for any kind of product from an armchair to a fountain pen.

In 1974, Ital Design moved from Torino to a large, new plant in the south-side suburb of Moncalieri. About 150 designers, technicians, modellers and skilled craftsmen work there.

Chapter 11

Amidships-Mounted Engines

A small group of American auto writers was in Sicily, to test the Fiat X-1/9, in the spring of 1972 (Fig. 11-1). Everybody had a lot to say about the Fiat, and about cars in general. I remember one colleague remarking, "At least, from now on, we won't have any difficulty when someone asks us for the definition of a sports car. It's a road car with its engine in the middle."

The point is well taken. But not absolutely correct, because the proposed definition doesn't make the Fiat 124 Spider, with its front-mounted engine, any less of a sports car. On the other hand, it recognizes that *only* sports cars can carry their engines amidships.

The amidships position for the engine usually precludes a normal-length rear seat. An amidships engine is one that is located ahead of the rear wheels, just inside the wheelbase, as opposed to a rear engine that is carried in the rear overhang, outside the wheelbase.

While some companies still build sports cars with rear engines (Porsche 911, Renault Alpine A-310, and DeLorean DMC-12), a growing number of sports car manufacturers have chosen to place the engines in the middle; those cars are mostly two- or three-seaters. The 1982 production models include the AC ME-3000, De Tomaso Pantera, Ferrari 308 GTB, Mondial, and Berlinetta Boxer, Fiat X-1/9, Lamborghini Countach, Lancia Beta Montecarlo, Lotus Espirit, Maserati Merak, and Talbot-Matra Murena. Engines range from small, transversely mounted four-cylinder units to a giant flat-twelve.

The amidships-mounted engine was common in the days of 19th-century auto pioneers. As engines developed, buyers demanded more power, and vehicle architecture fell into the stereotype we know. The amidships position was revived for single-seater racing cars such as the Auto Union 1934-40, the Cooper 1946-65, and came into gen-

Fig. 11-1. The Fiat X 1/9 was designed by Bertone in coordination with Fiat engineers. The car is assembled by Bertone.

eral use in Grand Prix racing in 1961, sports car racing in 1964, and for Indy cars in 1965.

A number of prototypes and special models appeared in earlier years, and their designs are in many ways noteworthy. The first was Edmund Rumpler, a Viennese engineer who designed aircraft, during World War I, after years of experience with various car companies in central Europe. You could say his idea, which he patented in 1919, was to put an airplane fuselage on wheels and run it on the road.

He built both open and closed prototypes in 1921 (Fig. 11-2). The bodies had stubby noses but long, tear-drop-shaped tails that indicated the logic of a rear-end engine installation. Rumpler opted for a central engine location, and he obtained it by moving the rear wheels back until there was only minimal overhang. The engine was a Siemens W-6; it had three banks of two cylinders (one vertical and one spaced at 60 degrees each way) on a common crankcase. Behind the power plant, a gearbox separated the flywheel from the final drive unit.

Rumpler's design was the first of a whole new breed. Benz bought a license to use the Rumpler patent for a new racing car project that became known as the Tropfenwagen (teardrop car). Three such Benz cars were built, in 1923, to designs by Max Wagner and Hans Nibel. After long delays, they were entered in the Italian Grand Prix at Monza. They were slow due to a lack of power from the 2-liter, in-line six. One retired due to engine trouble and the other two finished fourth and fifth. They had no immediate impact on racing car design or engineering trends in general.

In 1924, Benz converted one Tropfenwagen into a sports car by fitting headlamps and fenders. It was quite successful in minor hill-climb events in Germany. In 1926, Benz merged with Mercedes, where Dr. Porsche was chief engineer, and the Tropfenwagen was not developed further. Dr.

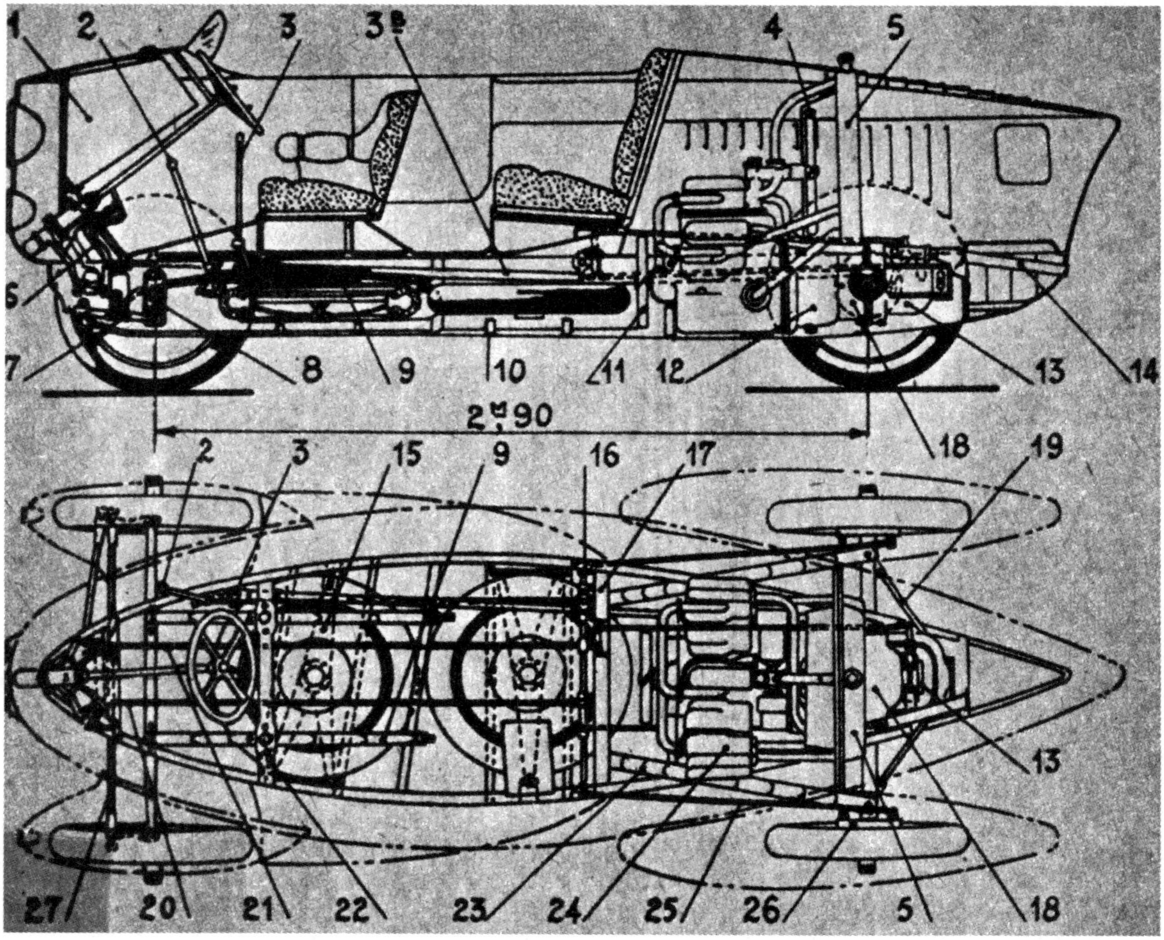

Fig. 11-2. A plan view and elevation of the 1922 Rumpler open touring car. Two spare wheels are stored below the floor.

Porsche revived it several years later for another company: Auto Union.

Auto Union was formed in 1932 by a merger of Horch, Audi, DKW and Wanderer. In 1933 the company decided to enter Grand Prix racing. Dr. Porsche had left Mercedes-Benz in 1929 and set up an independent engineering office, and Auto Union commissioned him to design its racing car. All earlier Porsche-designed cars had been equipped with front-mounted engines, but for this project, Porsche placed the engine in the middle, aft of the driver, but ahead of the rear wheel axis (Fig. 11-3). Did he remember the Benz Tropfenwagen? No doubt. As for evaluating it, Porsche's business partner was Adolf Rosenberger. Rosenberger had been one of the Tropfenwagen drivers, and he could provide first-hand information.

Porsche improved on the Rumpler configuration by placing the gearbox behind the final drive unit. This brought the engine closer to the rear wheels and thereby cut down the overall length requirement. It also helped put more weight on the driving wheels. The Benz engineers had designed the Tropfenwagen engine to the then-current Grand Prix formula that restricted displacement to 2 liters. For 1934, there was no displacement limit, but there was a weight-limit maximum 750 kilograms (1654 pounds.)

Porsche realized the weight-saving potential of the amidships engine, eliminating the propeller shaft, and decided on a 4.4-liter, V-16 engine. This engine weighed 540 pounds and delivered 295 hp. This was compared to 562 pounds and 80 hp in the Tropfenwagen. In 1935, the Auto Union engine was enlarged to 4.95 liters. This raised power output to 340 hp. The engine ran as a 6-liter in 1936 and 1937, and it delivered 520 hp (Fig. 11-4).

As the Grand Prix changed for 1938, new projects were started, but Auto Union decided to produce its new car design in-house rather than buy it ready-made from Dr. Porsche. The key man in charge were Dr. William Werner and Robert Eberan von Eberhorst. The new formula abolished the weight limit and reimposed displacement limits (3.0 liters supercharged and 4.5 liters unsupercharged). Auto Union chose a 3-liter, V-12 engine

Fig. 11-3. The 16-cylinder Auto Union chassis.

Fig. 11-4. The Auto Union in action, with Bernd Rosemeyer at the wheel.

with a supercharger. An all-new chassis was designed, but the engine remained in the same place. From an initial output of 420 hp in 1938, the engine was gradually developed to yield 485 hp at the end of 1939.

Of Auto Union's production cars, Horch and Wanderer were conventional; Audi and DKW used front-wheel drive for their most popular models. There was no spinoff of technology from the Grand Prix car into the ordinary road cars made by the same corporation, and all serious car companies regarded the central engine layout as a solution suitable only for racing cars.

Fortunately, not all were serious. A little amidships engined sports car was produced in Italy as early as 1922. Designed by Guido Ucelli, it was made by the San Giusto company in Trieste. The engine was an air-cooled, four-cylinder, in-line type mounted just ahead of the rear wheels. It represented a step ahead of Rumpler's ideas in that the transmission was located behind the final drive unit. This made for an extremely compact drive train. The chassis was ultra light—with a central box-section frame and all-independent suspension—weighing only 684 pounds. The complete car, with roadster coachwork, weighed 1500 pounds. The 750cc engine, delivering no more than 16 hp, was enough to give it livelier performance than many popular small cars of its day. But San Giusto was not a serious automaker. Only 10 of these little gems were produced. The marque disappeared in 1924.

Rumpler's idea of grounding the aircraft fuselage found followers everywhere. But they were independents seeking more to advance the state of the art than to get their particular creations into production. One such independent was Emile Claveau. Beginning in 1923, he devised a series of streamliners with advanced chassis engineering. In 1927, he exhibited a prototype at the Paris auto show. The engine was an air-cooled, flat-four of his own design, located just ahead of the rear wheel axis. He followed the Rumpler layout in placing the gearbox next to the engine, ahead of the final drive unit. Within an overall length of 10½ feet, Claveau provided a four-passenger compartment and a wheel at each corner with cycle-type fenders.

Another independent was Andre Dubonnet; he was the inventor of the "kneeaction" used by Chevrolet and Pontiac in 1934-35. With his engineer, Gustave Chevra, he built a large teardrop-shaped car in 1935, propelled by a Ford V-8 engine positioned just behind the passenger compartment. A semiautomatic transmission was interposed between the engine and the final drive unit.

William B. Stout tried many different designs before he built his Scarab prototype in 1935. He started off with an overhanging rear engine. Next he went to an amidships position only to find that it cost him too much interior room. He ended up placing the engine above the final drive unit (with stepdown gearing to a 3-speed transaxle).

Buckminster Fuller's 1933 Dymaxion was an admidships-engined design with a difference. The Ford V-8 was located forward of the rear axle. Its flywheel faced the front and the front wheels were driven in the same manner as Professor W.E. Miller's three-wheeler of the same period.

Although Andre Dubonnet had demonstrated his car to the Detroit auto companies, in 1936, his ideas were not appreciated there. Nor did the European industry display any real interest.

When Mercedes-Benz suddenly presented a sports car with an amidships-mounted engine, in 1935, it was a potential trend setter that was not given adequate recognition at the time. Very few were built, but why it came into being deserves some explanation.

While Dr. Porsche was working on the Volkswagen project with NSU and ZUndapp, Dr. Max Sailer, at Mercedes-Benz, thought it was time the Stuttgart-based company looked into the technology of small, rear-engined cars. This resulted in the 1934 Mercedes-Benz 130-H. It was an economy sedan powered by a 1.3-liter, water-cooled, four-cylinder engine standing vertically in the tail end of the chassis, driving the rear wheels. The engine was enlarged to 1.5 liters the following year, and Sailer decided to build a roadster version of the new 150-H. Keeping the full length would not produce the sport styling that Sailer was looking for. Shortening the chassis to make it a two-seater would give a ridiculously short wheelbase and create all sorts of stability problems.

Finally, he decided to turn the engine around to an amidships position. This more or less put the engine in place of the back seat and kept a reasonable wheelbase; maximum weight was concentration within the wheelbase. Small numbers of the 150-H roadsters (Fig. 11-5) were built in 1935 and 1936, but then it was replaced by the new front-engined, 170-V two-seater design in 1937.

A few amidships-engined racing projects emerged in the wake of the Auto Union. Harry Miller did not copy the Auto Union, though, for he had been contemplating four-wheel drive and a central engine position since 1932. The 1938 Gulf-Miller was as free of influence from Auto Union as the egg is from Columbus.

Miller's engine was an in-line six that was installed 45 degrees from vertical behind the driver's seat. It carried a centrifugal supercharger, at the extreme rear end, with a huge intercooler on the left side. From the crankshaft, a propeller shaft

Fig. 11-5. The Mercedes-Benz 150 H roadster.

ran *forward* to a transaxle at the front end of the chassis. From there another propeller shaft took the drive to the rear differential. Four Gulf-Millers were built. They were unsuccessful at Indianapolis, but later they succeeded in setting a number of international class records.

In 1938, Alfa Romeo began design work on its Tipo 512, an amidships-engined Grand Prix car for the 1940, 1.5-liter Formula (supercharged). The project was headed by a Spanish engineer, Wilfredo Ricart, who had come to Alfa Romeo in 1936. He chose a water-cooled, flat-12 engine, separated from the back of the driver's seat by a full-width fuel tank. From the clutch at the flywheel, the drive was taken via a 5-speed transmission located aft of the final drive, and then forward into the differential. A single prototype was built in 1940-41 and never raced.

Ricart also designed a sport-racing car with an amidships-mounted engine: Tipo 163. The chassis and drive line followed the broad lines of the Tipo 512, but the 163 engine was the 3-liter, V-16 developed for the Tipo 162, but running unsupercharged. It was built in 1941 but never developed.

A car that was probably intended for Auto Union was Porsche's project number 370. Instead it became a Cisitalia. It was a four-wheel drive Grand Prix car project with a 1500cc supercharged flat-12 engine. Two former Porsche engineers, Carlo Abarth and Rudolf Hruska, built the Cisitalia during 1946-1947 for industrialist Piero Dusio. By 1950, Dusio had moved to Argentina, and the car's racing career was over before it started. It made an occasional appearance up to 1953, but it was never in contention.

At the close of World War II, a number of automakers were toying with amidships-engine chassis designs. Olle Schjolin, after many years with GM, in Detroit, returned to Sweden about 1939 and went to work with Volvo. He came up with a design for a radical small sedan with a radial, four-cylinder engine placed, at an angle, fairly high up behind the back seat, but ahead of the rear axle. Apparently, a prototype was built, but the project was killed when Volvo needed all its resources and manpower to develop and produce the 444.

When Louis Renault visited the Berlin auto show in 1939 and saw the Volkswagen, he ordered his engineers to design a rear-engined economy car for the same market. A design was hurriedly produced, and a prototype was built before France collapsed at the start of WWII and Renault plants were put to work for the Wehrmacht.

Development work on the car that was to become the 4 CV went on secretly. It went into production in 1947, and by that time Fernand Picard was busy with the "108" project (a larger model). This was to be a longer car and, due to the heavier engine, Picard felt it was desirable to hoist it inboard. An amidships-engine "108" prototype was built, but interior space was unsatisfactory. The car was abandoned in favor of a still larger car, with a conventional layout, that went into production in 1941 as the Fregate.

French racing driver Jean-Pierre Wimille built a futuristic prototype sedan, in 1947, that was powered by a Ford V-8 engine positioned ahead of the rear wheel axis. The project died with its originator who was killed during practice for a race in Argentina in 1949.

Only one passenger car with an amidships-mounted engine was in production in 1947. That was the King Midget, built in Athens, Ohio, by Dale Orcutt and Claud Dry. It used a single-cylinder Wisconsin engine placed ahead of the rear axle. It revived the Olds-Ford-Cadillac layout that was nearly 50 years old. The King Midget was too small, both as a product and as a company, to have any effect on Detroit's thinking. It was through racing that the amidships engine would become fashionable.

It started in England, in 1946, when John Cooper built a tiny single-seater with a single-cylinder, Norton motorcycle engine mounted behind the seat, but ahead of the rear wheel axis, with chain drive to the wheels. It was an extremely agile car, of remarkable acceleration and speed, largely due to its lightweight construction. It became a trend setter and Cooper built more power cars using the same pattern. First came the 1000-cc, Cooper-JAP, V-twin for hill climbs, then came a 1100-cc, Cooper-Climax for Le Mans in 1955.

Two years later, the small Surbiton firm launched a Formula 2 Cooper-Climax (1500cc) and also used the larger engine in the Le Mans type. With the 4-cylinder Coventry Climax engine, Cooper switched from chain drive to monobloc construction, and placed the engines vertically in line behind the driver, with the final drive unit interposed between the engine and gearbox.

Under the 2½-liter Grand Prix formula, which lasted from 1954 through 1960, amidships-engined Coopers achieved the impossible. Giving away 750cc to the competition, Stirling Moss's 1750cc Cooper-Climax won the Argentine Grand Prix in January, 1958. At Monaco in May, 1958, Trintignant's 2200-cc Cooper-Climax defeated a whole field of cars with front-mounted engines and 300cc greater engine displacement. The central engine idea had demonstrated its superiority. Due to their light weight and ideal weight distribution, the Coopers had greater maneuverability for circuits with tight curves, burned less fuel (and therefore could run nonstop when others would stop to refuel), were easier on tires and brakes, and did not lag far behind in acceleration and top speed.

The competition learned fast. For 1960, Lotus and BRM also had amidships-mounted engines. Lotus used the Coventry Climax and BRM built its own, but both were four-cylinder units installed straight up and down and bolted to the transaxle. That same year, Ferrari built a Formula 2 prototype with an amidships-mounted Dino V-six in preparation for the 1.5-liter Grand Prix formula of the

1961-through-1965 era. The 1961 Formula One Ferrari carried the engine centrally, and a Ferrari 246 LM sports prototype with an amidships-mounted 2.4-liter engine ran at Le Mans. Actually, the first amidships-engined car to compete at Le Mans was a cobbled-up 1953 Porsche, entered by the factory and driven by Paul Frere.

The production-type Porsche of that era was Type 356; it had a flat-four, air-cooled engine mounted in the tail end of the chassis. These cars did well in both rallies and races, but not well enough for two drivers who were also mechanics; Helm and Walter Gloeckler. In 1951-52 they began to modify their Porsches by moving the engine inboard and greatly improving their stability.

Porsche reacted by trying to outdo the Gloecklers, and they took the approach of turning the whole tail end of the chassis around. This gave the same result in terms of weight distribution, and it was a simpler job mechanically except for one thing: the torsion bar suspension. Instead of running forwards, the torsion bars now ran backwards; this meant stretching the body and providing an anchorage. Also, it upset the suspension geometry, so that the Porsche with a central engine actually handled worse than the rear-engined model.

By 1954, Porsche had designed its 550 model with an amidships-mounted engine. And it was a winner. Designed by Ernst Fuhrmann, it was a superb piece of engineering with flawless drive train layout, weight distribution, and suspension.

In 1954, the small French DB (Deutsch-Bonnet) switched from flat-twin Panhard engines and front-wheel drive to Renault power, amidships-mounted. The little DB streamliners won great acclaim, year after year, at Le Mans where they showed remarkable speed and fuel economy. To compete, Jean Redele of Alpine, whose production models were using the rear-engine Renault 4 CV chassis, moved the engine inboard of the wheelbase for his racing specials. They too were quite successful.

After the Deutsch-Bonnet partnership broke up in 1963, Rene Bonnet built a car on his own, again with an amidships-mounted Renault engine. The big aerospace firm, Engins Matra, became interested in his car, which evolved into Matra-Bonnet Djet (Fig. 11-6), and finally Matra. By 1965, Rene Bonnet was out of the picture and Matra switched to Ford power, using the little V-4, 1.5-liter then made in Germany. The next step from Matra came in 1973 with the Bagheera, using the Simca 1100 power train with transverse engine, repositioned from the front to the middle. It was replaced in 1980 by the Murena.

Charles Deutsch also did his own car. He took the engine from the Peugeot 204 and moved it back to an amidships position in a newly designed streamliner he called the CD (first appearing at Le Mans in 1966). Several prototypes followed, but the CD never became a production car.

Also in 1966, the Neri & Bonacini garage in Modena was building a sports car using the Lancia Fulvia power train that was installed transversely behind the seats. Only one was built and its racing career stopped short before any serious engagements.

On the fringes of the racing scene, and more interesting for its design features than for its performance, was the Bugatti Type 251 of 1956. Designed by the great Gioachino Colombo, who had worked for both Alfa Romeo and Ferrari, it was a semistreamlined Grand Prix car with a straight-eight engine installed transversely ahead of the rear wheels. A five-speed transaxle was placed at the center of the crankshaft and extended backwards into the differential casing and final drive unit. Maurice Trintignant drove the car in the Grand Prix of the Automobile Club de France at Reims in 1956 but did not finish. It went back to Molsheim but never became raceworthly and had no influence on design trends (which were to be dictated by Cooper).

Cooper began using Maserati engines in some

Fig. 11-6. The Matra Djet, 1965.

sports/racing cars in 1959. Maserati built its first car with an amidships-mounted engine in 1961: the Tipo 63. It was powered by a 325-hp, V-12 installed longitudinally in the chassis, with the 5-speed transmission extending backwards from the final drive unit. It is significant mainly because it marks the start of Maserati's experience with this type of construction. Maserati is, of course, one of those companies that manufactures amidships-engined cars today.

Rover's chief engineer, Spencer King, became interested in the possibility of producing an amidships-engined sport car. When Rover began to manufacture the Buick V-8, he had the power unit he needed. A prototype (Fig. 11-7) with a rough, squared-off body was built in 1967, but the project was soon abandoned. Previously, Rover had installed a gas turbine in a BRM chassis in 1963, provided a sports car body, and raced it *hors concours* at Le Mans. The turbine was placed in the middle (driving the rear wheels). A second Rover-BRM turbine coupe was built and raced in 1965 (Fig. 11-8).

It was the first turbine car to actually compete

Fig. 11-7. A Leyland/Rover prototype, built in 1967, with body design by David Bache.

Fig. 11-8. A Rover gas turbine racing prototype, using BRM chassis, from 1965.

in racing, but several turbine-powered cars with amidships-mounted engines had existed for years. GM's first turbine car, the Firebird I of 1953, had a centrally positioned power plant, and so did the Firebird III of 1958 (Fig. 11-9). A GM engineer named E. C. Conklin actually patented this power train configuration (application 1954, granted 1962). Fiat used the amidships-engine configuration for its Tipo 8002 gas turbine car of 1954 (Fig. 11-10) and Renault chose a similar arrangement for its Etoile Filante, built in 1956. These cars were used only for demonstrations, but they might also have played some role in the research area.

Powered by an aviation-type Turbomeca engine, the Etoile Filante was sent to Bonneville Salt Flats; there it set a record at 191.8 mph over the flying mile. Amidships-mounted engines were nothing new for cars built specifically to attack the world land speed record, starting with Ray Keech's White Triplex of 1927 that combined two amidships mounted Liberty V-12 aircraft engines with a third one mounted in front—driving the rear axle.

Captain G. E. T. Eyston's Thunderbolt of 1937 had two Rolls-Royce, V-12 Merlin engines installed centrally in the chassis, driving two rear axles. John Cobb's Napier-Railton of 1938 had two Napier Lion engines installed in diagonal formation between the axles. The front one drove the rear wheels and the rear one drove the front wheels.

When Ken Norris laid out the turbine-powered Bluebird for Donald Campbell in 1956, he placed the giant Bristol Proteus engine in the middle. Carlo Abarth built a streamliner for setting international class records in 1957, using an Alfa Romeo Giulietta engine positioned centrally. When Sidney Enever created the MG EX-181 the same year, he

Fig. 11-9. The GM Firebird III gas turbine experimental car from 1959.

Fig. 11-10. The experimental gas turbine powered streamliner that was built and tested by Fiat in 1954. The body was designed by Fabio Luigi Rapi.

placed the engine aft of the driver's seat. MG has not followed up on this with a production car, but Abarth built a 5-liter, V-8 road racer with an amidships engine in 1969. All straight-line speed-record machinery is, of course, marginal in its relationship to everyday sports cars.

Consider what the manufacturers put into production. News of the Coopers and Porsches had considerable impact among Detroit engineers and designers. At Chevrolet, Zora Arkus-Duntov built, in 1959, a prototype with a V-8 located amidships. They called it CERV-1 (Chevrolet Engineering Research Vehicle No. 1). It served mainly to develop the rear suspension for the 1963 Corvette Sting Ray. In 1964, Arkus-Duntov served up the CERV-11, a four-wheel-drive prototype, with an amidships-mounted V-8. It led in turn to a true roadcar prototype that appeared in 1970 as a study for a future production model.

In all of these cars, the engine was disposed longitudinally in the chassis. When Chevrolet displayed two Corvette-based experimental cars with rotary engines in 1973, the installations were transverse. Much of what Chevrolet was doing in these years was influenced by what was going on at Ford. Therefore it's necessary to look at cars from Dearborn to explain some that came out of Warren.

Ford's first amidships-engine car was an experimental roadster built, in 1962, to designs by Roy Lunn (an English engineer who had formerly worked for Jowett and Aston Martin). It was named Mustang, and it was powered by a small, V-4 engine (built by Ford of Germany) used as standard equipment in the Taunus 12 M. When Ford decided to go after the Le Mans crown, Roy Lunn was put in charge of the team that designed and developed the cars. The initial Ford GT-40 was patterned after the Ford-powered Lola that Eric Broadley built in the summer of 1962. It led to the Mark II and the J-car; all of them used V-8 Ford engines installed longitudinally. Ford built a small series of GT-40 road cars, in 1965, but there were no real production plans at the time. Driven by a desire to cash in on its racing successes and take over part of the Corvette market with more modern machinery, Ford found its answer in Italy in the form of the small De Tomaso operation.

De Tomaso had built cars with amidships-mounted engines since 1961. The first was a Grand Prix car powered by a 4-cylinder OSCA engine. Then there was an Indy car with a Ford V-8 power, at the end of 1963, and the Vallenlunga sports coupe with a British-made, 4-cylinder Ford engine in 1966.

Also in 1966, De Tomaso built a sports car prototype with a 5-liter, Ford V-8. In 1967, he introduced the Mangusta, a clean coupe with an amidships-mounted Ford V-8. Ford and De Tomaso made an agreement for joint production of a new car to be based on the Mangusta; this resulted in the Pantera of 1970. The Pantera was the reason why Chevrolet disclosed its Corvette with a central engine that year. Ford's failure to make a success of the Pantera in the U.S. market also explains why Chevrolet is not yet producing a Corvette with an amidships-mounted engine. Ford eventually sold its De Tomaso holdings to Rowan Controllers (which is the property of the De Tomaso family). The Pantera is still in production.

A few amidships-engined cars worthy of note had been announced by other manufacturers in the meantime. A transversely installed V-12 was the dramatic new idea behind the Lamborghini Miura of 1965. It was an extremely compact installation that permitted considerable weight saving by shortening the car (Fig. 11-11).

The body was a masterly design by Giorgetto Giugiaro and produced by Bertone (Fig. 11-12). When Lamborghini replaced the Miura, in 1971, it was with the Countach, carrying its V-12 longitudinally with the transmission at the front end, and an exterior shaft leading back to the final drive.

In 1964, Porsche began production of the 904 that was equipped with a flat-four located ahead of the rear wheels. It was soon superseded by the rear-engined 901 that led to the 911. Chassis layout had nothing to do with the demise of the 904. But the 904 had used a steel frame with a plastic body and that gave problems.

Shortly afterwards, Porsche began design work on a low-priced sports car, with an amidships-mounted engine, under a contract with Volkswagen. This joint venture resulted in the Porsche 914, that was available with either a flat-tour VW engine or a Porsche flat-six.

Under the same contract, Porsche also developed a small economy car under the code name EA 266. It was to be a two-door, four passenger sedan, a possible successor to the VW Beetle, with a four-cylinder, watercooled engine placed ahead of the rear wheel axis. This put the engine under the back seat, added accessibility problems for the engine heated up the back seat, raised the back seat to an abnormal height, and pushed up the roofline. The EA-266 was cancelled when VW decided to use Audi engineering for its new generation of cars and go to front-wheel drive across the board.

Curiously, Pontiac was toying with a similar project at the same time. The car was to be smaller and lighter than the Vega, and there was nothing conventional about it. The prototype weighed about 750 kgs for a complete, two-passenger, commuter-type car with an all-steel sedan body.

The engine was a very small radial four that was hardly bigger than the automatic transmission created for it. Both the engine and the transmission were bolted to the rear axle and carried an unsprung weight (bad for both ride and handling in a larger car). Pontiac felt that for driving short distances on good roads, this could be a good way to control manufacturing cost. Complete prototypes were built and tested, but the project was killed, in part, because the commuter car market had not yet been established, and partly because GM president Edward N. Cole's plans for the GM rotary (Wankel) engine pre-empted it.

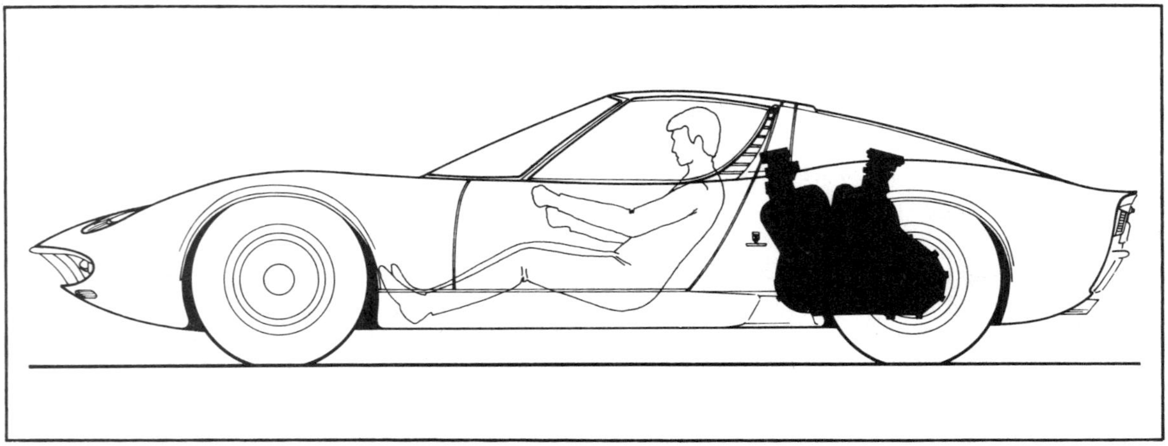

Fig. 11-11. The 1965 Lamborghini Miura had a V-12 engine mounted transversely.

Fig. 11-12. The Lamborghini Miura. The body was designed by Giugiaro and built by Bertone.

Colin Chapman went to the amidships engine layout for his 1960 Lotus Formula One and Formula Junior cars. The following year the Lotus 19 Monte Carlo was introduced after adapting the same layout for a sports/racing car. Lotus 23 was a Cosworth-Ford-powered sports car with an amidships engine, built at the end of 1961. It lead to the Renault-powered Lotus Europa, in 1966. In turn, it was replaced by the Esprit in 1976. The Esprit engine is installed longitudinally.

Alfa Romeo has yet to build its first amidships-engined production car, but the Tipo 33 competition car from 1967 had an experimental 2-liter, V-8 (that was being developed for the Montreal) located centrally in the chassis. It was designed by Carlo Chiti (formerly of ATS and Ferrari). A 3-liter version appeared in 1969. In 1971, the new 33 TT was equipped with a 4-liter, flat-12. Alfa Romeo's racing program, since 1976, has concentrated on a Grand Prix engine and the Tipo 179 Formula One car.

Ford of England prepared a sports coupe with amidships-mounted engine that came close to being produced in 1971. Called the GT-70, it was a half-way step between the original Mustang and the GT-40 (using a V-six Ford of 2.6 or 3.0 liters). It was designed by Len Bailey with manufacturing cost in mind, but the parent company killed it. It was left to AC, where the Cobra originated in 1963, to put a Ford-powered amidships-engined sports car into production in England. After years of Cobra experience, and developing the luxurious 428 GT, AC Cars Ltd. introduced the ME-3000, in 1974, using a British Ford, V-six engine.

When Jaguar began to develop its V-12 engine, it was first installed centrally in an experimental chassis in 1964. This F-type Jaguar was conceived as a possible Le Mans contender, but by the time it was race-ready, in 1967, it was no longer competitive and the project was dropped. The engine went into production for the E-Type in 1961.

It was certainly no thought of going back into racing that led Mercedes-Benz to develop its C-111 in 1969. But Rudolf Uhlenhaut and his engineering team needed a very high-speed chassis for development work, and that's how it came about. The first version was powered by a 3-rotor Wankel engine (placed longitudinally amidships). The second version had a four-rotor unit in the same place. The C-111 was kept in service as a mobile test bed after the Wankel engine project was put on the back burner in 1973. In 1976 it was used to demonstrate the speed capabilities of the turbocharged 5-cylinder diesel engine; it set a series of records (Fig. 11-13). But the company has no plans to put the C-111 into production—with any kind of engine.

The Maserati Bora of 1971—designed by the same Giulo Alfieri who had done the Tipo 63 10 years earlier—was the company's first production model with an amidships mounted engine. It had a four-cam, V-8 installed longitudinally ahead of the rear wheel axis. It was followed, in 1973, by the Merak that uses a V-six engine in the same arrangement.

Eager to preserve its sports-car image, BMW decided, in 1975, to create a modern sports car. Using a 3.3- or 3.5- six-cylinder engine, mounted centrally, vertically, and longitudinally, it was developed under the code name E-26 during 1976-77. The body was designed by Giugiaro. A few racing outings brought mediocre results, and the project was then given up. This allowed BMW to concentrate on a Grand Prix engine for Brabham.

Carrozzeria Bertone created its Stratos styling exercise, in 1970, using a four-cylinder engine installed transversely ahead of the rear wheels. Lancia's chief engineer, Sergio Camuffo, liked the car and stuffed a Dino V-six engine into it. The Stratos became an unbeatable rally car; it won the world championship three consecutive years (Fig. 11-14).

While this was going on, Fiat realized the potential of its front-wheel-drive 128 models as the basis for an amidships-engined sports car.

Fig. 11-13. The Mercedes-Benz C-III. The first version built in 1971 had a three-rotor Wankel engine. A redesigned model had a four-rotor engine. This one was built, in 1976, around a five-cylinder turbocharged diesel engine.

Giuseppe Puleo designed the X-1/9 around the 128 power train while Bertone developed the body for it. A more ambitious project was the X-1/20 that used the twin-cam 124 engine in a car of the same type. It developed into the Lancia Beta Monte Carlo that went into production in 1975. A turbocharged

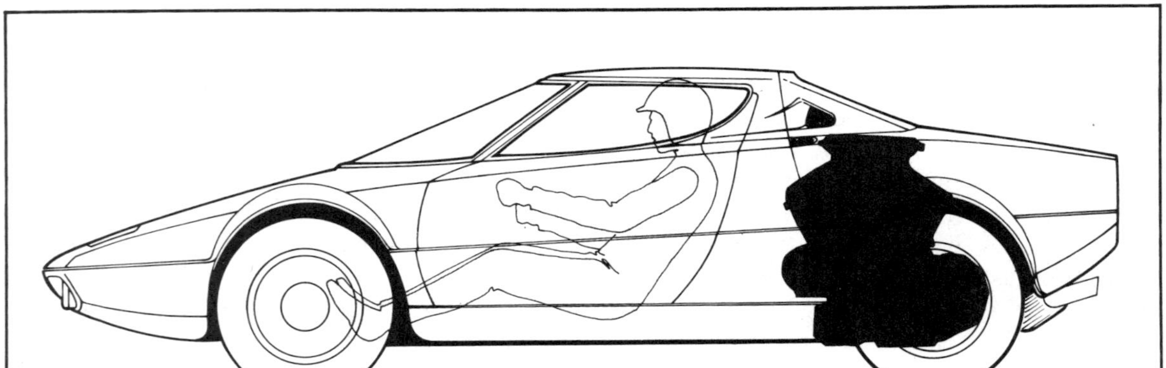

Fig. 11-14. The second edition, 1971, of the Lancia Stratos by Bertone. The engine is a Fiat Dino V-six, mounted transversely.

Monte Carlo was Lancia's main weapon in international racing for 1982.

The Renault R-5, new in 1972, is a front-wheel drive car with the engine mounted longitudinally aft of the front wheel axis. To create the R-5 Turbo in 1977, Renault turned the whole power package around and took the back seat out of the little car. The Turbo coupe became an impressive rally car for two or three years, but then it was surpassed.

Any front-wheel drive power train is suitable for transplanting to an amidships installation. But a number of makers of front-wheel drive family cars also build their own sports cars and prefer to stay with the front-wheel drive. Such is the case with the Volkswagen Scirocco. Sports cars will continue, I believe, to be available in a variety of configurations, including the traditional front-engine and rear-wheel drive concepts.

As for the future of the amidships-engine layout, its engineering advantages secure it a firm place in future two-seaters and competition cars. These advantages can be set out in ten concise points.

☐ **Improved Traction.** By positioning the engine close to the driving wheels, the unit load on the tires is raised and traction is improved. The center of gravity will be close to the rear wheels. During acceleration, weight transfer takes place (from front to rear). This further adds to the unit load on the rear tires. This has the effect of further reducing the risk of wheelspin on acceleration.

☐ **Better Braking Balance.** Under braking, severe weight transfer takes place—in a rear-to-front direction—with a front-mounted engine. This leads to a severe loss of braking effect in the rear wheels (because the unit load on the rear tires has been reduced). With the amidships engine design, the weight transfer will be the same, but because the center of gravity is located closer to the rear wheels, the loss of braking capability in the rear wheels will be less severe. This means more favorable weight distribution during braking to enable the rear wheels to perform a larger proporation of the braking duties. This provides the driver with better balanced braking for an easier control situation, with less risk of rear wheel locking and skidding (spinout).

☐ **Higher Cornering Capacity.** The amidships engine location permits a slight lowering of the engine's position in the chassis. More important, it permits a general lowering of the car floor, the seats, and all other mechanical components. This results in a lowering of the center of gravity. The car will be able to sustain higher lateral accelerations. This is of ultimate importance for a sports car that is likely to be used in racing.

☐ **Ride Comfort.** For any car, even the starkest sports car, ride comfort counts as long as it carries at least one passenger in addition to the driver. With an amidships-mounted engine, the lower center of gravity will reduce body roll on curves. Roll depends on the vertical distance between the center of gravity and the car's roll axis. With retention of the same suspension geometry, a lowering of the center of gravity will reduce body roll under lateral accelerations. There is great opportunity for weight saving in the mid-engine chassis (elimination of propeller shaft, more compact power train, shorter wheelbase, more compact body). This means softer springs can be used. A lower spring rate will lower the ride frequency, and the human body generally finds lower ride frequencies more comfortable (with proper damping—otherwise low spring rates will give rise to large amplitude; that can produce "seasickness").

☐ **More Responsive Handling**. With a front-mounted engine, there is a longer distance between the engine's center of gravity and the car's center of gravity than will be the case for the car with an amidships-mounted engine. This distance is very important because it dictates the car's polar moment of inertia (along with the disposition of other weight concentrations such as fuel tank, radiator, battery, and spare wheel).

A long distance between the engine's center of gravity and car's center of gravity means a high polar moment of inertia. This gives high directional stability and a resistance to change of direction. A short such distance gives low polar moment of inertia. That means less reluctance to a change in course, and that equals more responsive handling.

☐ **Improved Visibility.** With most front-mounted engines, a high hood line is unavoidable. That is deleterious to the forward view that is so vital to a sports car driver. The amidships engine location gives the designers full freedom to lower the hood line.

☐ **Lower Noise Levels.** Placing the engine behind the driver will result in less mechanical noise floating into the car from the power train. There can be more effective insulation on the wall between the seats and the amidships-mounted engine than on the firewall of a car with a front-mounted engine.

☐ **Passive Safety.** With the engine removed from the front of the car, the structural engineers have a much easier task of building controlled front-end crash into the body and chassis, and to provide the necessary padding in front of the passengers. Because the car will be lighter, the impact loads in a collision will be reduced, and the energy management task will be less enormous.

☐ **No Need for Power Steering.** Because the engine mass has been moved back to the rear wheels, the weight on the front wheels has been significantly reduced, and the effort needed to turn them is lower. This means the need for power steering has been eliminated. That's important to the sports car driver who prefers manual steering for its better "road feel."

☐ **Lower Production Cost.** Lower cost might not seem like a benefit for the driver, but it concerns him at the moment of making his purchase. Lower cost helps the maker keep the price competitive. The cost-saving potential of the car with the amidships-mounted engine lies in the fact that it can be made shorter and lighter without any loss in carrying capacity. Any weight saving will show up as a cost saving. The elimination of components (propeller shaft with two universal joints—and, with a transverse engine, ring gear and pinion) produces a direct cost cut.

Chapter 12

Bantamweights and Compacts

WHILE MILLIONS OF AMERICAN MOTORISTS ARE learning to live with small cars, the industry is exercising its brains to find ways of making them more satisfactory. In a market that has been attuned to powerful, roomy, and heavy cars until recent years, sudden and total change can be ruled out. The industry cannot change gears fast enough and the market needs a period of preparation.

The boom in European imports, beginning in 1956, followed by the arrival of the domestic compacts in 1960 and the subcompacts in 1971, have amply prepared the public and substantially altered the size mix in the national car park.

Europe and Japan have been building mainly small cars for the past 50 years. At the start of the 20th century, many American cars were small, too, but the distances to be traveled, loads to be carried, comfort levels expected—assisted by the tremendous economic strength of the nation—pushed car sizes up and up.

WHO BUILT THE FIRST SMALL CAR?

Ford's first car was a quadricycle, with seating for two, with no weather protection whatsoever. In Europe, some of the most popular personal transport vehicles of the pre-1900 era were Leon Bollee and de Dion-Bouton motor tricycles. Examples of small cars abounded in the years that followed.

What was the first small car? Isaac Watt Boulton, a railroad-locomotive repair man of Ashton-under-Lyme, England, made six small steam cars, probably for carrying only one person, and almost certainly three-wheelers, in 1848.

Before the arrival of the gasoline-driven cars in 1885-86, small, lightweight steam cars were built in England by Louch & Messenger, John H. Knight, and L. J. Todd. In America similar cars were built by Elijah Ware of Bayonne, New Jersey, Sylvester Hayward Roper of Roxbury, Massachusetts, and Lucius D. Copeland of Camden, New Jersey.

Peugeot built quadricycles as well as carriage-size cars in the 1890s. Riley, Wolseley, and Humber started with motor tricycles. Louis Renault began production of a small car in 1899. His first model laid the foundation for a giant of industry.

After 1901, small-car designers copied the Mercedes layout (just as the makers of bigger machinery were doing). Among early examples that traced the course of evolution for the small car were the 1905 Liliput, designed by Willy Seck, the 1909 Opel 8 PS (nicknamed *Doktorwagen*), Jules Salomon's Le Zebre from 1909, and the Wanderer 5/12 PS of 1911 that was nicknamed *Puppchen* (little doll). It had seating for two, in tandem formation, while the others had two seats side by side.

These and other examples of small cars were, by and large, designed and built like large cars scaled down. About 1910 some designers began to think of new small-car concepts that were basically different. Sometimes the layouts were not even suitable for larger cars.

THE CYCLECAR WAVE

It was Robert Bourbeau of Paris who invented the cyclecar concept that was first used on his Bedelia in 1911. It became a trend setter. Everywhere there seemed to be a market for a kind of car that bridged the gap between the motorcycle-and-sidecar and the "real" automobile.

The Bedelia had a motorcycle engine mounted in front, below a torpedo-shaped fuel tank, with an exterior belt drive to the rear wheels. Seats were arranged in tandem fashion (with the driver *behind* the passenger). Among the first French followers of the Bedelia example were La Ponette, Automobilette, Violette, and Lurquin-Coudert.

Some British designers shortened the narrow cyclecar to a single-seater (Carden, Rollo Pony, AV Monocar) while others widened it to carry the passenger alongside the driver. They weighed 500 to 650 pounds; the single-seaters could be as light as 300 pounds. In the side-by-side seating category were the Humberette—with a front-mounted V-twin engine, seamless steel-tube frame, and shaft drive—Swift, Coventry Premier, Walcycar, and others.

The Coventry Premier was built on a 72-inch wheelbase and used a tubular steel frame (extended from motorcycle practice). It was powered by a V-twin motorcycle engine driving the rear wheels by chain. The Swift was longer, with an 84-inch wheelbase, and used shaft drive. The 1912-1914 Walcycar was a stylish roadster, made in Birmingham, featuring a pressed channel-steel frame. Among British cyclecars still in production in 1921 were the AV Bicar, Bleriot-Whippet, Carden and Tamplin.

In America, the cyclecar was an outgrowth of the motorized flivver. The trend started in 1902 with the Orient Buckboard. It had a simple design with a wooden platform carrying a bench seat for two and a single-cylinder engine mounted between the rear wheels. It had tiller steering, a total absence of weather protection, and weighed 350 pounds. It was produced until 1909 in Waltham, Massachusetts and a copy made in France was sold as the Reginette. The Orient Buckboard evolved into the Metz; it had a two-cylinder engine and wheel steering, and it could be purchased as a kit for home assembly.

Tandem-seat cyclecars also existed in America. Among the first was the 1914 Woods Mobilette from Harvey, Illinois. The same year Walter Marr, former chief engineer of Buick, also built a tandem-seat cyclecar in Detroit. The Imp from Auburn, Indiana, also had tandem seating and followed the Bedelia layout with the exception of seating the driver in front.

Benjam Briscoe produced the Argo cyclecar in Jackson, Michigan, from 1914 to 1916. It was a convertible two-seater of conventional construction. The American cyclecar was dead, but the small-car idea was not.

On the European continent, the cyclecar trend declined after Peugeot began producing its Type BP 1 (famous as the Bebe or baby Peugeot) in 1913 (Fig. 12-1). Peugeot bought the design from Ettore Bugatti. Bugatti had drawn it up in 1911 as an answer to Le Zebre and first tried to sell it to Wanderer. The Bebe Peugeot has seating for two, side by side, and a four-cylinder engine with shaft drive to the rear axle. With its 71-inch wheelbase, it weighed only 772 pounds including the collapsible top. It was ahead of its time.

Strangely, in 1920 Peugeot replaced it with a narrow-track design having two seats in tandem formation. Why? Because the Bebe was too heavy to qualify for the lower tax applied to cyclecars. The Quadrilette, despite its 91-inch wheelbase, weighed less than 750 pounds. The body widened, in 1921, to permit an offset passenger's seat halfway alongside the driver's seat. It gained great popularity, and continued unchanged in 1922. But for 1923, Peugeot redesigned the Quadrilette with seating for two, side by side, and it became taxed as a car (Fig. 12-2). It was now competing against the 5 CV Citroen (a modernized Le Zebre, due to Jules Salomon) that also proved highly successful.

What finally killed the cyclecar in Great Britain was the arrival of the Austin Seven in 1922. To some extent inspired by the Quadrilette, it was an entirely original design by Sir Herbert Austin, Karl Brozyna, J. C. Haefeli, and J. R. Rix.

Fig. 12-1. The 1913 Peugeot Bebe Type BP I.

Fig. 12-2. The 1923 Peugeot Quadrilette.

Instead of being merely a two-seater, the Austin Seven offered essential motoring for four persons (Fig. 12-3). The engine was a small, in-line four—mounted conventionally in front—with a radiator ahead of it, and drive to the rear axle. The wheelbase was 75 inches, top speed was 45 mph, and fuel economy was about 36 mpg.

It had a simple chassis frame, made up like a tapered ladder (narrow at the front and wide at the rear), with straight, channel-section side members, and only two main cross members—one behind the engine and another just ahead of the rear springs. The front cross member supported the gearbox and provided anchorage points for the radius rods attached to the front axle. The rear cross member carried the thrust bearing for the short torque tube enclosing the rearmost half of the propeller shaft. The springs were mounted in brackets on the outside of the frame side members.

ARCHITECTURAL EXPLORATION

The first body style was a two-door convertible, and a coach followed in 1923. The bodies consisted of steel panels on a wooden framework. The convertible weighed 850 pounds.

The Austin Seven's layout and dimensions were quickly copied by Standard, Rhode, and Swift. By 1928, a modernized version was built under license by Dixi (later BMW) in Eisenach, Rosengart, in Paris. Steps to produce it in America were taken by a small company in Bulter Pennsylvania that began operations in 1931. In 1933, the Austin formed the basis for the first Datsun from Tokyo.

The Austin Seven could be criticized on the grounds of lacking small-car architecture, because it did not differ conceptually from the Austin Nine or Twelve, but fault could not be found with its success.

As a counterpoint, the Trojan can be cited as a thoroughly unconventional car that never caught on with the public. It became more popular as a light van or utility vehicle than as a passenger car (Fig. 12-4).

Leslie Hounsfield built the first Trojan prototype in 1913. His ambition was to build a simple car that was inexpensive to manufacture and inexpensive to run. The car had a hood, but it covered only the fuel tank and radiator. The engine was a two-stroke pancake engine located under the front passenger's seat, driving the rear axle by chain. The Trojan chassis was a pressed-steel tray, and the body was steel-paneled.

A four-seater, all-weather touring car, built on a 95.5-inch wheelbase, it was 135 inches long overall and weighed about 1700 pounds. It was tall and narrow. Some would say it was ugly. But there is nothing laughable about it. It is a valuable lesson in economy-car design. The Trojan engine delivered 10 hp, which gave the car a top speed of 40 mph, with 40-mpg fuel economy.

YOUR BREAKFAST ROLL, HERR SCHIRRMEISTER

Germany's economy under the Weimar Republic was more conducive to developing marginal cars than in the rest of Europe. The first radical attempt was probably the S.B. (Slaby & Beringer). This single-seater was little more than a battery-powered soap-box on wheels. Then came the as-

Fig. 12-3. The Austin Seven was the first of a new breed.

Fig. 12-4. The Trojan engine was mounted under the left front seat.

tonishing Hanomag 2/10 PS that was designed by Carl Pollich and Fiedelis Böhler. It was made in the giant Hannover locomotive works from 1924 to 1928.

Hanomag's tiny economy car became famous under the name *Kommissbrot* (because its shape reminded the Germans of the hardtack bread served in military canteers in World War I). Fenders were integral with the sheet-steel body. A single headlamp was mounted in the center, low down, while each fender carried a small marker light.

It was built on a 75.6-inch wheelbase and had a track of 42 inches, front and rear. Curb weight was 816 pounds. The seat was a bench-seat for two placed midway inside the wheelbase. The front end was mainly reserved for legroom, but there was also a small trunk for minor tote items. The engine was a vertical single-cylinder unit with water cooling. Radiator and fuel tank were accommodated inside the engine compartment between the rear wheels. The Hanomag was priced at 2300 Reichsmark while the best-selling car, Opel's Laubfrosch, sold for RM 4000. It could exceed 40 mph on good roads and would cruise at 25 to 30 mph.

ORTHODOXY AND ADVANCE

In 1931, the DKW company in Zschopau, Saxony, took a great leap forward in small-car construction.

The F-1 (Fig. 12-5) had a very low and narrow ladder-type frame, and most unstressed parts of the bodywork were made of plywood with imitation-leather covering. The engine was a two-cylinder, two-stroke unit mounted transversely in front for driving the front wheels. Two-seater roadster and convertible models came first. They were followed in 1933, by a low, two-door coach with seating for four people.

Built on a 94.5-inch wheelbase, it weighed 1136 pounds. Its 14-hp engine gave a top speed of 50 mph and the average fuel economy was 30-33 mpg. The same basic concept lasted until 1940. Then DKW moved the engine from within the wheelbase into the front overhang and streamlined the body.

The Morris Minor of 1928, on a 78-inch wheelbase (Fig. 12-8), became a big seller, but what it had over the Austin Seven was mainly a superior engine. When Ford began building small cars in England, in 1932, the Model Y (marketed as the Ford Eight in Britain and Ford Junior on the Continent) was an elegant little sedan that foreshadowed the styling of the American 1934-model V-8. It was designed in Dearborn by Eugene T. Gregorie. But its architecture represented no advance over the Model A.

Thoughts of a radically different people's car began occurring to several engineerers in Germany as early as 1926-27. One was Josef Ganz who was editor of a magazine called *Motor Kritik*. Another was a design consultant by the name of Bela Barenyi. Both had the idea of mounting the engine in the rear, fuel tank and spare wheel in front, with a chassis having all-independent suspension. Both published drawings and applied for patents on specific features. But neither one got as far as building a prototype.

The first company to make the attempt was ZUndapp in Nuremberg. ZUndapp came into being as a maker of detonators. The name is a contraction of the corporate title: ZUnder and Apparatebau

Fig. 12-5. The DKW Reichsklasse Type F-2 from 1933.

Fig. 12-6. The Morris Minor, 1929.

GmbH. ZUndapp began motorcycle production in 1921, and 10 years later was ready to tackle a car project.

PLANNING THE PEOPLE'S CAR

The President of ZUnder and Apparatebau GmbH, Fritz Neumayer, said what he wanted was a "Volks-Auto." For the design, he went to Professor Porsche who had established an independent engineering bureau in Stuttgart in 1930. The car, Porsche Type 12, was designed between December, 1931 and April, 1932. It had a tubular backbone frame and a rear-mounted engine. Porsche actually designed three engines: first, a five-cylinder radial, then a flat-twin, and also a three-cylinder radial unit.

The first ZUndapp prototype was a boxy car in contemporary style, but the second one had a fastback roofline, faired-in headlamps, and skirted fenders. The bodies were built by Reutter.

Type 12 never got into production because Neumayer realized that ZUndapp lacked the financial strength to develop the car and tool up a whole new plant at that time. But the design exercise was to prove valuable to Porsche early in 1933, when Fritz von Falkenhayn, president of NSU, offered him a contract for designing a small car. NSU had built cars from 1906 to 1929, but then sold its auto plant to Fiat. Feeling that motorcycles was not enough to plan the company's future on, Fritz von Falkenhayn wanted to get back into car production. But he had no product ready.

Professor Porsche revised his ideas and began on Type 32. The frame became a platform including a central tube, and the engine evolved into an air-cooled flat-four. Three prototypes with steel bodies were built (one by Reutter and two by Drauz). They had a short, sloping hood and a fastback roofline.

When it came to making a decision to go into production, von Falkenhayn hesitated. Finally he decided the financial burden would be too much for NSU to undertake.

Professor Porsche began talking with a government official who had Adolf Hitler's ear in automotive matters. Out of that came a request to the National Society of the Automobile Industry for recommendations on specifications for a mass-produced economy car. That was the start of the Volkswagen. The car was to have seating for four, weigh less than 1433 pounds, reach a top speed of 62 mph, and offer a fuel economy of 30 mpg or better.

OPEL'S REALITY VERSUS HILTER'S DREAM

Using type 32 as a base, Porsche made plans for Type 60, and submitted them to the Society. He was offered a contract and, on June 22, 1934, Porsche undertook to design and construct three prototype cars within 10 months. The work went far behind the deadline. It was not until October 12, 1936, that the three prototypes were presented to the Nazi authorities (Figs. 12-7 and 12-8).

These were the first of the beetle-shaped cars (with body design by Erwin Komenda). The Porsche organization was told to give each one a 30,000-mile test, which was completed in less than two months. After a partial redesign based on the test results, the government wanted 30 new test cars. That was too large an order for Porsche to handle, but Mercedes-Benz volunteered to make them. They were ready in the spring of 1937, and SS personnel thrashed them for 12 million miles.

Fig. 12-7. A Volkswagen prototype from 1937.

Fig. 12-8. A Volkswagen prototype from 1937.

Germany's auto and supplier industries were informed of the project and its progress. The Volkswagen was no secret. That led to some difficult moments such as the meeting between Hilter and Wilhelm von Opel at the Berlin auto show in 1936. When Hilter finished raving about the VW, Wilhelm von Opel pointed to the old Opel P4 (Fig. 12-9) and said, "There you see my Volkswagen." It was selling within 33 percent of the target price for the VW, which would need State subsidies to be held. It was already in mass production, and Opel was not selling it below cost. Hilter never forgave him.

The VW was not a good example of efficient space utilization (Fig. 12-10). Its whole architecture was full of compromises and it had insufficient luggage space and cramped rear seating. Weight distribution had a severe rearward bias, and the body shape led to instability in crosswinds. In addition, the engine was noisy.

British industry representatives examined the car and the factory ruins in 1945. The reaction from the chief engineer of Rootes was that it held no technical interest.

RADICAL ALPINE CONTRIBUTIONS

Another beetle-shaped car went into production while the VW was still under development. It was the Steyr 50, made in Austria from 1936 to 1938. It was succeeded by the Steyr 55, that was the same vehicle with a slightly more powerful engine.

Like the Volkswagen, the Steyr 50 had a flat-four engine. Unlike the Volkswagen, the Steyr engine was water-cooled and mounted in front. Though the drive was taken to the rear wheels, the car was just as low as the VW thanks to the use of a unit-construction body shell and all-independent suspension.

Shortcomings in the Steyr were not due to its layout. The fastback roofline made for poor back seat headroom because it was designed to be visually pleasing and look streamlined. Also, there were no side windows behind the doors. Rear-seat passengers were kept pretty much in the dark about the scenery they were traveling through.

Still, the Steyr 50 must be judged as architectually superior to the VW. Unfortunately, Steyr was a small car maker, and the production cost was just too high to make it competitive. It was listed at double the price of the Opel P4. Its mediocre performance and failure to offer substantially lower fuel consumption also played parts in driving the buyers away.

Two Italian cars that went into production during the VW's development period attracted world-wide acclaim. One was smaller and the other

Fig. 12-9. Opel's lowest-priced model was the P-4 of 1936.

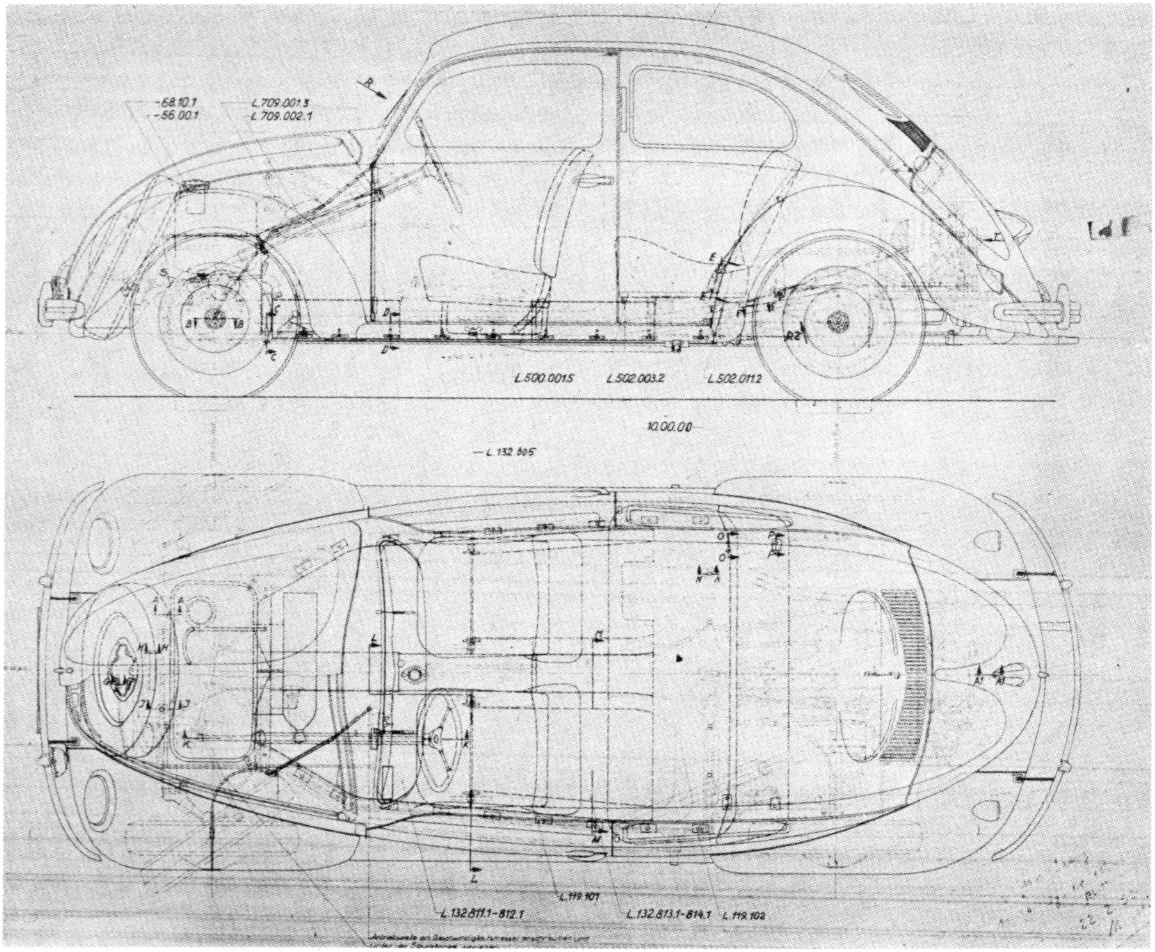

Fig. 12-10. An elevation and plan view of the production-model Volkswagen, 1939.

was bigger. I am referring to the Fiat 500 (Fig. 12-11) and the Lancia Aprilia (Fig. 9-7). Both had front-mounted engines and rear-wheel drive. The Fiat was a two-seater coupe and the Lancia was a four-door, four-seater sedan.

The Fiat 500 was the design of Dante Giacosa. He chose a lightweight, ladder-type frame that was just long enough to carry attachments for the front and rear suspension systems. The engine was a tiny, 13-hp inline four, that he mounted in the nose, ahead of the front suspension—even ahead of the radiator. This freed space for the pedals next to the gearbox, and the floorboards came right up to the wheel housings. This way it became possible to keep the wheelbase extremely short, only 79 inches, even with next to no overhang at either front or rear. With an all-steel body, the 500 weighed 1180 pounds. It had a top speed of 50 to 55 mph, and ran with an average fuel economy of 40 to 44 mpg.

A station wagon with four seats was added in 1948, and in 1949 the Fiat 500 was restyled for a more modern look (Fig. 12-12). With this change, the former wood-paneled wagon was replaced by an all-metal one. It was kept in production until 1954.

The Aprilia was the last masterpiece from the drawing board of Vincenzo Lancia. It was designed with the objective of carrying five persons at 80 mph, with low fuel consumption, using lightweight construction methods to obtain reasonable performance levels in a car having a small power unit.

The unit-construction body helped reduce the weight. The engine was an extremely compact, narrow-angle V-4 mounted between the front wheels with the gearbox built in unit with the final drive (transaxle) so that the passenger compartment could dispose of the total space between the wheels. The wheelbase was stretched to 108.3 inches, for an overall length of only 156 inches, so that the rear seat passengers would not have to sit on top of the rear axle area. For good measure, the Aprilia had no rear axle, but an intricate system of independent rear suspension.

Truck space was somewhat restricted due to

Fig. 12-11. The Fiat 500 set new fuel-economy records.

Fig. 12-12. A comparison of the Fiat 500 (center) with the 1922-model Austin Seven (top). The Fiat 500 C (bottom) appeared in 1948.

the "roundback" design of the body. It was a compromise between headroom requirement for the back seat, aerodynamic considerations, and the need for keeping it short overall to meet the weight target. It had a dry weight of 1787 pounds, and the 47-hp engine gave it the desired speed plus 26-mpg average fuel economy.

THE AMERICAN IDEA OF A SMALL CAR

Small cars did not have much of a market in the United States, even in the depths of the Great Depression that started under the Hoover administration and lasted through Roosevelt's first term.

The American Austin gave way to the Bantam, basically the same car with modern American styling, but sales remained very low. Powel Crosley Jr. started making the little Crosley economy car in Cincinnati, in 1939. It was notable for its small wheels—less space was stolen from the interior—and proportions that blended well with contemporary styling.

The size of car that was later classified as "compact" enjoyed greater support from the motoring public. Willys-Overland of Toledo, Ohio, was among the first to mass-produce a car about eight-tenths of the normal scale with the 1926 Whippet. Architecturally, however, it did not differ from the full-size Willys models.

Then came the Willys 77 of 1933. Stubbornly, sticking to a four-cylinder engine, even though Ford had standardized the V-8 and Chevrolet and Plymouth used only sixes, the Willys 77 was built on a 100-inch wheelbase. It had a boxy, practical sedan body with up-to-date styling. It was capable of 60 mph, and it averaged 22 to 25 mph fuel economy. It was replaced in 1937 by an airplane-nosed, semi-streamliner (using the same four-cylinder engine, good for 70 mph and 25 to 30 mpg). It was restyled again, in 1940, and renamed Willys Americar.

Chrysler Corporation did extensive experimental work during the 1930s with compact and smaller-than-compact cars. Included was a design for a radical Airflow-style prototype with a front-mounted, five-cylinder radial engine and front-wheel drive (Fig. 12-13). But none of its features were retained for the production models.

The Studebaker Champion of 1939 was a compact when compared with the Commander and President, but it was a roomy, four-door sedan with

Fig. 12-13. An experimental small Chrysler from 1936.

a six-cylinder engine that was not much smaller in physical dimensions than the Ford and Chevrolet of that time.

During the war, both Chevrolet and Ford had large programs for small post-war models. Ford's experiments were extremely broad; included was a car with a transversely mounted engine and front-wheel drive. But as the Light Car project advanced towards production, only conventional solutions were accepted.

At Chevrolet, light car studies crystallized into the Cadet; it came close to being cleared for production in 1947. It was a highly intelligent creation by Earle S. MacPherson who soon afterwards went to Ford and incorporated many of the same ideas into the British-made Consul and Zephyr of 1950.

For the American market, Ford killed its light car program when news leaked out about the Cadet's cancellation. Nevertheless a version of it was produced in France from 1949 to 1953 as the Ford Vedette.

COMMUTER-CAR PROPOSALS

After 1945, the American public was readier than before to accept small cars, not only in the cities of the Eastern states and the Midwest, but also in the suburbs in California. The redesigned postwar Crosley rode on a wave of new popularity. More than a score of minor companies put small cars in production. Examples were the Playboy, Keller, Davis, Rocket, Bobbi-Kar, and King Midget.

The 1946 Bobbi-Kar was a four-wheeler on an 80-inch wheelbase that was built by the Bobbi Motor Car Corp. of San Diego. The outfit moved to Birmingham, Alabama in 1947 and was reorganized as the Dixie Motor Car Co. Then George D. Keller set up as Keller Motors of Huntsville, Alabama, took over the Bobbi-Kar design, and built the same car as a three-passenger convertible and five-passenger station wagon until 1949.

The 1946 Playboy was a stylish miniature roadster made near Buffalo, New York by Louis Horowitz and Charles D. Thomas. Glenn Gordon Davis of Van Nuys, California designed an egg-shaped, three-wheeler intended to sit four persons abreast on one bench seat.

The 1947 Rocket was made by the Hewson Pacific Corp. in Los Angeles. It had three seats side by side in an aluminum body.

The Hoppenstand was a sleek roadster on a 90-inch wheelbase that stemmed from Greenville, Pennsylvania. It had a flat-twin engine under the seat.

Claud Dry and Dale Orcutt set out to build a miniature Jeep and came up with the King Midget that was powered by a single-cylinder rear engine. It went into production in Athens, Ohio, in 1948, and remained in production for 21 years.

THE BIRTH OF THE COMPACTS

When Henry J. Kaiser decided to put some of his great fortune into setting up as a major car manufacturer, his first plan was to build a medium-sized, front-wheel drive car with an abundance of aluminum parts designed by J. A. Gregoire. But that design got lost on the way to the fabled Willow Run production line. The Kaiser that appeared in the showrooms was as conventional as Ford and GM products.

The 1949 models from Chrysler could well be described as compacts, when compared with the Plymouth, Dodge and De Soto of the 1958-59 period. They were reduced in size and weight, roomy, and practical, but they were short on styling. With regard to layout and engineering, they were devoid of innovation. In the market place, they were duds.

The first modern compact was the Nash Rambler of 1950. It was built on a 100-inch wheelbase, and it was powered by a small six that gave 25 to 30 mpg fuel economy. Nash also displayed a truly small-car prototype called NXI (Nash Experimental International) in 1950. Public reaction was found to be encouraging, and the car went into production in 1956 as the Nash Metropolitan. It was built by Austin of Birmingham, England, and used the Austin A-40 chassis.

Kaiser Motors began building the Henry J. compact in 1951; it was an overstyled sedan on a 100-inch wheelbase. Two years later came the Aero Willys and the Hudson Jet. The Rambler was the best-seller among the compacts, and its success brought the Big Three to decide to make their own compacts.

Except for Chevrolet, the automakers concentrated on a styling approach, and turned their backs on novel concepts. Chevrolet chose a layout with a rear-mounted engine and ran into an "unsafe at any speed" charge from consumer advocate Ralph Nader.

The Ford Falcon (Fig. 12-14), Plymouth Valiant (Fig. 12-15) and Chevrolet Corvair came out as 1960 models. Studebaker tried for a shortcut, in the literal sense, in creating the Lark by chopping off the front and rear ends of the Commander.

The subcompacts, Ford Pinto and Chevrolet Vega, appeared as 1971 models. Both are case studies in how *not* to design and engineer a small car. With their front-mounted engines and rear-axle drive, their space utilization was poor. The Pinto was overstyled and lacking in practicality. The Vega went on to build up the worst callback record for any American car. American Motors kept them company with the Gremlin; it was simply a long-hooded Hornet with a chopped-off rear end. Only after 1975 did Ford, Chrysler, and GM begin to develop sensible small cars. Then it was done in close cooperation with their European branches.

NEW DEFINITIONS OF BASIC TRANSPORT

What happened in European small-car design after 1945 had its roots in prewar projects. Prototypes of the Citroen 2 CV, which was first shown in Paris in October, 1948, had been undergoing tests since 1936. The 1939 version (Fig. 12-16) was close to the final design in most respects.

Rustic in its simplicity, archaic in its equipment, and primitive in its appearance, it was designed as a marvelous balance of qualities that are difficult to combine. It was light at only 1160 pounds. Despite its 95-inch wheelbase, it was a roomy, four-door sedan. Despite a cheap interior and low-priced seating materials, it was comforta-

Fig. 12-14. A styling mockup for the original Ford Falcon, approved in May, 1958.

Fig. 12-15. One of the first scale models made at Chrysler for a small postwar car in 1947. Some of the design went into the 1949 models, and then it was dusted off when the Valiant study began.

Fig. 12-16. In 1939, the prototype for the Citroen 2 CV look like this.

able on or off the road. It was capable of cruising at 40 mph for indefinite periods, with close to 40-mpg fuel economy. If it was ugly to look at, it was beautiful to own.

Another solution with an air-cooled, flat-twin engine and front wheel drive was the 1946 Dyna Panhard . Based on a Gregoire-designed prototype (Fig. 12-17) built in 1943 for Aluminium Francais, it was a brilliant study in lightweight construction with heavy emphasis on operating economy. In contrast with the 2 CV Citroen, Panhard's solutions were costly.

The AF-Gregoire prototype was a two-door convertible with a weight of 1310 pounds that was capable of 55 to 60 mph and 50-mpg fuel economy with its 15-hp engine. The production-model Dyna Panhard was a four-door sedan (Fig. 12-21) with an all-aluminum, unit-construction body and a 24-hp engine. Its top speed was 60 to 65 mph, and average fuel economy 40 to 45 mpg.

Fig. 12-17. The 1943 prototype built by Aluminium Francais to J. A. Gregoire designs. The author is at the wheel.

Fig. 12-18. The 1946 Dyna Panhard.

Tiny cars with seating for one or two occupants appeared frequently in postwar France and Italy, A few of these cars, such as the Julien, de Rovin, and Bernadet, got into production. Not so with the Dolo, Mathis 333, Vulografo, Rolux, or OPES Ninfea.

Gabriel Voisin designed a tiny car in 1947. He called it the Biscuter. It was really like two motor scooters, placed side by side, with a low bench seat and a car-type steering wheel. It has a 9-hp single-cylinder engine mounted behind the seat. A Spanish company based in Barcelona built it from 1949 to 1961.

While Fiat resumed production of the 500, BMW and Rover borrowed its profile for new baby-car projects of their own. After building prototypes and making cost analyses, both companies decided not to go into production.

Wilhelm Gutbrod was the owner of a farm machinery works, in Plochingen, for whom Hans Scherenberg designed a small car that went into production in 1950. The first Gutbrod Superior was a two-seater coupe with a front engine and front-wheel drive, a 20-hp engine, and a top speed of 60 to 62 mph. Later a four-seater station wagon was added.

Hanomag exhibited a new small car resembling the Fiat 500 C at the Frankfurt auto show in 1951 (the Partner two-seater coupe), but production plans were canceled at the last moment.

BRITISH INNOVATION AT OXFORD

In Great Britain, Austin revived the Eight of 1939. Morris and Ford also produced Eights of prewar design, while Standard's Nine of 1939 evolved into and Eight but retained the old body. The Austin was built on a semiplatform frame; Morris, Standard and Ford used ladder-type frames. Morris was first out with a postwar replacement in the Morris Minor (Fig. 12-19) introduced in October, 1948. A mod-

Fig. 12-19. The 1949 Morris Minor.

Fig. 12-20. The 1949/50 Jowett Javelin was inspired by the Lancia Aprilia.

ern, small Ford was not seen until the new Anglia came out in 1953. It was preceded by the Austin A-30 in 1952 and followed by the Standard Eight of 1954.

The Morris Minor stemmed from a 1943 prototype, the Mosquito, designed by Alec Issigonis. It had a front-mounted, flat-four engine and rear-axle drive in a pretty and curvaceous two-door, unit-construction body shell. Before it was cleared for development, the engine was changed to a vertical, four-cylinder in-line (that Morris had in mass production), and the width was cut by 4 inches. The production model was an elegant little four-seater sedan built on an 86-inch wheelbase. And it weighed 1550 pounds. With its 27.5 hp engine, it was capable of reaching 65 mph and gave an average fuel consumption of 30 to 35 mpg.

It was a direct competitor to the Volkswagen, and clearly a superior design. It had far better space utilization, gave a quieter and more comfortable ride, and its roadholding was completely free of the VW's vices. A four-door Minor became available in 1956.

A somewhat bigger car of the same period was full of small-car engineering principles. It was built from 1947 to 1954 by Jowett at Idle, near Bradford in Yorkshire, and called the Javelin (Fig. 12-20).

The Javelin was designed by Gerald Palmer (who admired the Lancia Aprilia). He had also closely examined the Steyr 50. Jowett made its own flat-four, aluminum-block engine that was mounted ahead of the front-wheel axis. The transmission and drive line were conventional, with rear axle drive.

It weighed 2215 pounds with an overall length of 168 inches. It was well balanced on its 102-inch wheelbase. The passenger compartment was pulled so far forward that the wheel housings intruded in the pedal area, and the back seat was comfortably clear of the rear axle, with a spacious trunk buried in the fastback tail end.

Its unit-construction body had a front subframe to support the engine and help isolate its vibrations from the passenger compartment. The subframe combination with a unit-construction body was also developed by Mercedes-Benz; they used it for the 180 series in 1953 (Fig. 12-21).

The 50-hp, water-cooled, flat-four engine was enough to give it very lively performance. Zero-to-60 mph acceleration was accomplished in 22.5 seconds, top speed was close to 80 mph, and, in normal driving, its average fuel economy was 30 mpg. The Javelin did well in endurance racing; it averaged 65.5 mph in the 24-hour race at Spa-Francorchamps in 1949.

MEANWHILE, BACK IN TORINO . . .

In the Jowett class, of course, the Lancia Aprilia was unbeatable in long-distance racing as well as from a viewpoint of architectural efficiency. Lancia even created a smaller car, the Ardea, in 1939, by scaling down the Aprilia. This four-door sedan, with its 30-hp engine, weighed 1665 pounds, had a top speed of 67 to 70 mph, and gave an average fuel economy of 30 mpg. A true Lancia, bristling with technical refinements, it was too expensive to be a direct competitor to either the Fiat 1100 or the Volkswagen, but there can be no question about the superiority of its design.

In 1953, the Ardea was replaced by the Appia. It had a restyled body with a bigger trunk, but remained faithful to the same disposition of the mechanical elements as its predecessor. With a 99-inch wheelbase, it was 158 inches long overall and weighed 2100 pounds. The 43.5 hp, V-4 engine gave it a top speed of 80 mph with 28 to 31 mpg fuel mileage. As for the Aprilia, in 1950 it was replaced by the Aurelia—a heavier car with a V-six engine—in accordance with Lancia's interpretation of the trends in the market.

Fiat arrived at different conclusions; they saw a need for two separate classes of small car covering the price bracket of the 500 and widening it both upwards and downwards. Dante Giacosa, who had been promoted to technical director of Fiat, spent the war years letting his thoughts on small car design ripen. After 1945, Giacosa flooded the patent office with applications for novel layouts and clever features, including one for a 500-re-

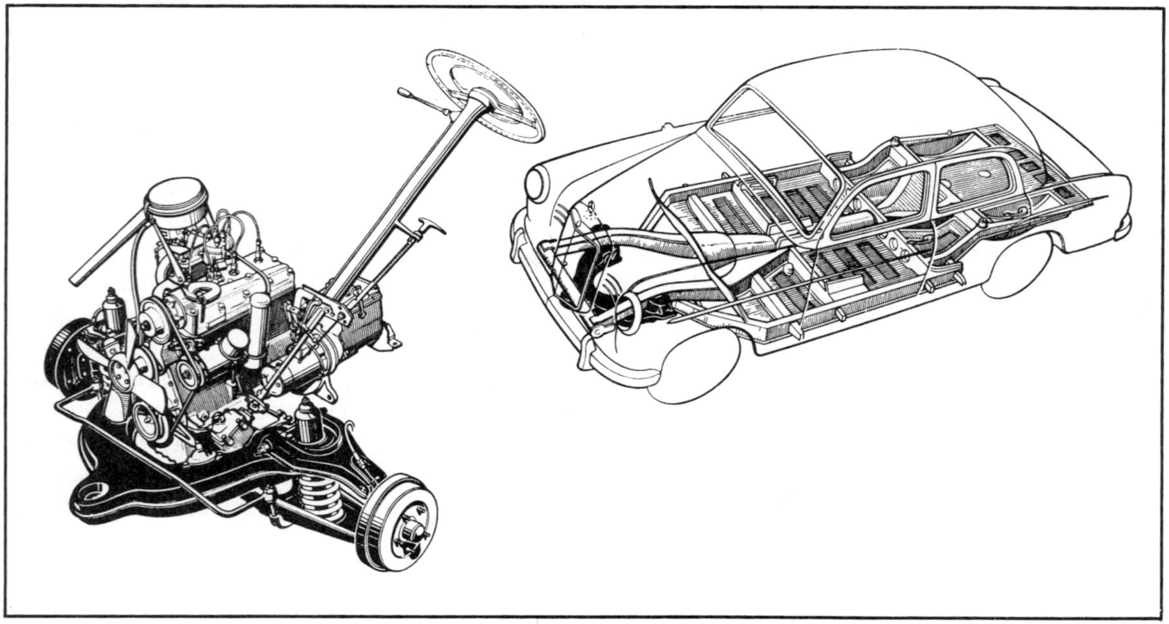

Fig. 12-21. The 1953 Mercedes-Benz 180 had a platform frame, semimonocoque body shell, and a subframe for the engine and front suspension.

placement with transverse engine and front wheel drive. But Fiat's production models came out with rear-mounted engines. Cost considerations weighed heavily in this decision. Giacosa might also have been influenced by seeing Renault's success with the 4 CV.

When Louis Renault saw the Volkswagen at the Berlin auto show in 1939, he told his engineers to prepare a rear-engine Renault with the same carrying capacity. It was developed during the war, and went into production in 1947, with several notable differences from the VW concept. The Renault had smaller wheels, and that saved a lot of interior space. It was a four-door sedan despite the smaller overall dimensions and lower weight. The 4 CV weighed 1320 pounds and was 143.3 inches long overall.

The engine was a vertical, in-line four (the only type of small engine Renault could manufacture without major new tooling investments) placed in the rear overhang. It delivered 21 hp and the car gave 33 to 36 mpg in normal use.

Fig. 12-22. The Isetta is the epitome of the bubble car.

Renault built the 4 CV until 1956. Then it was replaced by the Dauphine, that was roomier, more powerful, and very stylish (but it shared the 4 CV layout). It was supplemented by the boxy R-8, in 1959, and the stretched-body R-10 in 1962. Though they remained in production until the 1970s, they were the last Renault designs fitted with rear-mounted engines. The company began switching to front-wheel drive, in 1961, with the introduction of the R-4.

The R-4 had nothing new to offer architecturally. It had the boxy profile of cars made 30 years previously, and the engine was positioned lengthwise, inside the wheelbase, which pushed back the passenger compartment, leaving a long front containing machinery that could have been accommodated in a more compact and lighter package.

BUBBLE CARS

Fiat's solutions for the rear engine have also been outmoded, but not yet abandoned (though their days are numbered). But before looking at their detailed description, there is a phenomenon irreverently described as "the bubble car" that merits attention.

The bubble-car boom started in Italy with the creation of the Isetta in 1953 by Count Rivolta's ISO company in Milano. The Isetta was only 90 inches long overall, with a 58-inch wheelbase, and weighed about 700 pounds. Tire size was 4.80-10.

It was not a three-wheeler, as many assumed, but a four-wheeler. Nevertheless the rear track was very narrow (Fig. 12-22). It had seating for two, side by side, with entry and exit through the single front door, (that opened the whole front panel, including the windshield, instrument panel, and swung the articulated steering column out of the way).

The engine was an air-cooled, single-cylinder unit of 15.5 hp mounted behind the seat. The Isetta, incredibly, was capable of speeds up to 50 mph in standard form, with 55 to 60 mpg fuel economy.

In 1955, BMW began producing the Isetta (under license), and in 1957 added a bigger model with a full-width rear track and a flat-twin motorcycle engine. The wheelbase was extended to 67 inches and a rear seat was installed, with access through a curbside door.

NEW FIELD FOR AIRCRAFT DESIGNERS

A number of bubble-car concepts flowed from small companies, in Germany, such as the Heinkel, Goggomobil, ZUndapp, Brütsch, Kleinschnittger, Fuldamobil, Maico Champion, and Messerschmitt.

Egon Brütsch of Stuttgart offered a line of single-seater, three-wheelers, powered by 10-hp, air-cooled engines. Brütsch had no prejudice with regard to layout. He offered the Mopetta with a single front wheel and the Zwerg with two front wheels. The Mopetta carried the engine in front and had chain drive to the left rear wheel. The Zwerg (Dwarf) had a rear-mounted engine driving the single rear wheel. Brütsch sold his designs to the Victoria motorcycle works, in Nuremberg, where a four-wheeled Spatz 250 was built from 1956 to 1958.

The Kleinschnittger was made at Arnsberg, in Westfalia, as an all-steel two-door coupe or convertible. With two seats and very low construction, it weighed 706 pounds for the coupe and only 650 pounds for the convertible. It was driven by a two-cylinder, 15-hp engine mounted in front, with front-wheel drive.

The bubble car built by the Heinkel aircraft works in Stuttgart from 1955 to 1958. It was a three-wheeler, with a single, driving wheel in the back. In profile, it could easily be mistaken for an Isetta (Fig. 12-23). Like the Isetta, the front panel

Fig. 12-23. The Heinkei Kabinenroller (left) and the BMW-Isetta (right).

Fig. 12-24. Messerschmitt (nearest camera) and Fuldamobil (left).

opened like a door to give access to the interior. Unlike the Isetta, the steering wheel remained fixed in its position, mounted on an arm extending from the left body side.

The Fuldamobil, from Fulda, was a long-wheelbase three-wheeler. It had a balloon-shaped body and side doors (Fig. 12-24).

Aircraft designer Claudius Dornier came up with a bubble car design that was taken over by ZUndapp and produced as the Janus (Fig. 12-25). The name was chosen because it had a rear bench seat facing backwards, back to back with the front seat, calling to memory the two-faced Roman god.

It was a four-wheeler without side doors but an Isetta-like front door and a similar tailgate. The engine was mounted centrally under the seats (Fig. 12-26) to give adequate legroom both front and rear within a 72-inch wheelbase and an overall length of 125 inches. It weighed 880 pounds. With its wide track, it had excellent ride comfort and roadholding.

Hermann Holbein created the Champion rear-engine roadster, in 1948, and later added closed models to compete with the bubble cars. He sold the designs to Maico of Pfaffingen who adopted bigger engines and added a four-seater sedan (Fig. 12-27) and a station wagon.

Fritz Fend of Rosenheim in Bavaria invented an open three-wheeled, single-seater with two wheels in front and a rear engine, using motorcycle controls, such as handlebar steering, hand clutch, and throttle.

This idea appealed to Willy Messerschmitt in

Fig. 12-25. ZUndapp Janus.

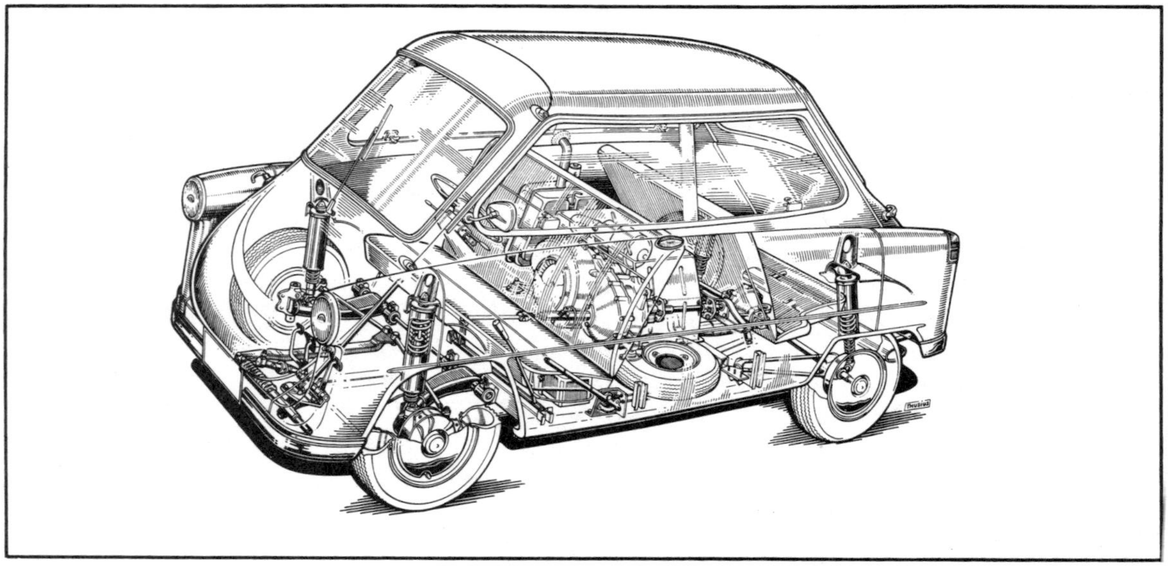

Fig. 12-26. ZUndapp Janus: The centrally mounted engine was flanked by fuel tank (right) and spare wheel (left).

Fig. 12-27. The Maico Champion two-door sedan with a rear-mounted engine.

Regensburg. He adopted the layout but added a second seat in tandem formation (Fig. 12-24). It had a plastic canopy top hinged on one side. With a 10-hp, single-cylinder engine, its top speed exceeded 50 mph. Messerschmitt later added a four-wheeled model with a two-cylinder, 20-hp engine. Its top speed was 90 mph with 50 to 55 mpg fuel economy.

SIZE 36 OUTSIDE; SIZE 39 INSIDE

Two manufacturers, Hans Glas and NSU Motor Works, narrowly missed getting into bubble-car production, but developed very small cars of advanced design. When the Fiats killed the bubble cars, Glas and NSU went on.

Fiat's 600 went into production, in 1957, and was followed by the new 500 two years later. The 600 was a two-door, four-seater sedan (Fig. 12-28) with a unit-construction body. The small tires (5.20-12) made an important contribution to efficient space utilization. The engine was a water-cooled, in-line four—mounted vertically and longitudinally in the rear end—leaving room for a small trunk under the hood. Built on a 78.8-inch wheel-

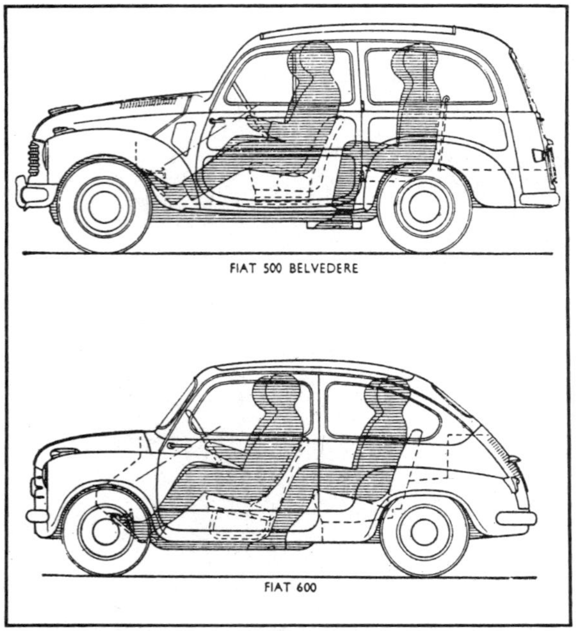

Fig. 12-28. A comparison of the new Fiat 600 (bottom) with the Fiat 500 Belvedere station wagon. The newcomer was less expensive to make, faster, less expensive to run, but had less luggage space.

base, it was 130 inches long overall and weighed 1301 pounds. The 24.5-hp engine was sufficient to give it a top speed of 62 mph, with 36 to 40 mpg average fuel economy.

Fiat even built a six-seater version of the 600 called Multipla (Fig. 12-29). It was created by the same formula by which Volkswagen had made its Microbus from the basic beetle. That consisted of tilting the steering column towards a vertical position and moving the front seats up above the front wheel housings. The hood disappeared, and the windshield base was moved up to the front. Thus there was no trunk at all, but ample space inside for luggage and passengers. The 600 Multipla actually looked like an overblown bubble car. It became popular as a taxicab in the large Italian cities, and it was bought by a variety of business enterprises as well as private families.

The 600 sedan was not a direct attack on the bubble cars, but it became a dangerous rival for VW and Renault. When the new 500 came out, the bubble cars were doomed. It was almost the same size as the bubble cars (Fig. 12-30). It had a 72.4-

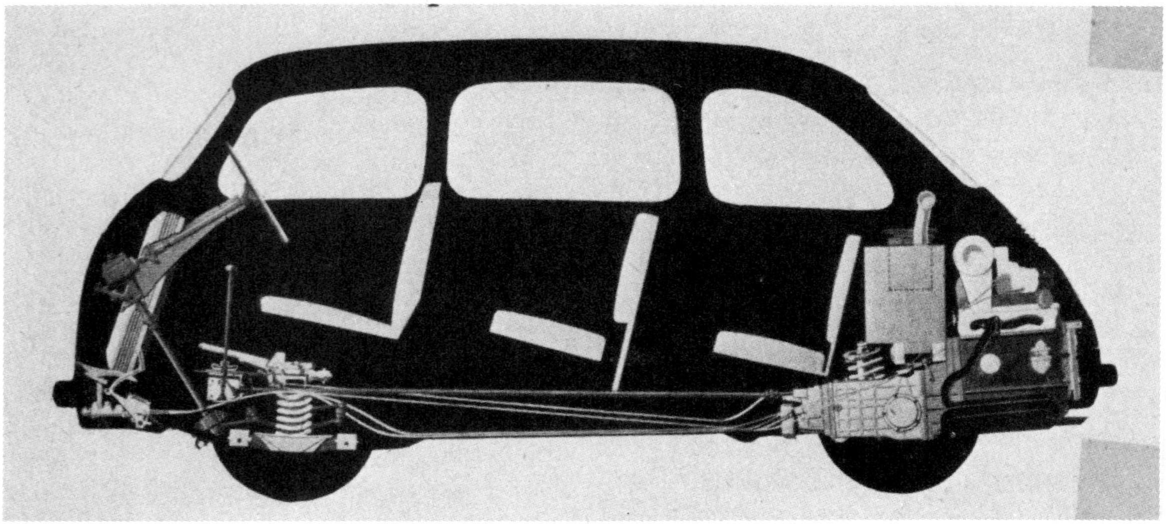

Fig. 12-29. Fiat 600 Multipla had seating for six occupants, all facing forward. Compare it with the 1921 Rumpler.

Fig. 12-30. Fiat-Nuova 500 with a rear-mounted, two-cylinder engine, 1957.

inch wheelbase and was 117 inches long overall. It had four seats and weighed 1147 pounds. Front seating dimensions were acceptable for all normal uses—including long trips.

While it was possible for two adults to sit in the back, it was difficult; and to sit there for any length of time was impossible. It was a back seat for small children, pets and parcels. But it is important to realize that the 500 was more than a two-seater (which is what it looked like from the outside).

The engine was an air-cooled, parallel-twin that was fitted vertically in the rear. Trunk space under the hood was almost as good as in the 600. The new 500 had a top speed of 63 mph, and fuel economy averaged between 45 and 50 mpg in normal use. No other mass producer of cars dared come into this market segment, and Fiat has cleverly followed up in recent times with such designs as the 126 and Panda (more about each one later on).

WIRTSCHAFTSWUNDER (UP THE GERMAN PRICE LADDER)

Hans Glas was a small manufacturer of farm equipment and motor scooters in Dingolfing. In the fall of 1954, he began building the Goggomobil, a small sedan on four 10-inch tires, that looked like a bubble car with a front hood and a rear deck. It had room for two children in the rear seat, and it had two bucket seats in front. The air-cooled, parallel-twin engine was mounted in the rear. There was no front trunk under the very short front hood. Any luggage had to be carried on available floor space inside or on vacant seats.

Wheelbase was 71 inches, and the base model weighed 960 pounds. The 13.6-hp engine gave it a top speed of 50 mph and a fuel economy of 40 mpg. After the shock of seeing the Fiat 500, Glas's chief engineer, Karl Dompert, reacted quickly and produced new designs—moving up-market. The Goggomobil name was dropped when the Isar came out in 1958 (available as two-door sedan and station wagon, both with front-mounted engines and rear-wheel drive). By 1962, Glas was offering four-cylinder engines and body designs by the Italian stylist Frua. Glas almost succeeded too well. The range expanded, and by 1965 the firm was taken over by BMW.

BMW had made a strong recovery based on its motorcycles and the Isetta. But when the bubble-car burst, BMW was ready with a real car to take its place. It was a fine little vehicle called the 700 (Fig. 12-31). It had an elegant body, designed by Michelotti, mounted on a platform with 83.5-inch wheelbase.

An air-cooled, flat-twin motorcycle engine was mounted in the rear (so low that there was room for a real trunk under the deck). Additional storage space was found in front (under the hood). Four seats, with acceptable dimensions even in the rear, were provided. The complete car weighed 1411 pounds. It was built until 1964 when BMW introduced the middle-size 1500, with a four-cylinder, front-mounted engine.

NSU had enjoyed great success with its mopeds and motorcycles after 1945, and was the

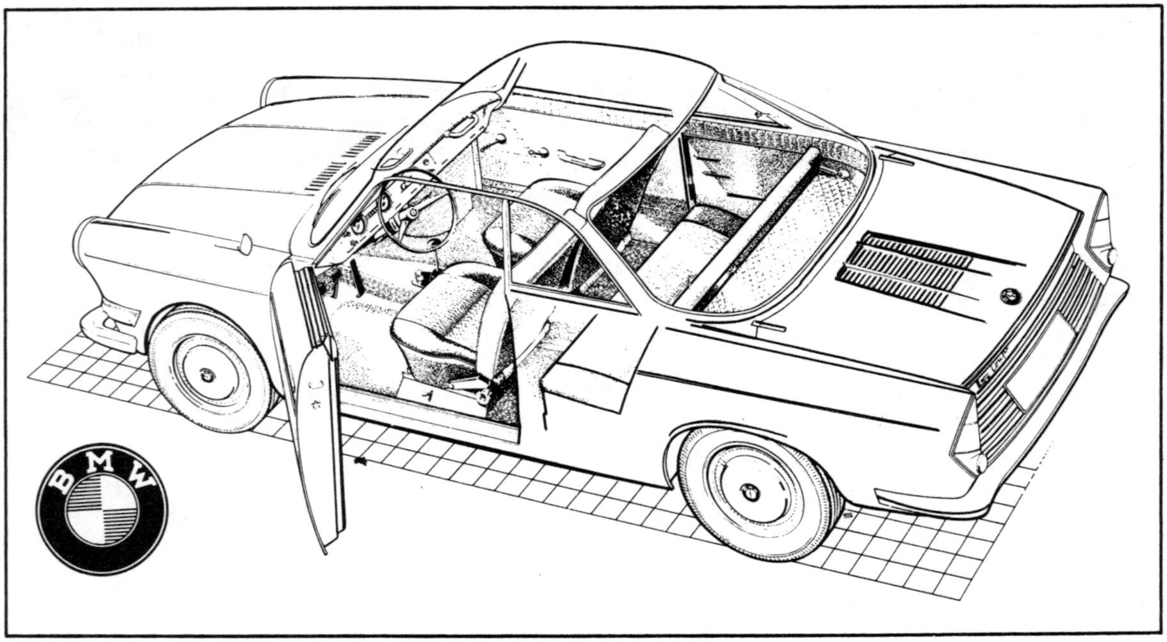

Fig. 12-31. The BMW 700 had a flat-twin motorcycle engine mounted in the rear.

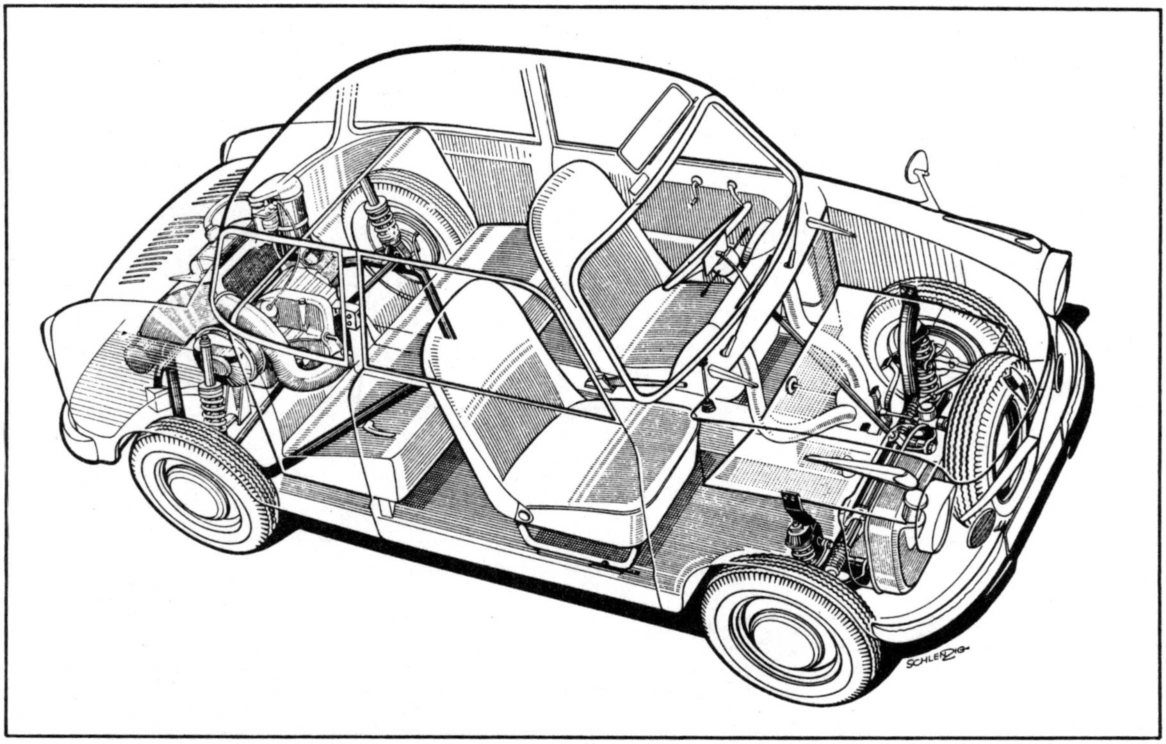

Fig. 12-32. The NSU Prinz was one of the best thought-out designs in its class.

world's biggest motorcycle manufacturer in 1955. Then the market went sour, and Gerd Stieler von Heydekampf, the NSU president, decided that the best recipe for the survival of the firm was to start building little economy cars. He had seen the Fiat 600, but knew nothing about the coming 500 when he gave the engineering department orders to prepare a four-seater, steel-bodied sedan with a weight of about 1100 pounds and a top speed of more than 62 mph.

NSU spent its last money reserves in developing the Prinz minicar and borrowed $7.5 million to erect and equip a new plant for producing it. Lengthy prototype testing delayed production until September, 1957. The first Prinz cars (Fig. 12-32) rolled off the line late in 1957.

The Prinz was designed with the lowest possible production costs in mind. That led to the decision to install the engine in the rear, driving the rear wheels. The air-cooled, parallel-twin design delivered 30 hp and was mounted transversely to permit the use of an existing motorcycle transmission. Built on a wheelbase of 78.75 inches, with a front and rear track of 47 inches, the NSU Prinz had an overall length of 123.75 inches. Curb weight was only 881 pounds.

But it was not a true four-seater. The back seat was better than in the Fiat 500, but not by much. Then Fiat added a station wagon body for the 500, giving ample room for four and their luggage. The Prinz had lively performance, but its 33-mpg fuel economy was not as good as the Fiat's.

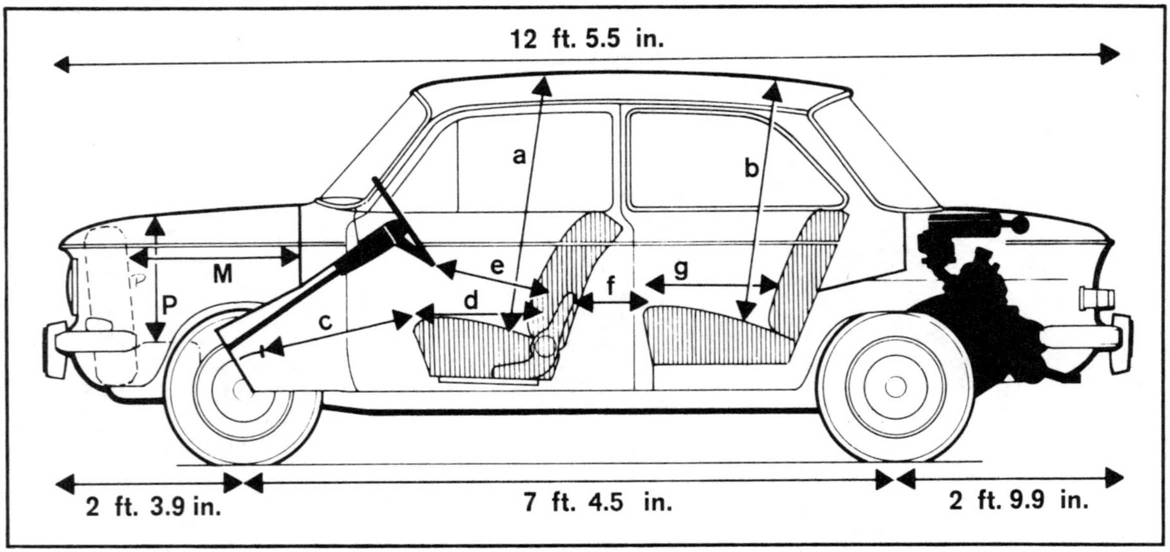

Fig. 12-33. The NSU Prinz 4 had more space inside than many larger and heavier cars.

Fig. 12-34. Corvair styling translated to European size: The NSU Prinz 4.

Von Heydekampf decided to move out of the bottom end of the market pyramid. He gave orders to design a roomier car with more performance. The NSU Prinz-4 two-door sedan arrived in 1961, with Corvair-like styling, foreshortened to fit its 80.3-inch wheelbase. See Figs. 12-33 and 12-34. It was followed by the Prinz-1000 in 1964, with a four-cylinder engine mounted transversely in the rear. The engine was so close to the rear wheel axis that there was room for a trunk in the back. The front hood was lengthened for the Type 110, in 1965, and in 1967 NSU began production of the Ro-80 compact luxury car with a twin-rotor Wankel engine and front-wheel drive.

A BOMBSHELL FROM BIRMINGHAM

A severe jolt shook the European small car market to its foundations in 1959 when Austin launched the Mini. Alec Issigonis went to work on preparations for a successor to both the Morris Minor and the Austin A-35 as early as 1952. An experimental car with transverse engine and front-wheel drive was built before Issigonis left for a 3-year stint with Alvis. In his absence, the engineering staff wasted

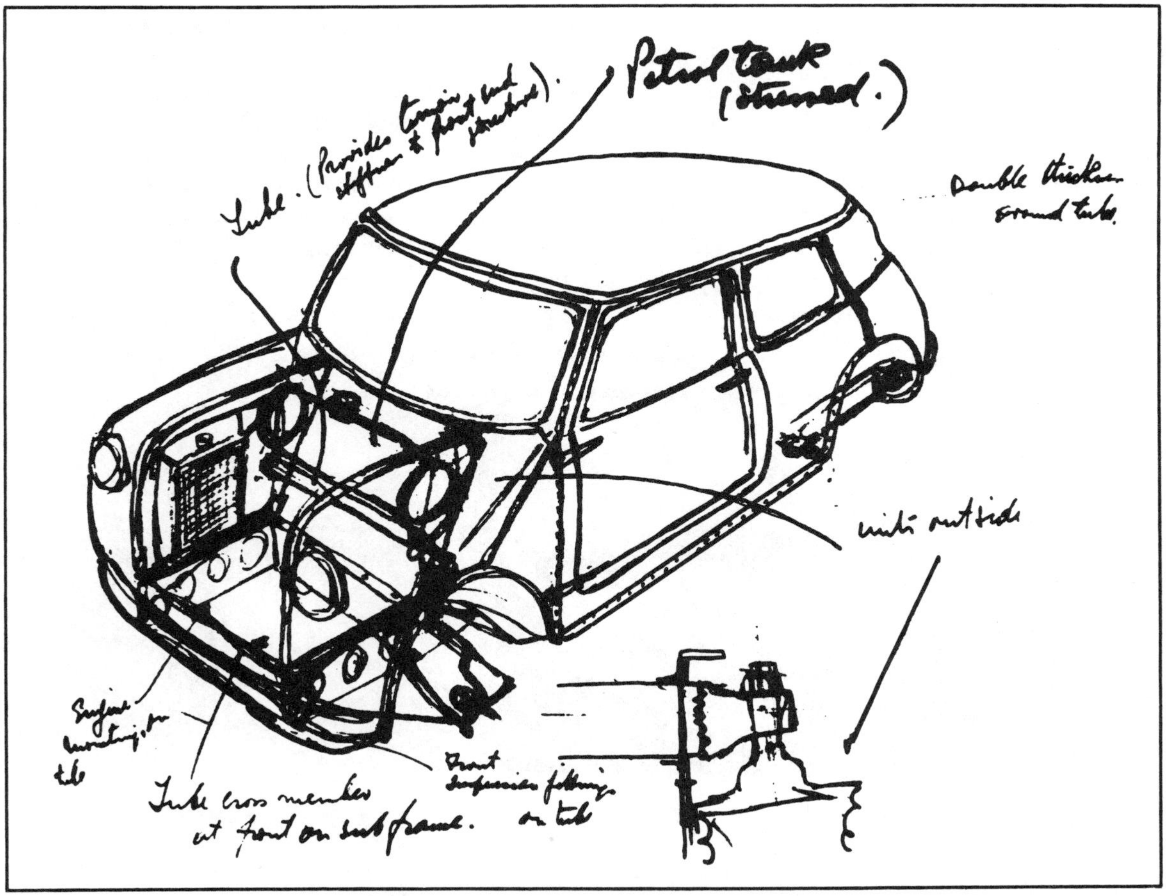

Fig. 12-35. The sketch by Issigonis shows the Mini layout with the radiator in the right-hand front fender.

their time on more conventional projects that were quickly ditched when Issigonis returned in 1956. In 1957, he presented plans (Fig. 12-35) for the ADO-15 to the president, L. P. Lord, and it was given full priority.

With its 80-inch wheelbase and 10-inch wheels, the Mini was small, light, and surprisingly roomy. It weighed 1230 pounds (Figs. 12-36 and 12-37). Styling freedom was so restricted, however, as to rule out anything but a wagon-type rear end. But it did not have a full tailgate. The window was fixed in its frame (which was part of the body structure). Only the lower part could be opened to give access to a small trunk. The miracle was that there was a trunk at all, within an overall length just over 120 inches.

The engine was inherited from the Austin A-35. Delivering 37.5 hp, it gave the Mini a top speed of 75 mph. Fuel economy varied between 27 and 33 mpg, according to driving methods and load.

An enlarged version, the ADO-16, followed in 1962, and the Mini architecture was then applied to bigger and bigger cars—including six-cylinder models—for about 10 years.

The Mini is still in production despite a search for a potential Mini-successor that has been going on for 15 years. Issigonis drew up a hatchback model, code named 9X in 1968, but it was never developed. He retired in 1971, and the new top engineers began work on the ADO-88 in 1974, In 1977 it was found that it failed to take full account of changes in the market, and of what consumers' expectations would grow into in the years ahead.

A new project, code named LC 8, was undertaken and approved for production in 1978. It was planned as an up-market car from the Mini (not a direct replacement) with more space, more style, and greater comfort. The LC 8 became the Metro. It was built on an 88.6-inch wheelbase and had a curb weight of 1638 pounds (Figs. 12-38 and 12-39). If the market ever gets weary of the Mini, Austin can quickly make a stripped down Metro.

Fig. 12-36. Incredibly clever packaging within minimum dimensions: the Austin Mini.

Fig. 12-37. Stylists never got to work on the Mini shape. Issigonis gave it a most purposeful look.

MORE SMALL-CAR DESIGNS

The Rootes Group, whose smallest car in 1946 was the Hillman Minx, felt the need to compete in the lowest price brackets in order to maintain market share. Design work began in 1938 on a project known as Little Jim, a two-door, four-seater sedan with styling that was aged before it was drawn. It was not revived after the war. In 1949, a little car with a two-cylinder engine, nicknamed Slug, was proposed. Prototypes were built, tested, and rejected.

Rootes waited till 1956 before tackling another small-car project. Known internally as the Apex, it had a 25-hp, rear-mounted, air-cooled engine, and was built on a 75-inch wheelbase. Tested in 1958, it was rejected. A new Apex two-door sedan was designed, in 1959, with a 79-inch wheelbase, wider body, crisp styling, and a 36-hp water-cooled engine, mounted at a 45-degree slant in the tail end. That project developed into the Hillman Imp (Fig. 12-40); it was cleared for production in 1961, and was introduced in May, 1963. It was aimed right at the Mini and it failed miserably.

With a 42-hp engine, 82-inch wheelbase, and

Fig. 12-38. The 1980 Austin Metro.

two-door sedan body, it weighed 1501 pounds. It was claimed to be capable of reaching 80 mph, but that figure is probably optimistic. Fuel consumption was not unreasonable, but the car had other defects that were largely unconnected with body design. Its handling was capricious, and it was unstable at speed. For a few years, a station wagon Imp was built. Due to the engine position, it had a high cargo floor and could not carry much payload. Unladen, the rear wheels already carried 66 percent of the 1603-pound weight.

While the Austin Mini certainly set new standards in space utilization, it was not the ultimate in small car design. Aerodynamically, it is poor, it's noisy, and it's short on comfort for anything but short trips.

An interesting evolution took place, in France, where the Dyna-Panhard had started as one of the leading examples of lightweight construction. The original body was taken out of production in 1953, and the factory retooled for a bigger vehicle, the Dyna-54, designed by Louis Bionier.

Panhard's aerodynamic studies date back to the 1930s, but they were pursued more intensively from 1948 onwards. The body shape for the Dyna-54 was evolved in the wind tunnel. It was low and wide and had smooth lines. Architecturally, it was a masterpiece (Fig. 12-41). The engine was the air-cooled, flat-twin inherited from the preceding models, mounted in the nose and driving the front wheels. The battery was housed under the hood. The fuel tank and spare wheel were behind the rear suspension, leaving room for a very big trunk.

The chassis frame (See Fig. 12-42) was a platform fabricated from aluminum pressings, with a tubular steel cross-member at each end. This platform did not extend outside the passenger compartment. It was located entirely within the wheelbase, with a front subframe for the power package, and the rear suspension bolted to it.

A low, flat floor was obtained; this allowed a low roofline. Two bench seats were standard, intended for two persons side by side front and rear, but wide enough for three individuals of normal roundness on trips of short duration. Because of the wide rear seat and the big trunk, the Dyna-54 became popular as a taxicab in France. What Panhard had done was to create the world's first small, light, and streamlined six-seater family car.

The four-door body was an all-aluminum structure that helped keep the body weight down to a mere 1610 pounds. Built on a 101.2-inch wheelbase, it was 180 inches long and 63 inches wide.

When Citroen took control of Panhard & Levassor in 1955, a big change was made to lower the cost of the 1957 models. Frame and body were redesigned for production in steel, and the weight was increased to 1874 pounds.

Citroen designers also did some hollowing out of the doors to make for more elbow room (turning the passenger compartment into a true six-seater). The weight gain caused some loss of performance and fuel economy, but the Dyna-54 remained quite lively with 0 to 60 mph acceleration in 20 seconds

Fig. 12-39. Not as stark as the Mini, the Metro has more comfort and space.

Fig. 12-40. The Hillman Imp also borrowed something from the Corvair's styling.

Fig. 12-41. The Panhard Dyna 54 had brilliant space utilization, excellent aerodynamics, and used light alloys extensively.

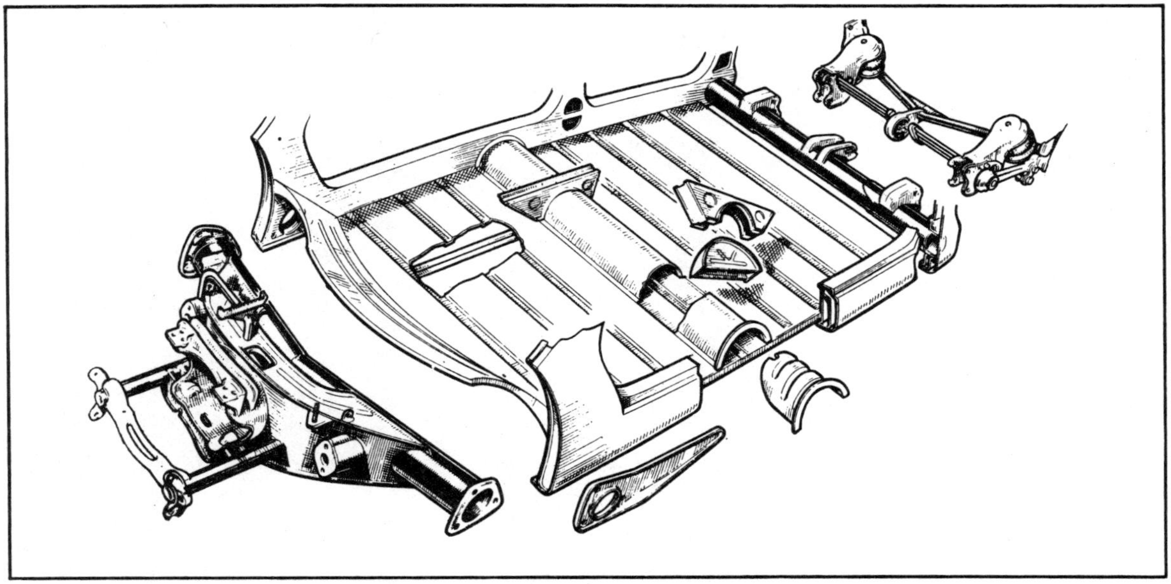

Fig. 12-42. The platform frame of the Dyna 54 was an aluminum casting reinforced by tubular steel cross-members.

(and that from a 42-hp, flat-twin!). The top speed was a little above 90 mph, and average fuel consumption was 33 to 39 mpg.

With the facelift of 1961 (the name changed to PL-17), the output climbed to 50 hp and a two-door convertible and a four-door station wagon were added. Production ceased in 1968. But the Dyna-54 and its derivatives remain milestones of design.

Part 3
The State of the Art

Chapter 13

GM's Commuter-Car Proposals

WHILE CHEVROLET WAS DEVELOPING THE Vera, and Pontiac was still pursuing a rival project (using a four-cylinder radial engine mounted on the rear axle), the GM engineering staff was thinking further ahead. Several prototypes were demonstrated to the press in May, 1969. This showed that awareness of unconventional solutions was not lacking in GM's technical ranks.

Three different versions of the smallest vehicle, XP-512, were on hand. Reminiscent of bubble cars, they were described by GM spokesmen as being "urban" cars intended only for short trips in central city areas. The three versions differed in styling and power systems. One was battery-electric, another had a small gasoline engine, and the third was equipped with a hybrid-electric system.

The XP-512 concept was a simple, four-wheeled two-seater, with no trunk (open or closed), a top speed of 30 to 40 mph, and limited acceleration. Their wheelbases ranged from 50 to 52 inches, height was from 52 to 56 inches, and length was from 66 to a little more than 86 inches. All had front entry a la bubble cars. The hybrid and electric cars had canopylike, top-and-windshield combinations. The gasoline-powered 512 was an open roadster.

The battery-electric vehicle (Figs. 13-1 and 13-2) was built on a steel platform chassis with integral roll bar and had a plastic body. It weighed 1250 pounds (including the 84-volt battery pack). The coaxial drive motor, integrally mounted on the rear axle, made it possible to use a compact planetary gear drive that passed one axle shaft through the center of the drive motor.

The canopy was designed so that, in fair weather, the car could be driven with the canopy front raised (and the rear window retracted) to serve as a convertible (Fig. 13-3). The canopy could also be removed so that the car became an open

Fig. 13-1. The battery-electric version of the XP-512.

roadster. A small package area behind the seat was accessible either through the open rear window or by folding down the right half of the seat.

The car had a range of 58 miles at a speed of 25 mph. Going faster used up the energy faster; the range shrank to 47 miles at its top speed of 30 mph. Acceleration from standstill to 30 mph took 12 seconds.

The hybrid-electric vehicle was a boxier design (Fig. 13-4). Apparently intended only for daytime operation, it had no lights. Its power system consisted of a tiny, single-cylinder gasoline engine coupled with a direct-current electric motor through an electro-magnetic clutch. With the clutch energized, the gasoline and electric motors operated at the same speed. The electrical energy was produced by a 72-volt power battery pack. The car operated either in all-electric or hybrid mode, and battery recharging was done with a built-in charging unit connected with a standard 115-volt household electrical outlet. In the hybrid mode, top speed was 35 mph. Its range in the all-electric mode at 30 mph was 5.2 miles. In the hybrid mode, with three gallons of gasoline, it's range was approximately 150 miles.

The gasoline-driven XP-512 was designed as a sports roadster (Fig. 13-5) with a unit-construction plastic body shell. Built on a 52-inch wheelbase, it was 86.3 inches long overall, 55 inches wide, and 51.9 inches high—to the top of the roll-bar. A 12-hp, two-cylinder engine was mounted horizontally, in the rear, coupled to an automatic transmission using variable-ratio belt drive and a centrifugal clutch. That gave the car a top speed of 45 mph. Acceleration from standstill to 30 mph took 18 seconds. With a four-gallon fuel tank, its range was approximately 280 miles (equivalent to 70 mpg).

POWER SYSTEM COMPLICATIONS

A slightly larger car, designed to fit into the exist-

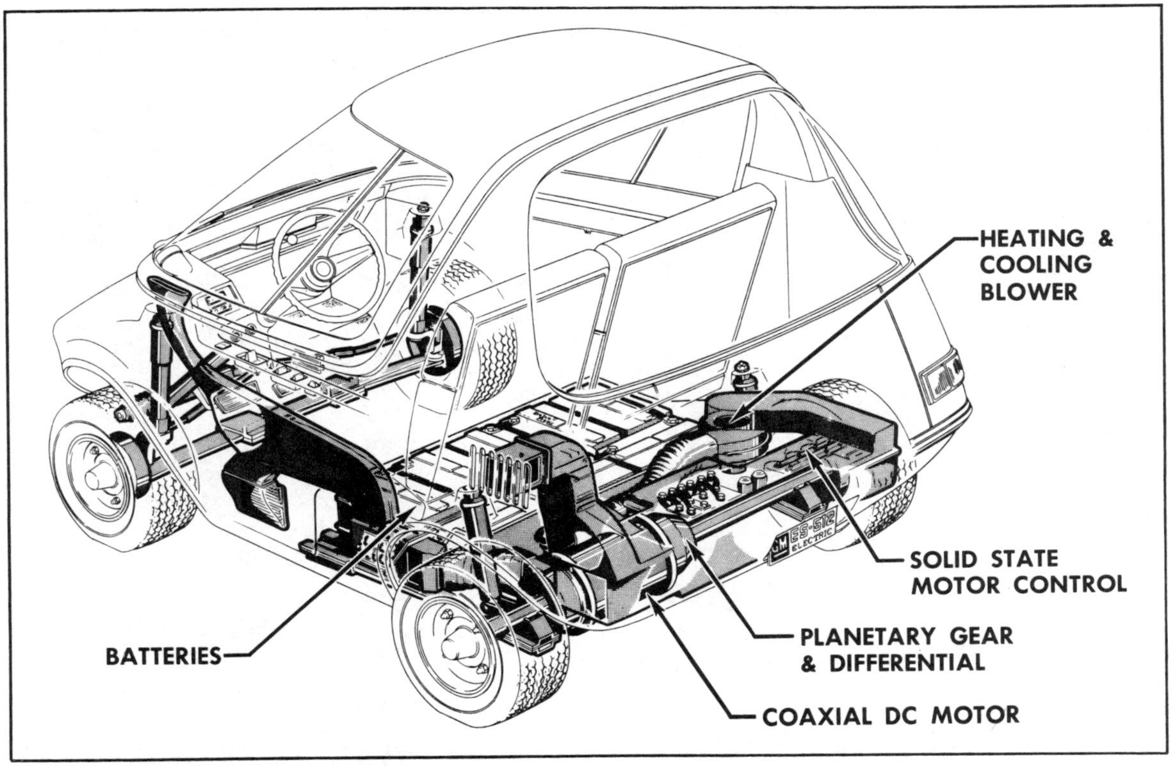

Fig. 13-2. Architecture of the battery-electric XP-512.

Fig. 13-3. The door and top arrangement on the XP-512.

ing traffic picture, was also shown, but only in mockup form (designated XP-883). It was a two-door, four-passenger sedan with a steel frame and plastic body (Fig. 13-6). The rear seat faced backwards and could not accommodate adults, but the backs of the passenger seats could be folded down to provide a cargo space 84 inches in length. Access to the rear seat area was provided by a tailgate hinged at the edge of the roof. The XP-883 had a 68-inch wheelbase with an overall length of 122.2 inches. It was 57.3 inches wide and 46.3 inches high. Both front and rear tread were 49 inches. It had a front overhang of 27.5 inches and a rear overhang of 26.7 inches.

The basic concept for the XP-883 originated at the GM Styling Staff. Designers developed the

Fig. 13-4. A hybrid-electric version of the XP-512.

vehicle as a small commuter car—capable of receiving any of several kinds of power plants, including internal combustion engine, electric propulsion, and hybrid gasoline-electric—to evaluate their potential in meeting the car's design purpose.

Although all these power plants were found acceptable, the first selected for further study was the hybrid-electric power plant. A two-cylinder, engine-electric motor combination was designed specifically for XP-883. The design objectives for the car included a weight of approximately 2100 pounds, a maximum speed of 60 miles per hour, and acceleration of 0 to 60 miles per hour in 28 seconds and 0 to 40 in 12 seconds.

The hybrid-electric power system consisted of an 18-hp, flat-twin, water-cooled engine, a dc series wound electric motor, a flywheel alternator for recharging batteries, six 12-volt batteries to provide a 72-volt system, an electronic control system, and an onboard charger that could be connected to an external 115-volt ac source.

The transmission system consisted of two gear reduction units (both providing a 5 to 1 reduction to axle speed). A planetary system transmitted

Fig. 13-5. A gasoline-engine version of the XP-512.

power directly from the coaxial electric motor to the drive shafts on the front wheels. Power from the gasoline engine was transmitted through the second gear reduction unit by a worm gear arrangement.

The XP-883 could operate in either an all-electric or hybrid mode. In either mode, it would accelerate from 1 to 10 mph by the electric motor. At that speed in the hybrid mode, the gasoline engine would start and provide the power function for steady speed operation as well as power to recharge the car's batteries. Both power plants were automatically combined for maximum acceleration.

QUICK AND EFFICIENT ON THREE WHEELS

The designers of the GM engineering staff also searched for the most efficient way to carry two persons from the suburbs to their downtown offices at freeway speeds. Because it was in 1969, the goal was 70 to 75 mph instead of 55 mph. The proposed solution was a streamlined three-wheeler that was designated XP-511 (Fig. 13-7).

The three-wheel layout—one in front, two in rear—provided excellent stability and maneuverability. It reduced weight, simplified the steering and allowed for an uncomplicated "backbone" type of frame. The low center of gravity, only 13½ inches off the ground, offered exceptional lateral stability and cornering capability. Two steering systems were tested in the XP-511. One was a simple handlebar with a 2:1 ratio and the other a more conventional steering wheel system with a 9:1 ratio.

The rear-mounted engine was a 67-hp, four-cylinder unit made by Opel that was matched with a three-speed transmission (automatic). The commuter could accelerate from 0 to 60 mph in 16 seconds and had an 80 mph top speed. In city driving

Fig. 13-6. A mockup of the XP-883.

it averaged 30 to 35 mph.

The XP-511 weighed 1300 pounds and had an overall length of 149.0 inches, a wheelbase of 86 inches, a rear track of 54 inches (width of 63 inches overall). The height was 40 inches and ground clearance 4.5 inches.

The 511 chassis consisted basically of two major subassemblies, a fiberglass body, and a steel chassis. It had a single hinged and counterbalanced canopy for exit and entry (Fig. 13-8). Its contoured bucket seats were semireclining and built into the body, together with head restraints. Because the seats were fixed, control pedals were electrically adjustable.

The chassis consisted of a Y-form steel frame, the three wheels, and a rear power train. The power train and rear wheels were mounted solidly to the frame so that the drive axles were made as simple one-piece shafts; the whole power train mass was essentially unsprung. The frame backbone passed through the tunnel in the body to support the single front wheel in the tapered front section (Fig. 13-9).

These experiments have yet to bear fruit in terms of production-model derivations. GM continues to develop its technology, and in 1980 they

Fig. 13-7. The XP-511 high-speed commuter car.

Fig. 13-8. The door arrangement of the XP-511.

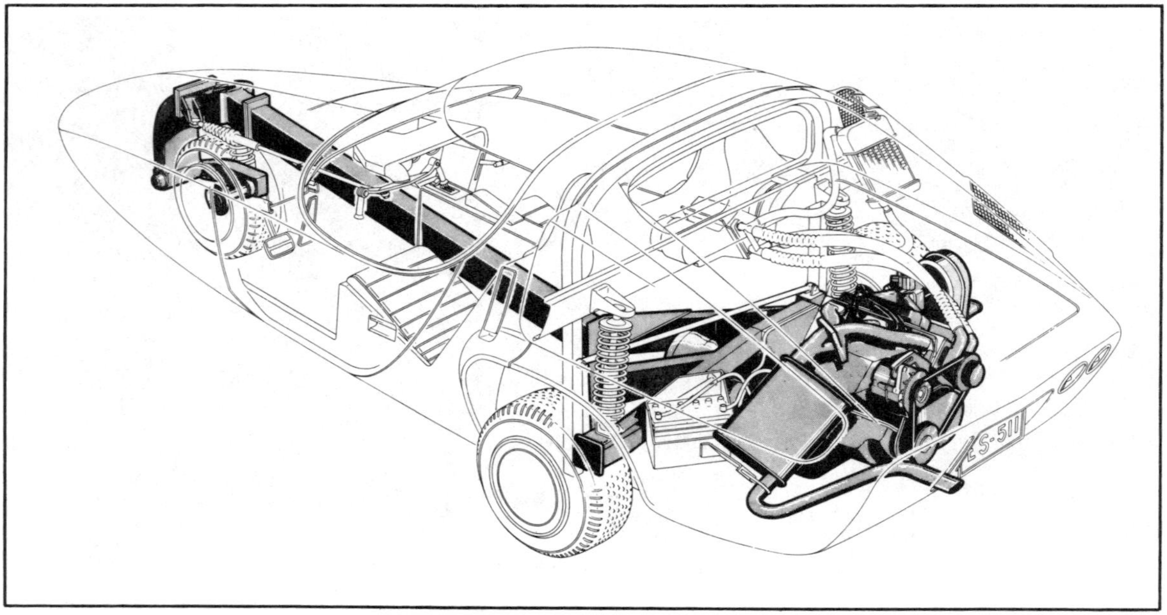

Fig. 13-9. Technical elements of the XP-511.

Fig. 13-10. A battery-electric proposal for 1985.

released details of an experimental battery-electric car proposed for production in 1984-85 (Fig. 13-10). The clay model shows a smart two-seater coupe. Lightweight batteries of high energy density were stated to be of the nickel-zinc type. The disposition of the power system elements was not given.

Another small car project (Fig. 13-11) was revealed at the same time. This short, four-wheeled commuter car was intended for either a 45-hp, three-cylinder engine or a 55-hp, V-four engine. The first version was calculated to achieve a fuel economy of 60 mpg at a constant 55 mph.

Research into unconventional power systems and alternative fuel is carried on relentlessly, but body design developments are occurring at a pace that has outstripped a lot of fundamental energy research. That is largely due to the emphasis GM now places on aerodynamics.

NOT JUST AROUND THE CORNER

Opel's Tech I coupe, revealed in September, 1981, is based on the production-model chassis components of the current front-wheel drive Kadett (with transversely mounted four-cylinder engine). In its development, as small a frontal area as possible was sought because this is a decisive factor. The result is a low vehicle profile, with soft lines, that funda-

Fig. 13-11. An experimental urban car built in 1980.

mentally corresponds to a new body concept. In order to reduce the frontal area, a seating arrangement approximating that of the Opel Manta coupe was chosen. This yielded not only a sporty appearance, but also a maximum of functionality. Opel designers therefore call it a four-door hatchback coupe. See Fig. 13-12.

The front of Tech I has a wedge-shaped upward slope, and the rest of this vehicle has what is known as "a flow-favorable optimally convex side area," and a high rear that discourages airflow separation. The all-around windows also favorably influence the airflow characteristics; the glass surfaces are flush with the outer skin, and also cover the roof pillars.

The air intake cross section can be varied according to the respective engine and transmission combination. Only as much air is permitted to enter under the engine hood as is required for the optimal cooling of the power plant. This resulted in an air drag coefficient "range" from 0.235 to 0.245.

Further measures to reduce the air drag include an underbody covering to quiet the airflow under the vehicle, a spoiler on the lower edge of the rear (which also counteracts soiling), and new aluminum-alloy wheels that largely preclude turbulence build-up without compromising the cooling of the brakes.

In the design of the rear body section, new aspects in the matter of safety were also taken into

Fig. 13-12. The Opel Tech I is called a "four-door coupe."

Fig. 13-13. The glass surfaces of the Tech I enclose B posts and C posts.

consideration. The area between the rear hatch and bumper is covered with white, rippled acrylic sheet (Fig. 13-13). The colors of the lamps are only discernible when a signal is given. This eliminates reflection effects.

A new electronic instrument panel was designed according to the motto, "less is often more"—so that the driver has in his field of vision only those indicators directly required in the driving situation of the moment. Opel regards the instrumentation developed here as trend-setting in the clarity of its display figures as well as in its versatility. Most of the switches can be operated with a fingertip, and without the driver taking his hand away from the steering wheel.

The Tech I more than hints at things to come at General Motors. But there are significant manufacturing questions that remain to be solved, such as the window glass, for instance. The concept of the four-door coupe does hold a lot of promise.

Chapter 14

France's Popular Marginal Motoring

GO TO A PROVINCIAL TOWN IN FRANCE AND you'll see a whole new generation of little boxes on wheels puttering around. Some of them can do 75 miles to the gallon. There are about 20 different makers of these machines. Different types can seat from one to four persons and carry varying amounts of cargo (from a briefcase and umbrella to over 600 pounds).

These vehicles offer marginal performance and comfort. They have been called motorized shopping carts. Their popularity is growing as the cost of car ownership continues to climb. France has about 50,000 of these little things on the road with an annual production run at the 20,000-unit level in the 1980-82 period.

Two things are happening that promise to make the marginal car a factor to be reckoned with in France's socioeconomic life. These vehicles were first retailed mostly through motorcycle dealers. Since 1980, established car dealers, who have previously derided this class of cars, have been applying for franchises.

In addition, "marginal" car construction has up to now been carried out mainly by small entrepreneurs operating in backyard shops. Now we find major industrial organizations outside the automotive area trying to diversify into this type of car production.

LIGHTWEIGHT AND LOW-SPEED CARS

The reasons for the popularity of the minicar goes beyond its 75-mpg fuel economy. Under French law, they come in two classes: those with engines under 50cc, and those up to 125cc. The first class is limited to a top speed of 28 mph and can be driven by anyone over 14 years of age (no registration, no driving license). The 125cc machines can be driven by anyone over 16 who has passed an oral exam on the highway code (this also entitles them to ride motorcycles up to 125cc). Registration is needed for this class of minicar. Some makers offer export models with 250cc engines.

The minicars compete in price range with used economy cars and overlap with the lowest bracket in the new-car market. Some account will have to be taken of this phenomenon whenever a car maker sets out to create a new ultrathrifty model. There could be design and engineering lessons in these marginal cars that will apply to far larger machinery.

A typical minicar weighs about 600 pounds. They are small, 70 to 78 inches long overall, and can be parked nose-to-the-curb in city streets.

Body structures are usually made of plastic, attached to metal frames, but some makers prefer steel bodies. It goes for all of them that maximum use of standard mass-produced parts from wherever suitable equipment can be found—motorcycles, cars, trailers, or motorboats—is sought with great ingenuity.

Bodies are usually two-seaters with a cargo area in back and a rear hatch. There is a diversification trend in body types towards the sporty as well as the utility car.

Suspension systems are usually all independent, steering is by rack and pinion, and brakes are four-wheel hydraulic with motorcycle drums or discs. They have car-type electrical systems meeting all basic safety requirements for lights, etc.

Most engines are single-cylinder, two-stroke units (the noisiest and the dirtiest, as critics say)

Fig. 14-1. The Mini-Comtesse Luxe.

Fig. 14-2. The Comtesse Sport.

installed in the rear end. But some makers favor front-wheel drive. Most transmissions have a centrifugal clutch and no gearing except for an automatic variable-ratio belt drive (of the snowmobile type). All have a reverse gear.

These engines are so light that their location makes little difference to the weight distribution. The seats are placed as centrally as possible, but legroom requirements often push them aft of the midway point of the wheelbase.

A RECENT PHENOMENON

The under-50cc class came into being in 1971 with the arrival of the Mini-Comtesse (Fig. 14-1), a product of Acoma in Angers, to designs by Henri Faucheux, encouraged by legislative measures. Several body styles, from a sports roadster to a buslike wagon, are in production; all have plastic bodies.

Engines are bought from Motobecane (France's second-largest maker of mopeds and light motorbikes). A three-wheel model was added in 1977, and the Super-Comtessa arrived in 1978. Acoma then introduced the Comtesse Sport (Fig. 14-2) and the Mini-Comtesse Break (Fig. 14-3) in 1980. The Break is a station wagon that is an almost perfect cube in its proportions.

The 125cc class has existed since 1969, but

Fig. 14-3. The Mini-Comtesse Break.

remained on the fringes of the market until the first of the recent petroleum shocks hit Europe. The Willam (Fig. 14-4) was the first to enjoy a boom in sales. It is built by Lambretta in Italy to designs by Henri Willam, the Paris-based importer of Lambretta motor scooters and three-wheeled mini-trucks. One basic model has led to a whole range of different models. The newest one, the Bretta coupe, is powered by a 49cc Minarelli engine. Willam also offers a van on the same chassis, a three-wheeled, two-seater Sulky, and a three-wheeled, single-seater—the Brio.

Willam's first major competitor was called Marden (Fig. 14-5). The Marden was created in 1967 by Teddy Marson (a former Citroen executive turned BMW dealer). Financed by Robert Denape, he got a small factory started. The name of the product was formed by the first three letters in the name of each partner (Mar-Den).

Prototypes were built with Citroen flat-twin and Sachs rotary Wankel engines. A 125cc Lombardini Beta engine was standardized in 1975, and a 49cc (Motobecane) version was added in 1978. Now, Marden cars are also available for export with a 250cc engine. There is even a battery-electric version. Marden is made by SEMA (Societe d'Exploitation Marden Automobile) in the industrial zone of Offranville, near Dieppe, on the Channel coast.

Plastic bodies come from SNPE of Dijon; they are one of France's largest makers of plastic panel constructions. Steel chassis frames are made by Mottez of Armentieres. A new suspension design, with lightweight arms and coil springs, was developed by automobile engineers from Alpine-Renault. Disc brakes from the Fiat 126 were adopted.

Marden production did not reach the industrial level until the enterprise engaged a new president, Robert Barret, in 1980, and was reorganized as SEMA. Barret is a former sales director at Mercedes-Benz of France, and he has some auto racing experience.

Fig. 14-4. The Lambretta Willam.

Fig. 14-5. The Marden coupe.

SMALL-MARKET CHOICES

By 1975-76, two important names had come into the field: Arola and Capucine. An enterprising mechanic named Daniel Manon built a prototype in 1975 and formed the Arola company in January 1976. One month later, he had an assembly line in operation at Corbas, an industrial suburb of Lyon.

Arola uses Peugeot engines only. Always conscious of the weight problem, Manon has stuck to extreme simplicity in his designs, from the starkest three-wheeler (with handlebar steering and open-sided cab) to the newest four-wheeled coupe (Fig. 14-6). In 1980, Manon signed a contract with former racing driver Maurice Trintignant for his assistance in developing new products.

The Capucine (Fig. 14-7) stands apart from the typical marginal car by having a front-mounted engine and front-wheel drive. The 49cc engine is a motobecane. The optional 125cc unit is made by BCB.

As a project, the Capucine had its start one day in 1974 when Daniel Renard—Citroen dealer and antique car collector at Aniche in northern France—decided on a new career. He sold every-

Fig. 14-6. A single-seater Arola.

thing and bought a disused Saint-Gobain glass plant. There he installed tooling to make his minicar and incorporated the business as ERAD (Etudes et Realisations Automobile du Douaisis). ERAD makes the body and all mechanical parts in-house. Only the engines and tires are purchased from outside suppliers.

In addition to the minicars, ERAD also offers some light commercial models equipped with a 125cc ACME diesel, capable of 40 mph with a load of 1000 pounds.

The large Jeanneau shipyards at Herbiers in Vendee are now building a vehicle called the Microcar. The AZ-49 is a town car with a rear engine,

Fig. 14-7. The Capucine 2 coupe has seating for two.

shaped like a cube with a sliced-off edge to make a very rakish windshield. Another model, intended for farms and villages, is called RJ-49. It has a front-mounted engine under a short, sloping hood, and front-wheel drive.

Etablissements B. Delsaux of Guesnain in northern France began production of the Modulo in 1975. Societe Bellier is in production with its Formule 85, equipped with a 49cc Motobecane engine, and is also planning a diesel version. The Garbo is a product of Vitrex Industrie of Paris. A rear-mounted 49cc Peugeot engine is used.

SINGLE-CYLINDER DIESEL ENGINES

Only one four-seater marginal car is built, and it's a diesel. The Duport Parco is a modern-looking sedan (Fig. 14-8). powered by a single-cylinder, water-cooled 10-hp Lombardini engine. Duport uses the gearbox from the Renault R-4, R-4 brakes, and R-4 steering. The chassis frame and suspension system are designed and manufactured by Duport. Duport also makes a lower-priced, two-seater called the Caddy (Fig. 14-9) that is also diesel

Fig. 14-8. The Duport Parco Diesel.

Fig. 14-9. The Duport Caddy has a single-cylinder diesel engine.

powered. The Duport cars are capable of highway travel, with a top speed of 50 mph, and fuel economy of up to 75 mpg.

Guy Duport is a small industrialist at St. Fereol near Faverges in the Savoy Alps. He got his start in business by making ski lifts. He began to develop the Caddy in 1977, and now building cars is his main business.

In both the Caddy and the Parco, the engine is mounted vertically in the right rear corner, driving the rear wheels. Despite being a single-cylinder engine, it is the heaviest part of the car. Everything else is made of lightweight materials. The frame is welded up from angle-iron sections, and the body is fiberglass-reinforced plastic with a fabric sunroof. Duport has his own aluminum foundry where he makes the suspension arms, wheels, and other light alloy castings. The whole Caddy weighs 842.6 pounds.

EXPERIENCE COUNTS IN MATERIALS

The Rivelaine is a creation of the GMT company (Generale de Mecanique et Thermique); it is a subsidiary of the coal giant Charbonnages de France. GMT found itself with lots of unused manufacturing

Fig. 14-10. Two versions of the Minoto.

capacity at the Wingles plant in northern France and looked to the minicar to fill the void.

The Rivelaine recalls the styling of the Isetta. The prototype was finished in March, 1980, and by July, testing and development work had proved the basic soundness of the design. A preseries was built for field testing and production started in January, 1981. Engines come from Fichtel & Sachs, and the transmission is GMT's own, patented, stepless-automatic drive.

Florival is a 49cc vehicle built by the Societe Mechanique de Florival at Guebwiller in Alsace. Production began in 1981.

An entrepreneur from southwestern France, Jacques Rochereau, turned his attention to marginal

Fig. 14-11. The Ligier coupe has an all-steel body.

Fig. 14-12. The Cicostar and the rear end of its chassis.

motoring in 1977 and designed the Minoto (Fig. 14-10) using a rear-mounted Motobecane engine in a steel frame with a two-seater plastic body. It weighs only 50.7 pounds. He formed Bel-Motors at Saint Reverend to start production. After the Minoto came the Veloto. Bel-Motors is now a subsidiary of Societe de Travaux Metallurgiques, which was founded in 1973 by Marius Moreau.

Teilhol of Ambert in the Allier district in the Massif Central is best known for producing the Rodeo recreational vehicle for Renault. Now Teilhol has a marginal car called Simply. It comes with 49cc or 125cc Motobecane engines, and is quite typical of current practice.

Guy Ligier, located on the outskirts of Vichy, is famous for his F-1 Grand Prix cars (now part of Talbot). The Ligier minicar borrows its styling from the 1966 prototype of Claudius Dornier (who had designed the ZUndapp Janus 13 years before).

The body is an all-steel structure (Fig. 14-11). Ligier has lots of experience with steel bodies after years of making tractor cabs for Renault. In chassis design, Ligier can draw on the topmost technology, and the minicar has MacPherson spring legs front and rear. Also, Ligier uses oversize brakes that are claimed to stop the car in 7 feet from its top speed of 27 mph. The 49cc engine is a Peugeot that is coupled with Mobymatic stepless-automatic drive.

During the past few years, other names such as Clayes-Flandria, KVS, Minicatsa, Fiam, Cicostar (Fig. 14-12), DIP, and Flipper have appeared. These are not built on an industrial scale.

Chapter 15

Modern Small Cars for Mass Production

DESIGNING FOR MASS PRODUCTION DEMANDS A great deal of coordination between the designers and the marketing people as well as between the product engineers and the manufacturing engineers. The following examples of mass-produced small cars, now in production, must be regarded in a different light than the cars described in the previous chapters as well as the design exercise in the chapters that follow. These cars are the present design reality.

SUCCESSOR TO THE TOPOLINO

The Fiat 126, which has been in production since late in 1972, can therefore hardly be described as "state-of-the-art" any longer. Yet it is only quite recently that a prominent Volkswagen engineer admitted to me that for efficient packaging, one could not do better than Fiat had done with the design of the 126.

A span of 36 years divides the Fiat 126 from the Fiat 500. In their separate ways, both models bear witness to the profound changes in our way of life, in our thinking, and in social structures; they are even more striking than those that have occurred in engineering design. Indeed, what distinguishes the modern utility car, over and above the obvious technical expertise involved, is the different role that it is called upon to play.

The design must not only be economical and bring the advantages of car ownership to an ever greater number of people, it must also be capable of providing maximum mobility in town traffic. Such a vehicle would be designed for a public already familiar with car ownership and looking for a reliable second car.

To produce an automobile with compact external dimensions at an economical price has always been a headache for manufacturers. It is not just a matter of scaling down the size of the car and its components; new design solutions are needed.

With the Fiat 126, Fiat has managed to find room for four seats and a trunk within an overall length of 120.5 inches. Its curb weight is a phenomenally low 1278 pounds. See Fig. 15-1.

Shortcuts are obvious. The rear seating dimensions are not generous, and the trunk is small. The engine—a little 23-hp, parallel-twin—occupies all the space behind the back seat. The spare wheel stands straight up behind the front bumper (setting the height of the hood line). The fuel tank is placed under the rear seat. The seating buck permits an ultrashort wheelbase of 72.3 inches, and the body is a unit-construction shell drawn tightly around the seating envelope (Fig. 15-2).

PANDA: A MODERN MODEL T

Three years after the introduction of the 126, it became apparent to Fiat that there was a demand for a more versatile car of approximately the same weight, but roomier and more practical. The rear engine position of the 126 is a drawback for space utilization as far as trunk room and the capacity to carry bulky objects is concerned.

Fiat's product planners reasoned their way through to a new type of car that became the Panda (Fig. 15-3). It is designed as a short-distance utility car, and no attempt was made to make it streamlined or suitable for high-speed motorway driving.

Design work on the Panda started, in the sum-

Fig. 15-1. The Fiat 126 Personal 650.

Fig. 15-2. Disposition of the mechanical components in the Fiat 126.

mer of 1976 when Fiat instituted a project code-named Zero, and commissioned Ital Design to create the body. In November, 1976, Giugiaro presented a schematic mockup of the interior and the first sketches for the exterior design. Two full-size models were built in January, 1977, and preliminary studies into weight and cost were completed. The project was then given the go-ahead. By September, 1977, the first complete Zero body was delivered to Fiat.

Crash tests were made and analyzed. No changes were needed as a result of the tests because the initial calculations had been absolutely correct. Tooling was ordered and production lines were cleared. A test fleet of pilot-production models were running by the end of 1978, and the first production models rolled off the assembly line at Desio in the first month of 1980.

Built on an 85-inch wheelbase, the Panda has an overall length of 133 inches and a curb weight of 1500 pounds (Fig. 15-4). Front wheel drive gives the Panda a low floor, right to the back bumper, and a full-depth hatchback.

Clever detail design stresses modernity and versatility. The rear seat, for instance, can be arranged in many positions to offer exceptional passenger and cargo accommodation. It can be folded forward (and removed altogether) or it can be laid down flat, giving a double bed when front seat backs are fully reclined. It also can be suspended as a cradle or hammock. A split rear seat is offered as an extra-cost option.

The Panda has full instrumentation, heating, ventilation, roll-down door windows, no-draft vents, rear window defroster and wiper, and all the safety features of other small Fiats. Impact-re-

Fig. 15-3. The Fiat Panda 45.

sistant front and rear bumpers match up with lower body sections coated with polyester for a distinctive appearance and extra protection in rough usage.

Where cost has been saved, it has been subtly concealed or combined with new mechanical elements that ensure longer life and reduced need for maintenance and repair. As an example, there is not a piece of curved glass anywhere in the Panda. Even the windshield is flat plate glass.

Two power trains are available in the Panda. By clever planning, both are shared with other models. The first is a 30-hp version of the vertical-twin used in the Fiat 126, mounted transversely in the front, so that the same basic transmission fits in perfectly. The second is the 45-hp, four-cylinder engine from the Fiat 127, complete with transmission and drive shafts.

The Panda 45 has a top speed of 87 mph and can go from standstill to 62 mph in 19.1 seconds. Fuel economy is 28 mpg in the ECE urban driving cycle, 40.5 mpg at 56 mph, and 31.3 mpg at 74.5 mph. With the two-cylinder engine, the Panda can reach and cruise indefinitely at 65 mph; at a steady 50 mph it will give 51 mpg. A four-wheel drive Panda will be produced by Steyr-Daimler-Puch in Austria.

A GALLIC WORLD-CAR CONCEPT

The Citroen Visa (Fig. 15-5) is not designed for the same purpose as the Panda. It is intended as an

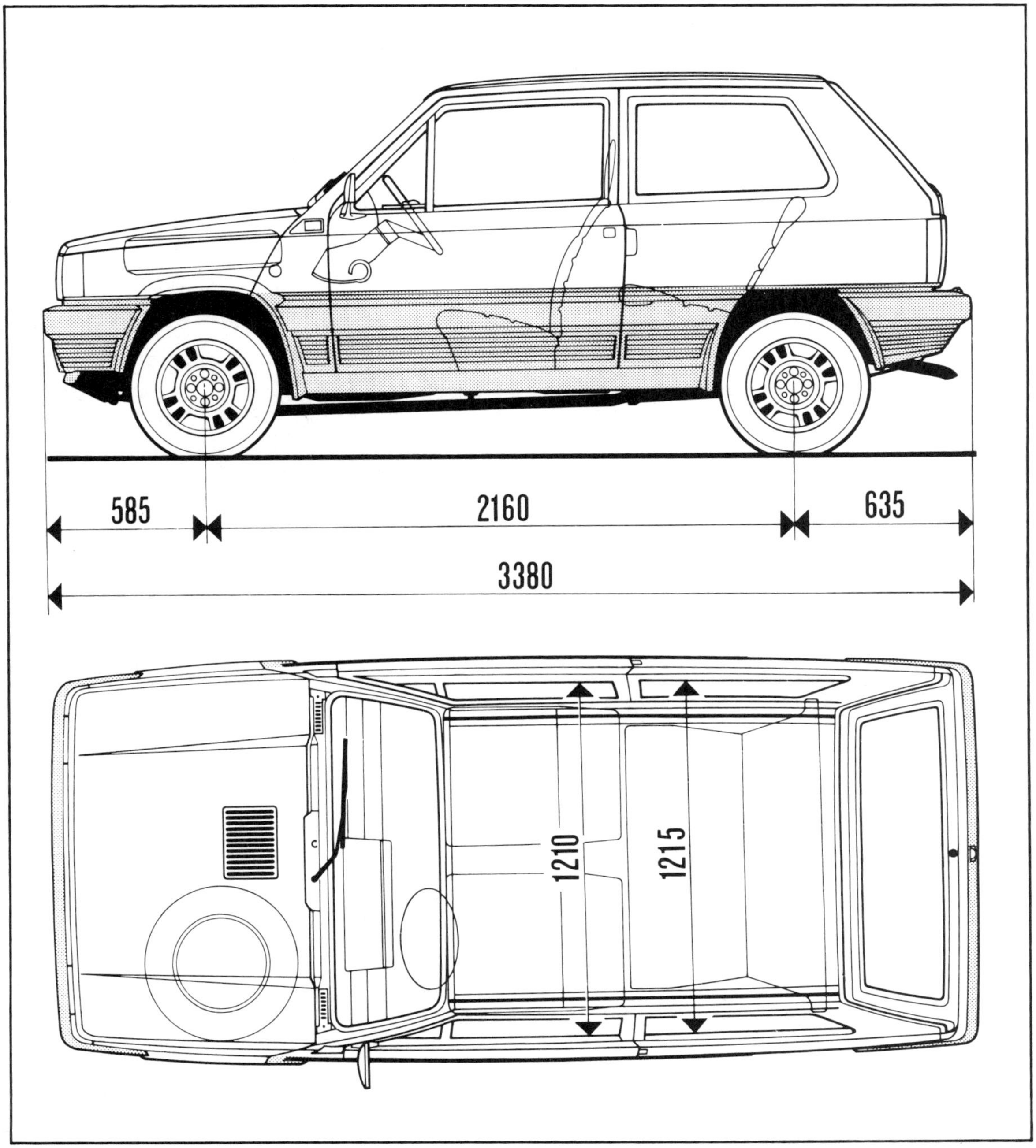

Fig. 15-4. Dimensions and layout of the Panda.

Fig. 15-5. The Citroen Visa.

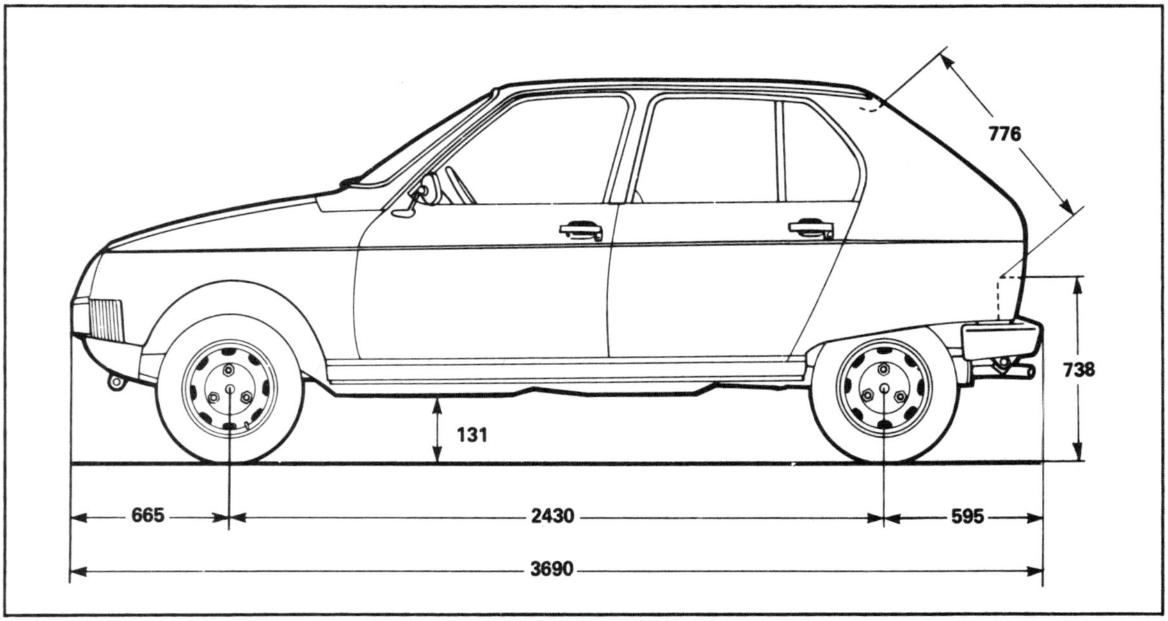

Fig. 15-6. A dimensional sketch of the Visa.

all-around family car, capable of sustained highway travel, with a low purchase price and low operating costs (Fig. 15-6). It weighs 1600 pounds.

In concept, it is a better streamlined, more modern version of the Peugeot 104 that was launched in 1972. The basic body shell is the same and the chassis is the same (Fig. 15-7). The Visa was introduced, in 1978, as a replacement for the Ami-8.

With its 95.3-inch wheelbase and 145.3-inch overall length, the Visa is the shortest four-door saloon built in Europe. It has a body of remarkable versatility. The hatchback gives access to a narrow but deep boot. It is narrow because the wheel housings are extended to enclose the spring legs. A light plastic shelf, easily removable, hides the boot's contents from view when the car is parked.

The entire rear seat can be folded and stowed away. The seat bottom goes against the front seat back, and the rear backrest is laid down forwards to form a metal platform. The car then becomes a two-seater utility van.

Citroen has been very clever about devising a choice of power trains for the Visa. Because it is based on the Peugeot 104, the 104 engine—a transversely mounted four-cylinder unit bent back to 18 degrees from horizontal—is available. It delivers 57 horsepower and gives the car a top speed of 85 mph with an average fuel economy of 33 mpg.

As an economy option, Citroen offers an air-cooled, flat-twin of 36 hp (lodged in the nose a la 2 CV). It can reach 80 mph and 0 to 60 mph acceleration takes 25.2 seconds. Its average fuel consumption is fractionally short of 40 mpg.

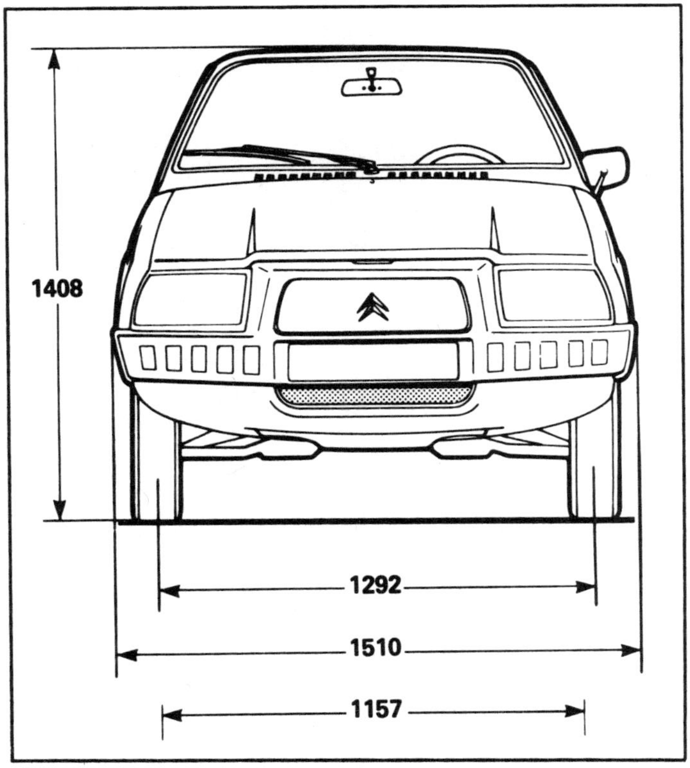

Fig. 15-7. The Visa combines a tall and narrow body with modern looks.

Fig. 15-8. The Honda City of 1981.

PROGRESS UNDER THE RISING SUN

Japan has a long history of building tiny cars, but, until recently, there has been little to learn from their design and technology. Among the first of the new generation stands the Suzuki Alto; it went into production in 1979. It is a four-door hatchback (Fig. 15-7) with front-wheel drive that is powered by a three-cylinder transversely mounted engine. Built on an 84.6-inch wheelbase, it is 130 inches long overall and weighs only 1456 pounds.

Inside this package, Suzuki designers have found space for normal seating dimensions for four occupants. Rear seat width is 45 inches, rear head room 35 inches, and rear legroom 32 to 37 inches (depending on front seat position).

As a packaging exercise, it outdoes the Mitsubishi Colt (Mirage) and the Daihatsu Charade. The Suzuki engine delivers 40 hp and gives a top speed of 83 mph. Fuel economy is 35.6 mpg in the urban driving cycle and 45 mpg in a steady 56 mph.

Now Honda has gone Suzuki one better with an announcement of the City in November, 1981. The City (Fig. 15-8) is a front-wheel drive, two-door hatchback with four seats, 133.8 inches long overall, and a weight of 1450 pounds. (Fig. 15-9).

The engine is a 63-hp, water-cooled four that is

mounted transversely under the short, sloping hood. The radiator is placed in front, the fuel tank is under the rear seat, and the spare wheel is lying down between the tank and the rear bumper (Fig. 15-10).

The body looks boxy. Having been developed in the wind tunnel, it has a drag coefficient of only 0.40, with zero-lift characteristics throughout its speed range (90-95 mph maximum). Fuel economy is 45 mpg in the Japanese urban driving cycle, and 68.4 at a constant 37 mph.

Inside the City, Honda has provided nearly 10 cubic feet of luggage space (plus a number of smaller pockets). All those areas that have conventionally been regarded as dead space have been given an individual going over to find a use for each of them—even the spaces under the seats. The result is a total of 16 useful storage pockets. Take for example the "cool" pocket in the instrument panel. When air conditioning is used, it keeps up to four beer cans drinkably cold. The two-level desk-type instrument panel abounds with trays and pockets. From the safety perspective, all essential areas are covered in soft padding.

The tops of the seat backs have been kept as low as possible for an air of interior roominess. The bottom ends are provided with openings that improve the passage of air for comfort. The hip point is raised for a nearly erect driving posture. The head-rest stays are integral with seat back in a distinctive construction that is simple, light, and sturdy.

To keep the City as light as possible, extensive use is made of high strength-steel. In some areas, plastics are used as an alternative to steel. Computer analyses of the body, wheels, etc., have been conducted, and a rational seating plan as well

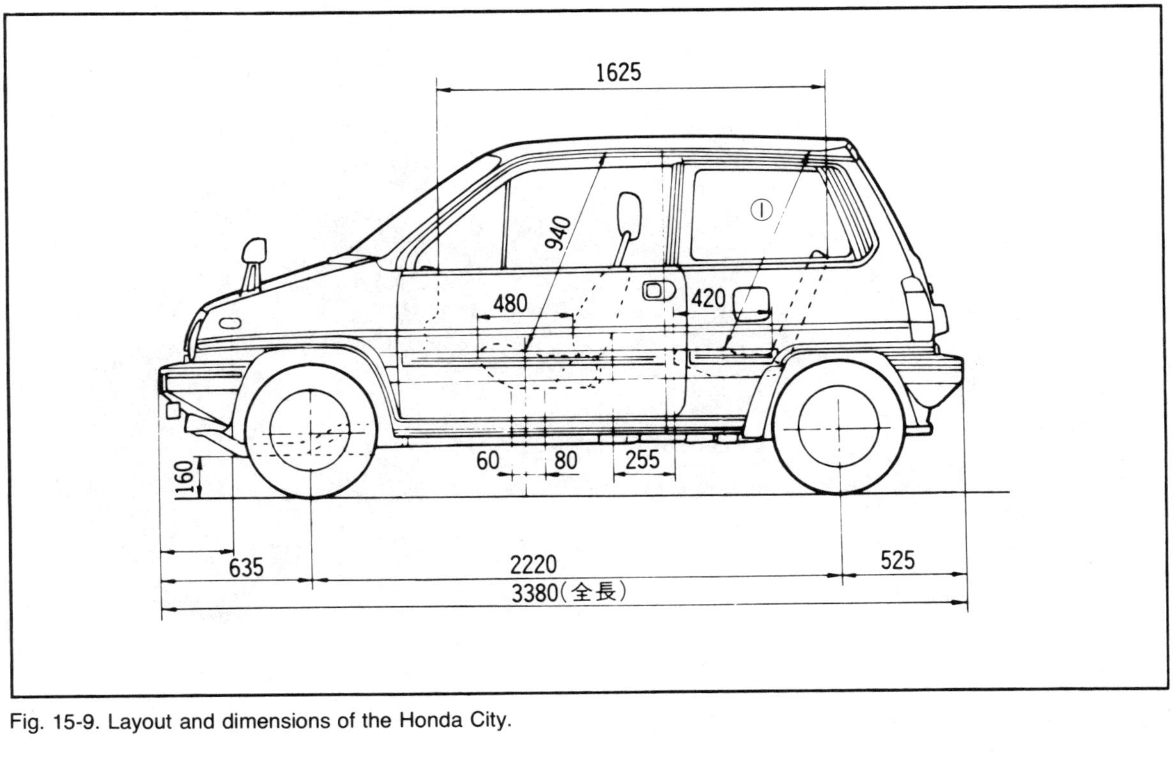

Fig. 15-9. Layout and dimensions of the Honda City.

as tidied-up, nonessential accessories and moldings have resulted from a thorough weight reduction program.

TOWARD THE STYLIZED SHOE BOX

Is the Honda City the ultimate small car? Probably not. As long as there is a hood, there is a chance to shorten and lighten the car some more. Consider the urban-car prototype that Giugiaro designed for Ghia in 1971 (Fig. 15-11). It has a sloping front but no real hood. And its length was only about 118 inches. Similar thinking shows up in the Minissima (Fig. 15-12) by Ogle Design, an urban car based on Austin Mini chassis components.

David Ogle was an aircraft pilot in the Royal Navy during the war. Afterwards he studied at the Central School of Arts and Crafts in London. He worked for several years as a designer in the radio industry, but in 1954 he set up his own design studio at Stevanage, a new town north of London.

His first sales were designs for household appliances. The first Ogle car, built in 1959, was a special body on the Riley 1.5 platform. Ogle Design Ltd. moved to Letchworth in 1960. Two years later David Ogle was killed in a traffic accident.

The company was saved, and a new chief designer, Tom Karen, continued Ogle's unfinished projects such as the SX 250 that was based on the

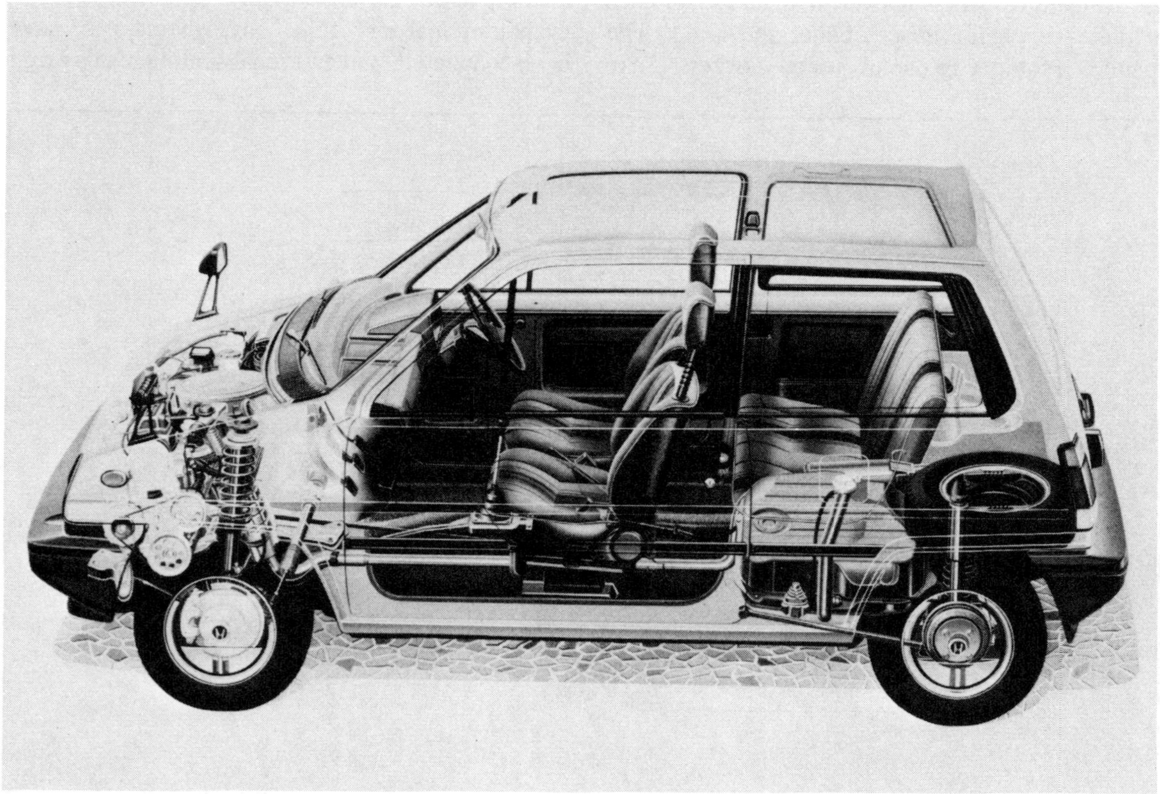

Fig. 15-10. Clever space utilization in the Honda City.

Fig. 15-11. The De Tomaso urban car from 1971.

Daimler SP 250 chassis. It evolved into the Reliant Scimitar in 1964.

Tom Karen was born in Vienna, Austria, in 1926 as the son of a well-to-do auto enthusiast. The family fled Austria in 1939 and Tom Karen landed in London in 1942. He went Loughborough College and graduated as an aircraft engineer. Next, he got a job in Ford's styling department, did a brief stint with David Ogle, and then worked on appliance designs for Hotpoint and Philips. In 1962 he returned to Ogle Design.

The Minissima created a lot of attention when it appeared, but it was ahead of its time. There was no reaction from British Leyland; they should have been exploring the market between the Mini as well as the next higher bracket. The ideas behind the Minissima are still valid, and could serve other mass producers of automobiles.

Fig. 15-12. The Minissima by Ogle Design.

Chapter 16

Italian Projects to Spark New Trends

WHEN STYLING AND ENGINEERING GO HAND IN hand, the participants sometimes stimulate each other to the extent where one good idea breeds others and results surpass original visions. Some recent examples from Italy are cases in point; they represent fresh solutions for cars of widely different character and purpose. This chapter describes one urban car, one taxicab, one family sedan, and one high-performance luxury car.

ECOS BY PININFARINA

An urban car with battery-electric power was created in 1978 by Pininfarina in collaboration with Fiat. They took a different approach from most earlier attempts that were usually two-seaters with small passenger compartments and little luggage space. The ECOS (Fig. 16-1) offers a solution that respects current safety laws, has four seats, has good accessibility, and has a lot of luggage capacity compared to the outside dimensions.

Among the main design objectives was ease of entry and exit; that meant a high roofline and high doors. Visibility is extremely important in the urban environment, and that led to the adoption of vast glass areas. The interior should be attractive and comfortable. That fixed the dimensions at the level of a middle-size family car rather than the smallest four-seater economy car (Fig. 16-2). Easily accessible luggage space with high capacity was regarded as a necessity. That led to the reservation of a large area in the tail end for use as a trunk.

The first idea was to form a kind of bench from a row of batteries on which passengers could sit. In effect, the height of the batteries, and that of the structures necessary for containing them, could be developed naturally into a seating surface set at an optimal height for passengers. The battery packs were split between front and rear seats. That gives even weight distribution and ready access to batteries and motor. Layout of the "mechanical parts

Fig. 16-1. The ECOS urban car was created jointly by Fiat and Pininfarina.

proper," including the electric motor and the regulation unit, did not present special problems.

Chassis components were taken from the Fiat Ritmo (Strada). With front-wheel drive, the drive motor and control unit were mounted between the front wheels on the ECOS. The spare wheel stands straight up behind the front bumper. The body was designed for easy adaptability to either hinged (swing-open) or sliding doors. The back seat can be easily folded (which triples the luggage space). A large tailgate is fitted, and liftover height is at floor level. The main roof support is the B-post (integral with a roll bar). Blind spots due to A-post and C-post obstruction has been minimized.

The shape and shortness of the hood also stem to a large extent from concern for the driver's view of what lies ahead. If architectural reasons only were to be taken into account, it could have been made even shorter, but safety considerations indicated that anything shorter would create problems with United States and European standards.

The front and rear lights are housed in a position protected by a molding that goes right around the vehicle. At the front is fitted an air intake designed to cool the electric motor and the regulation group by means of forced air circulation.

To stress its broad scientific approach, Pininfarina also stated that, "A special roof can be provided overlapping the production roof and forming an interspace with it, self-ventilated by means of solar cells which, at a given temperature, operate front and rear fans so as to create in hot seasons and zones an effect comparable to that of a car left in the shade."

The ECOS weighed 2756 pounds, when in running order. It was built on a 86.6-inch wheelbase and its overall length was 133.5 inches. The range was 40 miles in city-traffic conditions. Top speed with a full load was 50 mph.

MEGAGAMMA BY ITAL DESIGN

The taxicab proposal in this group comes from Ital Design. It was first exhibited in 1978. The designer, Giorgetto Giugiaro, says the study is linked to the taxicab program instituted in 1976 on behalf of the Museum of Modern Art in New York.

In setting his priorities for this design, Giugiaro put comfort, usefulness, and safety ahead of speed and aerodynamics. See Fig. 16-3. He chose a

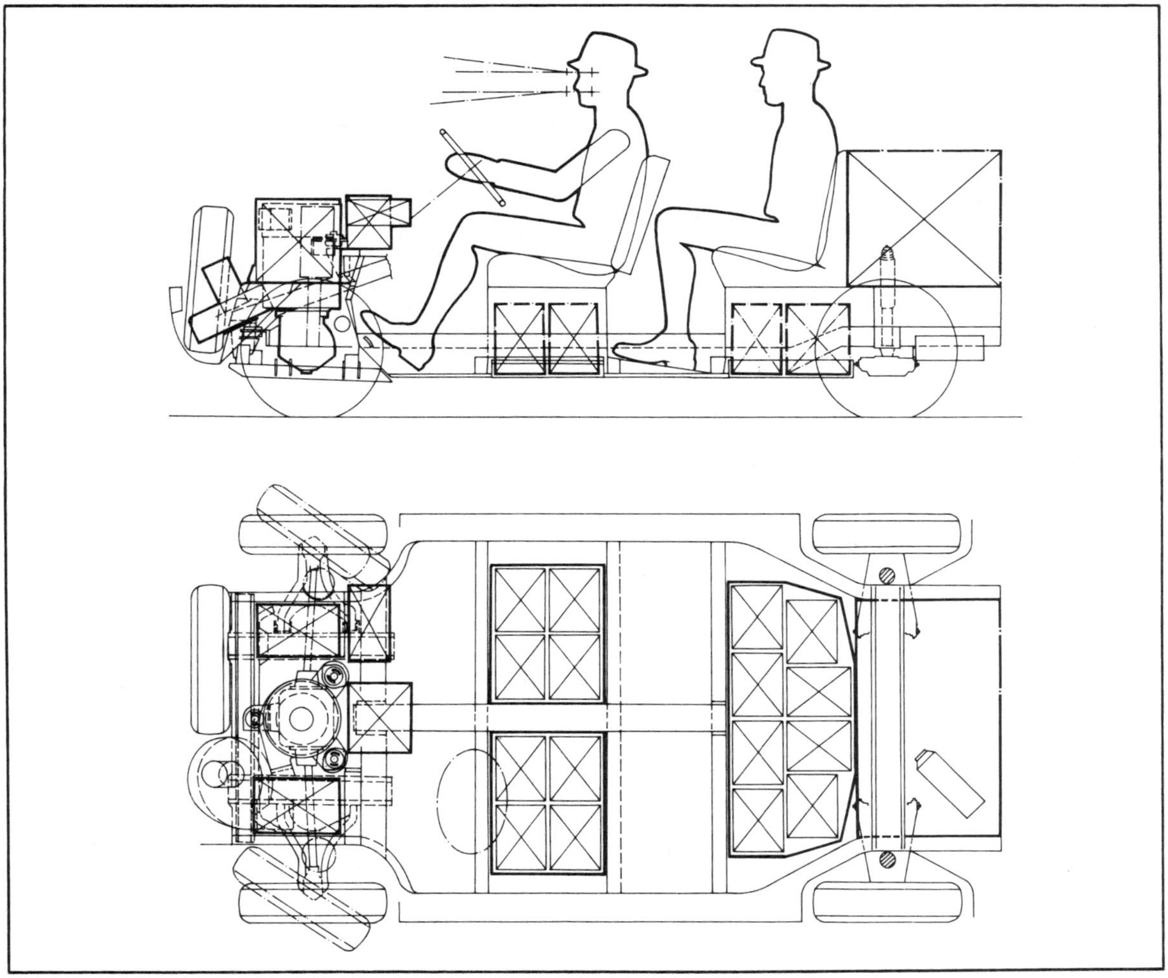

Fig. 16-2. Battery packs are located under the seats in the front-wheel drive ECOS.

Fig. 16-3. The Megagamma by Ital Design.

production-model chassis (the front-wheel drive Lancia Gamma with its flat-four engine). The prototype is called Megagamma; it combines certain van features with some ideas from the family sedan.

It has a short, sloping hood and very wide doors without wheelhousing intrusions. To provide proper seating and ease of entry within a short overall length, Giugiaro realized it was necessary to raise the roofline. Compared with the standard Gamma, the Megagamma is about 9.5 inches higher, but 12 inches shorter. The actual dimensions are 170 inches in overall length and 64 inches in height. See Fig. 16-4.

Raising the floor and roof levels have contributed to raising the center of gravity, but not to the extent of causing any problems in the dynamic behavior of the vehicle. With increased frontal area and shorter length, the aerodynamics have had to suffer. But at city-traffic speeds little or no rise in fuel consumption can be detected when compared with the Gamma sedan. Of course, taxicabs are not always crawling along congested streets. They also make brisk runs to airports, sports arenas, and convention centers in suburban areas. The Megagamma shape is so smooth that, up to 55 mph, the overall air drag is comparable to the Gamma sedan's.

The driver's seat has been pushed forward and moved up to meet a corresponding tilt of the steering column. The rear floor level has been raised to provide seating accommodations more or less as one finds in a luxury limousine. The underfloor space has been utilized to store the fuel tank and spare wheel. Trunk capacity is virtually unchanged, but the trunk configuration has changed from long and flat to short and high.

The hinged shelf covering the luggage compartment is removable if more space is required in this area. Seats are mounted on a horizontal floor, and access and exit maneuvers are extremely straightforward and convenient. The solution adopted for the rear seats makes it possible to remove the cushions and fold away the structure behind the back of the front seats. The flat-floored compartment thus obtained is perfect for goods transport.

The seats consist of a chromium-plated, tubular structure to which are hooked the cushions that form seat and back. Setting is by electric control. The front seats slide fore and aft and can be adjusted for height and rake. The rear seats recline and are also set on slides. If cushions are removed, the structure, which hinges at two points, can be folded away vertically behind the back of the front seats in order to take up a minimum of room.

The public transport adaptation is also of interest. In this case, the vehicle can be fitted with a sliding passenger door. Two fold-down seats can be fitted opposite the rear seat and the space alongside the driver can be used for large pieces of luggage. The Megagamma can also be fitted for other professional and industrial uses. By raising the roof further above the door frame, the volume obtained is ideal for a variety of uses (apart from goods transport, minibus, school bus, etc.).

MEDUSA BY ITAL DESIGN

Even more difficult to design than a taxicab/utility vehicle is the four-door family sedan that is capable of transporting four people comfortably and economically at high speeds. Giugiaro addressed himself to the task and, in 1980, revealed an adroit and original solution; the Medusa. See Fig. 16-5.

For this design, Giugiaro made up his list of priorities. Out of that came an emphasis on light weight, compact dimensions, and low air drag that overruled all ideas of conventional configurations and construction. He was ready to sacrifice some trunk space for increased creature comfort and to compromise some practicality if it would help make

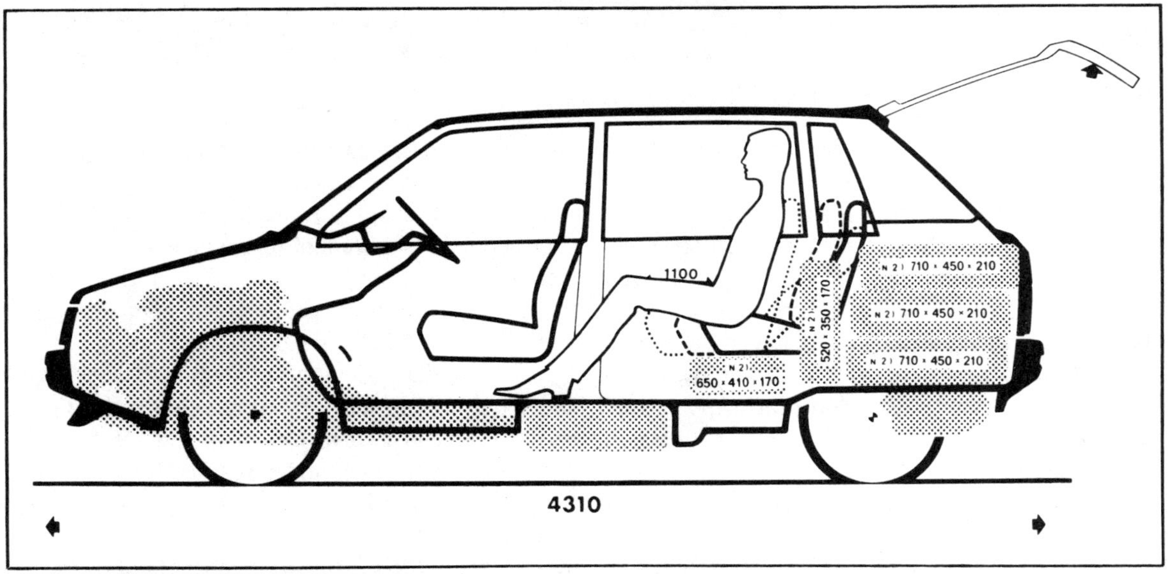

Fig. 16-4. Passenger and luggage accommodation in the Megagamma.

Fig. 16-5. The Medusa by Ital Design has a rear-mounted engine.

the car more efficient in its main purpose of carrying people from one place to another.

The passenger compartment and its envelope, as they came out of the initial studies, suggested a rear-mounted engine. That's where Ital Design stepped right out of styling and into engineering. Making a survey of what was available in the way of standard components for such a layout, Giugiaro settled on the Lancia Beta Montecarlo sports car, with its transverse four-cylinder engine installed on the rear wheel axis.

The Montecarlo power train was accommodated in the back of the Medusa (along with the rear suspension). Front suspension, steering, and brakes are also taken from the Montecarlo. And Giugiaro put a Lancia grille on the body to do justice to its mechanical elements.

The body shape is rounded, clean, and without sharp edges. Wind tunnel tests show a drag coefficient of 0.263. That puts the Medusa at the head of the line of low-drag prototypes using existing power trains and chassis elements.

The Medusa is built on a 109.8-inch wheelbase and is only 171 inches long. It has seating for three in the rear and two in front (with space for two normal-size suitcases under the hood).

The side windows are fixed and flush with the outside framework. The lighter, strong doors leave more room to the passenger compartment owing to the suppression of window raising mechanisms.

The openings have been obtained by electrically operated sliding panes.

The car has three windows at the side for greater visibility and extra lightness to the eye. The cut of the front door is flush with the A-post and the rear door finishes at the cut of the C-post. The drip channels on the roof have disappeared replaced by internal drain pipes. The door handles are flat and flush with the side panel.

A spoiler is applied to the base of the rear window to protect and mask air outlets from the engine compartment. The rear lights are scooped out of the form while the bumper, which houses the number plate, contains extra slits designed in the wind tunnel to optimize outlet of air from the engine compartment.

FERRARI PININ BY PININFARINA

It was in 1980 that Pininfarina displayed its Ferrari Pinin (Fig. 16-6). It's the first four-door Ferrari ever built, and it's not just a styling prototype, but a fully engineered proposal for a future production car.

Mechanically, the Pinin is pure Ferrari. The engine is the 360-hp, flat-twelve from the 512 Berlinetta Boxer, positioned in front and driving the rear wheels, with the clutch, 5-speed gearbox, and independent rear suspension taken from the Ferrari 400. Because of the BB engine's extra width, Pininfarina could not use the 400 front suspension, but designed a new system to fit within the confines of available space.

The Pinin is built on a 108.3-inch wheelbase and is only 190 inches long overall, with a 21.5-cubic-feet trunk in the tail end. The design was not made as an aerodynamic study, but as a car of greater distinction for the richest men in the world. It is flawless in its proportions, and bears witness to good taste from any angle it is viewed. With a discreet front spoiler, but without deflectors, wings, and other airflow-control aids, it has a drag coefficient of 0.36.

Fig. 16-6. The Ferrari Pinin has a front-mounted, flat-twelve engine and rear-wheel drive.

Bumpers are integrated with the general shape and have high energy-absorbing capacity. Windshield wipers are concealed under a flat when not in use. When the flap is raised, it serves to feed fresh air into the car's interior. Windshield pillars have been eliminated, and all glass fits flush with the body surface to improve the car's aerodynamics by boundary layer control. The Campagnolo wheels have spokes shaped as turbine blades to assist in cooling the brakes.

In designing the interior, much attention was paid to the problem of the rear passengers who are "the most neglected" in traditional sedans. Physical comfort for them has been obtained by careful study of available room, the design of the seats and the possibility of varying their rake. The front seats can be adjusted longitudinally and vertically, and the rake of the backrest is adjustable electrically. A special device makes it possible to "memorize" three positions so that those selected can be restored automatically in the event of their being altered.

Led (light-emitting diode) displays handle most of the information duties. A screen makes use of a luminous traveling spot to replace the traditional moving hand in speedometer and tachometer. An electronic route-guidance system is included, and there is a monitoring system for a service and repair needs. The latter stem from advanced electronic studies by Veglia-Borletti. Pinin's headlight system relies on multiparabolic reflectors developed in cooperation with Lucas and Carello.

Chapter 17

Ford and Ghia Prototypes

FORD HAS ITS OVERALL DESIGN CENTER IN Dearborn. The British and German car-manufacturing branches also have their own studios. In addition, there is Ghia, in Torino, that collaborates with all three and carries out a lot of show-car and prototype construction, on its own initiative, and on behalf of the Ford companies.

FROM REVERIE TO SUBSTANCE

For many years, under George Walker, Elwood P. Engel, and Gene Bordinat, most of the "idea" cars from Dearborn were pure styling exercises. They tended towards unrealistic reveries.

The first of a thinking-man's prototypes appeared, in 1968, in the shape of the Techna (Fig. 17-1). It was really an engineering project. The design staff was brought in during the final stages to give form to the substance.

In 1968, Ford was still thinking in terms of the six-seater sedan with a V-8 engine as the average car. Against that background, George H. Muller, a highly imaginative engineer, combined a number of innovative ideas into one car.

The basic theme of the Ford Techna was more effective use of space for increased functionality and greater occupant comfort (within a streamlined, 51-inch high profile). The sleek exterior was designed in the Corporate Projects Office and the interior was created in the Advanced Interior Department. The car was 209.7 inches long, and had a wheelbase of 120 inches and a curb weight of 4745 pounds.

The V-8 engine was pushed 9 inches further forward than in other Ford cars, and canted down and to one side to connect with an offset, triple-jointed driveline. This resulted in a small tunnel located between the left and center occupants rather than in the middle of the car. Thus providing ample footroom and improved seating comfort for six occupants (Fig. 17-2).

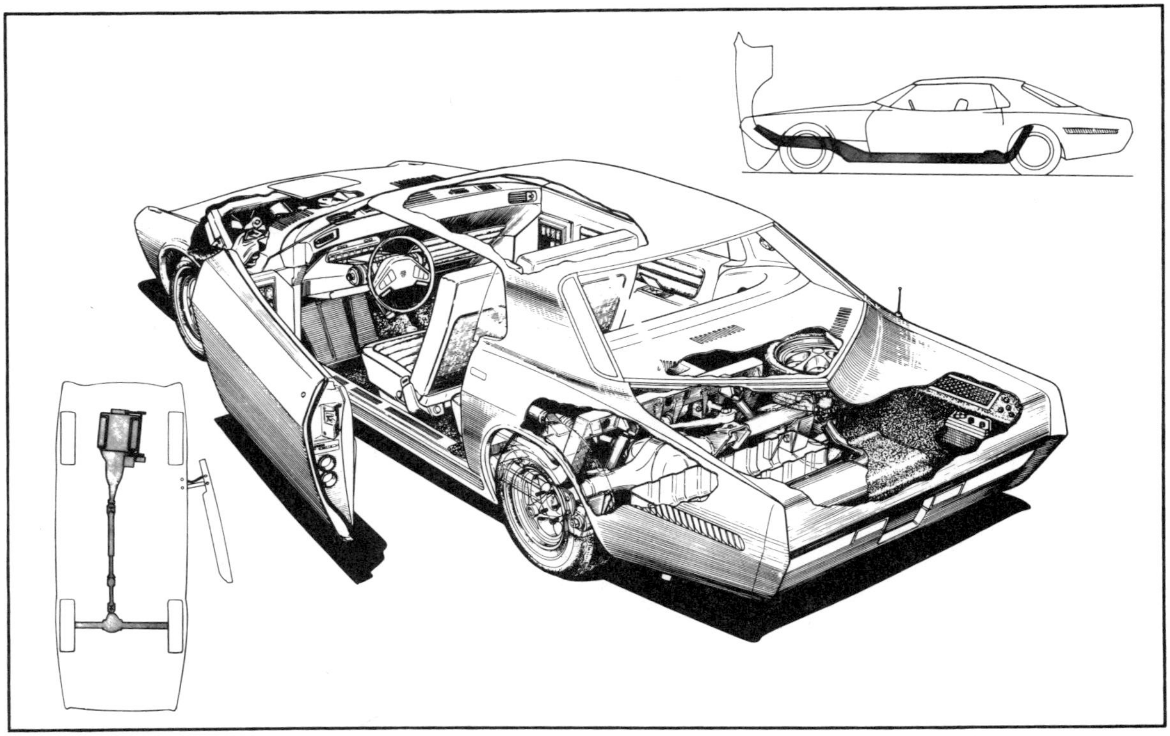

Fig. 17-1. The Ford Techna is experimental car from 1968.

Additional interior space was obtained through the use of bonding techniques rather than welding for many of the steel body panels, a honeycomb steel floorpan, a swept-away instrument panel, and a three-quarter length frame.

Improved access to the Techna passenger compartment was provided by a combination of smaller sills with revolutionary power-operated, parallel-hinged doors. By opening straight out from the body rather than swinging, the two six-foot long doors provided the same easy entry as with four doors (even when the car was parked within 18 inches of another vehicle).

The Techna was equipped with a structural glass windshield that eliminated the front corner posts, providing sweeping vision (Fig. 17-3). Forward vision was also improved by the low hoodline that was shaped so that reflections to the driver's eye position were reduced.

The entire front end skin—hood and fenders—was a one-piece construction, welded and bonded together, to be pivoted forward for complete access to the engine compartment. Twin aluminum radiators were mounted at a 50-degree angle relative to the ground. The alternator and air conditioning units were placed at the rear rather than at the front of the engine, and the battery was located in the trunk.

A FINE ITALIAN HAND

After the Techna had been built, all of Ford's engineering reserves were tapped to meet federal standards for safety and emission control. Future-concept cars had to wait. The advent of high-priced petroleum, however, might have hastened their return. It certainly caused a dramatic reordering of priorities.

Among the first of noteworthy Ghia creations for Ford was the Megastar of 1977 (Fig. 17-4). Designed for maximum energy conservation and at the same time to offer a new solution to the problem of providing modern styling with practical accommodation, it was a wedge-shaped prototype based on the German-built Granada running gear with unusual "tear-drop" side windows and very low aerodynamic drag.

Inside, a full five-seater saloon car package was retained. Folding rear seats more than doubled the luggage volume. The rear window lifted up for loading, and all other windows were sealed to allow smooth, turbulence-free body sides. Novel features included headlamps that are wound up electrically from low-mounted recesses, a quick-release fire extinguisher behind a pull-out panel in the facia, and a four-unit modular instrument layout.

A year later Ghia exhibited its Megastar II (Fig. 17-5). It was a five-seat, two-door hatchback based on the Taunus/Cortina platform. The Taunus four-cylinder engine, four-speed transmission, front and rear suspension, and the alloy road wheels were retained. But the overall length of the floor pan was reduced by 13.8 inches at the rear, the fuel

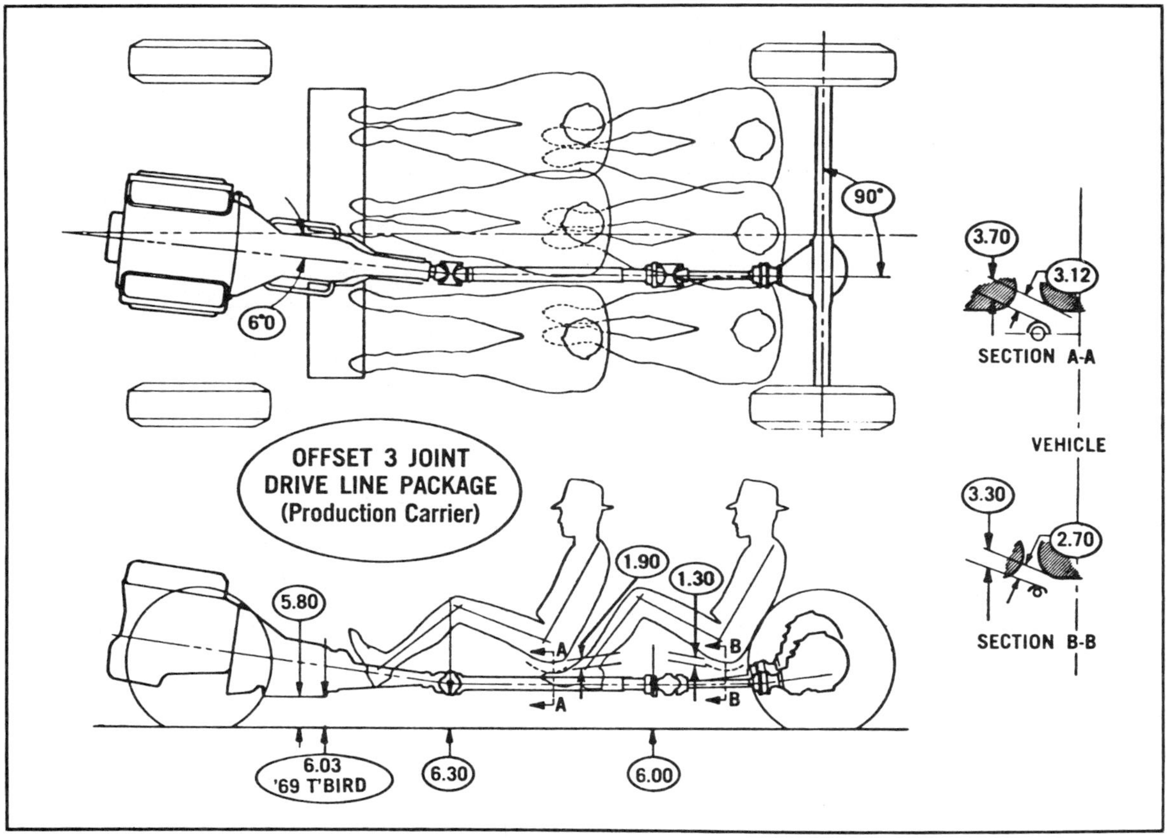

Fig. 17-2. The layout of the Techna drive line and seating.

Fig. 17-3. The Techna had an offset floor tunnel. The cantilevered roof allowed elimination of the A-posts.

tank was repositioned, and the spare wheel was moved to the engine compartment. The front bulkhead was also moved forward and the front end extended to accommodate a long chisel-shaped nose. The all-steel bodied Megastar II measured 48.4 inches high, 68.7 inches wide, and 165 inches long.

EXPLORING NEW BODY STYLES

In the spring of 1979, Ghia built the GTK (Fig. 17-6) using the Megastar styling theme for a station wagon body on the Fiesta platform. The front end of the GTK was extended forward to reduce its overall height and to incorporate a louvred air intake for the radiator. Although the position of the engine and front-wheel drive transmission was not altered, a now low-profile radiator was moved forward to let the spare wheel be housed directly above the transmission.

The Navarre was created jointly by Ghia and

Fig. 17-4. The Ford Megastar by Ghia.

the Dearborn studios. It was a sports coupe, with seating for five, powered by a V-8 engine, built on a 110-inch wheelbase.

Ghia was back in the spring of 1980 with a five-passenger urban prototype called the Pockar (Fig. 17-7); it was so-called because of its door pockets, accessible from outside. It was only 114 inches long overall, yet had the same passenger and luggage space as the Fiesta—which is 140 inches long. The Pockar was also considerably lighter. It was planned for a rear installation of the Fiesta power train with the engine tilted back at a 9-degree angle to clear the rear seat.

Ghia also unveiled the Ford Altair in 1980. It was a four-door sedan—based directly on the four-door Granada production model—2 inches longer, 3 inches lower, and 4 inches wider, with reduced frontal area and lower aerodynamic drag (its drag coefficient was 0.35, compared with 0.44 for the standard car).

Avant-Garde was the name of a Ghia prototype that appeared in 1981; it was based on the front-wheel-drive Escort (Fig. 17-8). The two-door sports coupe had a two-plus-two seating arrangement in a streamlined body shell (drag coefficient about 0.30). It was built on a 94-inch wheelbase, and it measured 155.5 inches in overall length. Overall width was 63.8 inches while the height was kept down to 49.7 inches.

At the front of Avant Garde, the bumper and spoiler were an integral unit. From the low front end, the body-work flowed smoothly to the sharply

Fig. 17-5. The Megastar II.

Fig. 17-6. The Ford GTK by Ghia.

Fig. 17-7. The Ford Pockar by Ghia.

raked windshield that was set at an angle of 67 degrees to the vertical. Twin cooling ducts for the engine were positioned between the front bumper and hood. The headlamps and front direction indicators were partly recessed behind rotating panels to improve airflow. The intake for the heater and ventilation system was positioned flush with the surface of the hood close to the windshield.

The rear wheel arches became more enclosed and underbody airflow was improved by the adoption of a smooth, flush-fitting undertray running from front to rear. Airflow around the front wheels and sides of the car was improved by increasing the depth of the front spoiler and lower sills. At the rear, an airfoil was added to the tailgate.

THREE WHEELS FOR SUPERIOR ECONOMY

Ghia's Cockpit three-wheeler (Fig. 17-9), with two seats in tandem formation, brought the Messerschmitt to mind. The idea came from Bob Lutz, chairman of Ford of Europe, who wanted to experiment with concepts for a car with superior economy. After preliminary engineering drawings had been made up, the project was handed over to Ghia for styling.

Access to the seats was provided via a framed, front-hinged and lift-assisted (by hydraulic struts) canopy. This design enabled the driver to make optimum use of confined spaces in congested city parking areas where there would not be sufficient clearance for entry or exit by driver or passenger through conventional side-hinged doors. The windshield is laminated glass while the whole canopy and rear window are made in tinted reinforced plastic. From the outside there is one curb-mounted lockable canopy release. The interior provides room for baggage alongside the bucket seats.

Fig. 17-8. The Ford Avant Garde.

Fig. 17-9. The Ford Cockpit—shades of Messerschmitt.

Miscellaneous items can also be stored on the rear parcel shelf, under the front hood (which also houses the horizontal spare wheel), and on either side of the central instrument binnacle.

The triangular all-steel chassis frame, with its additional side support bars, carries all the mechanical components. The transverse spring front suspension is controlled by hydraulic shock absorbers while the rear-mounted, air-cooled, single-cylinder engine and driving wheel are mounted at the apex of the triangular chassis and sprung by two integral coil spring shock absorber units.

"In building a car of unorthodox design, we have tried to retain as many of the regular features found on conventional cars as possible," says Filippo Sapino, managing director of Ghia operations. "For example, the individual front wheel housings gave us the opportunity to group the remotely controlled exterior rear view mirrors, direction indicators, air extraction louvres and rectangular headlamps neatly together."

The engine air intakes are located either side of the car on the rear bulkhead. Access to the engine and its ancillaries is via a one-piece removable rear panel. Also located above the engine is the 1⅓ gallon fuel tank with a flush-fitting external cap.

The complete Cockpit weighs 770 pounds and is 129.3 inches long, with an 87-inch wheelbase. Its front track is 48 inches. Calculated fuel consumption is 80 mpg.

PROBING THE FORD IN YOUR FUTURE

When Ford's Probe I (Fig. 17-10) appeared at the Frankfurt auto show in 1979, it was just a plastic shell with a dressy interior and simulated instrumentation. Two years later, it reappeared as a

Fig. 17-10. Ford's Probe I.

Fig. 17-11. The Probe II.

working prototype (with a Mustang engine and transmission).

Designed and built under the direction of Don Kopka, at the design center in Dearborn, it was built to demonstrate how a practical coupe of the next decade might be conceived to obtain the lowest possible aerodynamic drag.

Meanwhile, Kopka and his staff had designed the Probe II, aimed at the short-term future—a concept prototype for 1985, Ford said (Fig. 17-11). The body is a less futuristic derivative of Probe I, and the vehicle is tailored around the Ford Escort power train.

The five-door family car is designed for an optimum combination of functionality and aerodynamics. The radiator air intake is positioned below the bumper, headlamps are recessed behind clear plastic shields, a very fast windshield angle eases the airflow transition from hood to roof, and a miniature rear spoiler is built into the backlight frame.

Probe II specifications include the use of lightweight materials such as graphite fiber, plastics, and light metal alloys. It has a 100.4-inch wheelbase and is 179.1 inches long. Height is 52.5 inches and width 70.4 inches.

Ford's Probe III (Fig. 17-12) originated at Brentwood, in England, where Uwe Bahnsen led a design team for advanced concepts, assisted by Patrick Le Quemente for the exterior and Trevor

Creed for the interior. Heinz Ostendorf was in charge of the aerodynamics. The result is a four-door hatchback sedan with a drag coefficient of only 0.22.

The Probe III chassis components are taken from the Taunus/Cortina, with front-mounted engine and rear axle drive. It has the same 68-hp engine, but it is 10 mph faster than the production models (with a top speed of 110 mph). At top speed, its fuel mileage is 23.5 mpg. Cruising at 56 mph, it would give 52 mpg.

The body features flush-fitting glass and headlamps, faired-in mirrors, rear fender spats, and a double-decker rear spoiler. In addition, Probe III has a full undertray with an adjustable front section that permits a variable-venturi effect; this creates a suction force that counteracts aerodynamic lift as a function of speed.

Compared with the Taunus/Cortina production models, Probe III has lost some rear seat legroom and headroom. Front seating dimensions, however, are superior in Probe III. The streamliner is nearly 12 inches longer overall, and trunk space has been slightly improved.

The latest from Ghia is the Ford Quicksilver that was unveiled at the Geneva show in 1982. Taking a cue from the Ital Design Medusa, Filippo Sapino has created a four-door sedan with the en-

Fig. 17-12. The Probe III.

Fig. 17-13. The Ford Quicksilver by Ghia.

gine mounted behind the rear seat (Fig. 17-13). The engine is a transversely mounted Ford V-six; it shares the transmission and drive line with the AC 3000 ME. That leaves room for a trunk in the tail end (though of somewhat subnormal dimensions). Access to the engine is from the rear. In addition, the back seat can be folded to give access from inside (necessary to change plugs on the front bank, for instance).

The engine is well insulated for noise. The space under the short hood in front is used for storing the spare wheel and providing additional luggage capacity.

The Quicksilver provides seating for five occupants inside an overall length of 180 inches. The wheelbase is 116 inches. Body width is 69 inches and the height 49.8 inches. With its smooth surface and high windshield angle, the Quicksilver seems highly streamlined. Tests have shown a drag coefficient of 0.30. That is higher than expected, but Filippo Sapino says: "With some detail work, we can probably get that down a couple of points."

What about Ford's next-generation production cars? The Probe II will be bypassed. The essential elements of the Avant-Garde and Probe III will be adopted for use in mass-production cars of all sizes.

Chapter 18
The Fiat VSS

WHAT FIAT CALLS ITS VSS (VETTURA SPERImentale a Sottosistemi; Experimental Subsystems Vehicle) is a fully practical car, close to present technology, that could be mass produced in 3 to 4 years (see Fig. 18-1). It's an economy car intended for 20 years' durability in service. With low aerodynamic drag (its drag coefficient is 0.36) and a very sizeable weight saving, it is 70 kg lighter than the current production-model Fiat Ritmo (whose mechanical elements it shares). The body is expected to be absolutely rust free. The reduced weight and rust-resistance stem mainly from the use of plastic body panels made up as subassemblies and then bolted or bonded to a cagelike central support structure, welded up from special steel pressings (Fig. 18-2).

Rustproofing and lightening the car were not Fiat's only aims with the VSS. The main goal was to take some of the labor cost out of producing cars.

NEW APPROACH TO MAKING CARS

Rethinking the car to fit the industrial realities of our time has been on Fiat's mind for years. Their efforts have been spurred on by repeated fuel-cost shocks and the sharp rise in Japanese car penetration in some of its traditional markets. Productivity might be the name of the game, but there is no way Fiat can compete—in its Italian-based plants—against automobiles made in Japan. The reason is socioeconomic and has nothing to do with either automation or product quality. As long as the Japanese auto worker puts in 2300 hours a year, and his Italian counterpart about 1400 hours, Fiat will have a 36-percent, manpower-cost handicap.

Fiat's search for higher productivity leaves the human factor on the side of uncontrollable factors. The solutions must be found in the product and in manufacturing processes. Important advances have been made, such as the Robogate system of body-

Fig. 18-1. A general view of the Fiat VSS.

Fig. 18-2. The doors and hatch arrangement on the VSS.

shell assembly, and Fiat's Comau welding and spray-painting robots have been widely acclaimed.

With the VSS, Fiat has taken a very long step forward. The subsystems concept is the heart of the principle. It is more than a modular concept that is just putting building blocks together.

Inherent in this concept is a new approach to making cars. It involves decentralized production of subsystems joined to the central structure in a logical sequence. This gives a sharp reduction in the number of assembly operations and therefore shortens the assembly time.

The primary aim is simplification of the method of building cars without necessarily simplifying the product. A reduction in the number of assembly operations means changes in the vehicle's materials and construction.

During 1978, when Fiat devised its first schemes aimed in this direction, a local think-tank called The Idea Institute, was brought in to collaborate in a consultative capacity. In the later stages of conceptualizing the VSS, the Idea Institute's brainpower was enriched by securing the services of Rudolf Hruska, recently retired engineering director of Alfa Romeo. Hruska created the Alfasud—not just the car, but also the Pomigliano d'Arco plant layout and methods.

"The steel cage" (Fig. 18-3) "of the VSS weighs 30 percent less than the naked body shell for the Ritmo," explained Paolo Scolari, Fiat's technical director. The Ritmo body shell is light (only 406 pounds). But the steel cage for the VSS weighs a mere 269 pounds. The cage is the main structural member to which are attached all power train and chassis parts—engine, drive line, suspension, steering and brakes. It provides a safe, nondeformable cell for the occupants.

No basic difference exists with regard to the

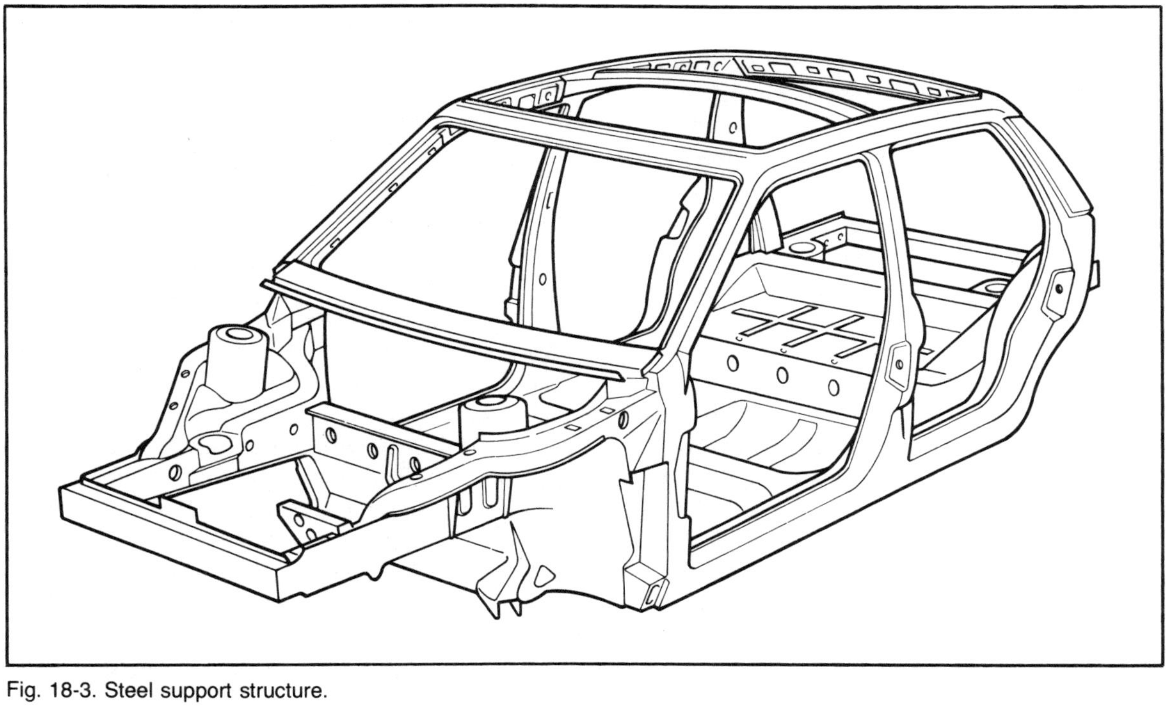

Fig. 18-3. Steel support structure.

mechanical subassemblies and their joining with the supporting structure. The difference lies in the body assembly.

SELECTIVE USE OF STEEL AND PLASTICS

All body elements are treated as subsystems or parts of subsystems. The front end—including an integral bumper—is one such subsystem. The hood and upper front fenders make up another subsystem. Cowl sides, front and rear doors, rear hatch, rear panel, and roof panel form individual subsystems.

There is a dramatic reduction in the number of welding points on the cage when compared with a body shell. Because the steel cage can be welded up from relatively small sheet-steel pressings, the large hydraulic presses used for large panels are no longer needed. They are replaced by molds for plastic components that are far smaller and less expensive.

The plastic components can be molded, finished, and made up into subassemblies in separate, specialized plants. The types of plastics are carefully selected for each application. Fiat invited four of the leading producers to assist in the project: Bayer, BASF, SNIA, and General Electric Plastics. The industrial implications here are enormous. It could mean re-planning entire factories with new layouts, new materials-flow setups, and new tools and machinery.

Power train, suspension systems, and other mechanical subassemblies are more easily mounted in the cage than in a unit-body shell. This is because the cage is open on all sides. Assembly processes can therefore be simplified, perhaps, automated, and certainly speeded up.

Even with a conventional paint system, the production cycle of the VSS takes 5 hours less than building the Ritmo (with robot carriers and robot welding of the body).

The basic tooling for the cage can be used for a longer-than-normal period. It need not be affected by the addition of new body styles or even a model change.

INCREASED VERSATILITY

A number of different body sections can be adapted to the same cage. Hatchback coupes, notchback sedans, and station wagons need to differ only in the C-post to rear panel section. Similarly, front ends can be exchanged, replaced, and restyled without affecting the basic structure (Fig. 18-4). That means drastic budget cuts for new-model development and tooling.

All components have been designed using a mathematical model and have undergone practical experiments and tests similar to those carried out on conventional car components (Fig. 18-5). Each subsystem of the VSS car has been built using the material that best suited its function according to the basic VSS principle of function separation. In the future, this method will allow the introduction of materials, other than those used on the VSS, if new ones permit greater freedom from the standpoint of both design and production.

CHOICE OF PLASTICS

At the front and rear ends, materials with elastic properties are used. In other areas, the plastics have greater rigidity. The front end of the vehicle consists of a unit made up of the front bumper, rigidly fastened to the structure through brackets, and of a multirole front section independently fastened to the structure. Both these elements are constructed in polycarbonate in order to ensure that, in the event of front impact, they deform without incurring permanent damage. They are suitable for injection molding.

The hood is designed to extend as far as the wheel arches. In a conventional car, four separate elements make up the engine compartment: outer hood panel, inner hood framework, right front fen-

Fig. 18-4. Body style versatility of the VSS.

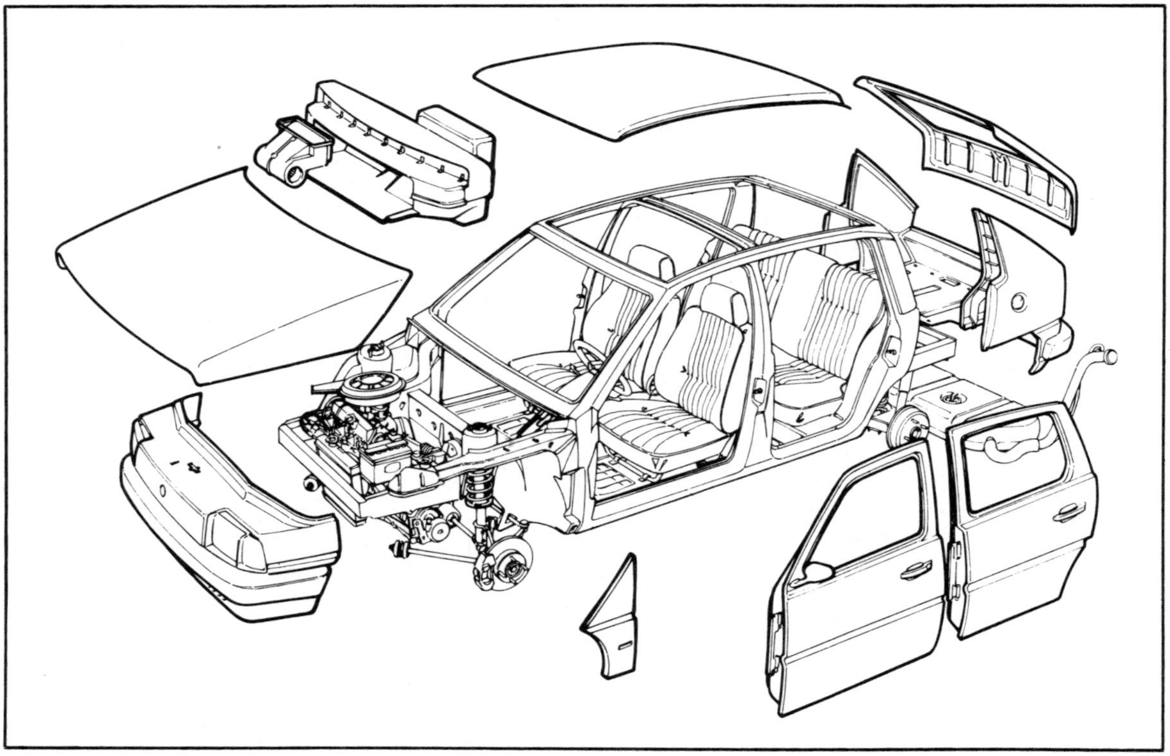

Fig. 18-5. The nine subassemblies of the VSS.

der, left front fender. On the VSS, these four elements are combined into one single component that includes the air scoop. It is made of polyester resin (reinforced with fiberglass) having a thickness of 0.118 of an inch.

Each door consists of two half-shells in sheet-molded unsaturated polyester, bonded at the edge, plus an inner cross member in high-strength, low-alloy steel. This has the dual function of load bearer and linkage to the body; the window mechanism, linkages, door hinge, and lock are all built into this inner cross member. The plastic components and cross member are fixed together using external plates on both the front and rear parts of the door.

The rear end of the VSS consists of elements that differ in terms of both shape and material employed. The luggage compartment floor and sides, rear cross member, and bumper are polycarbonate. The rear fenders are unsaturated polyester. The elements of the rear panel are bonded through off-the-line procedures. The complete subsystem is then fastened to the main structure with bolts.

The rear hatch consists of one large single shell of 0.1-inch-thick, sheet-molded unsaturated polyester with a central recess for the rear window. The whole subassembly is made torsionally stable and rigid by bonding the window to the lip. A reinforcement ribbing around the perimeter provides a groove all along for fitment of dust- and water-tight gaskets. The roof is made of 0.118-inch-thick polyester foam that is bonded to the steel structure.

The dash panel is composed of two subsystems. First is an instrument-and-control module carrying the steering wheel and upper column, electrical switches, and all instruments. Second is the dash panel that has interior ventilation and air-distribution functions, padding, and accessories. Noise insulation has been considerably improved due to the plastic materials used as well as to the adoption of dividing joints between steel and envelope.

COST, SUPPLY, AND ENERGY

The VSS prototype (Fig. 18-6) has been designed

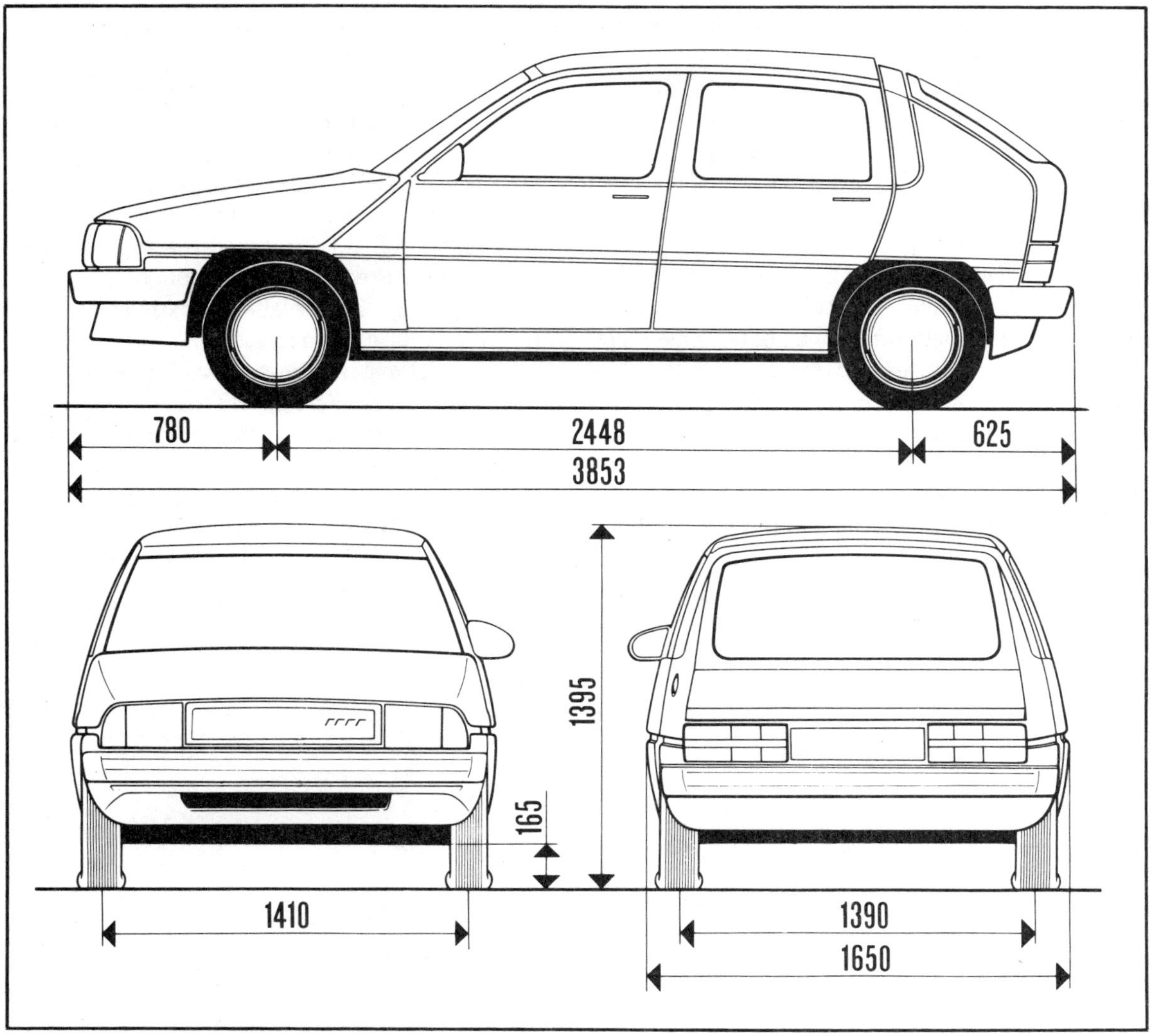

Fig. 18-6. Dimensions of the VSS.

for immediate feasibility on the basis of the technologies and experience acquired to date. Many of the solutions it provides will be feasible in the near future. Nevertheless, the full adoption of production and design methods, as demonstrated by the VSS, faces difficulties in terms of both economy and organization. The economic difficulties are due to the higher cost of materials and to the necessary investment to be made for new production tooling. Organizational difficulties stem from both the structure of the subcontractors' market and the current plant organization schemes.

Such sweeping changes, that would represent vast investments in new plant, would perhaps not make sense if they could not also help reduce the cost of, say, building a range of different models around the same mechanical components, and to achieve major weight reductions while simultaneously securing quality improvements, longer life, and a variety of features that translate into tangible, competitive advantages in the market place.

"We will be able to produce a VSS type of car after 1985," states Paolo Scolari, "but there are many questions that must be solved in the meantime. Fiat is already a leader in the use of plastic materials. The Ritmo has 15 percent plastic content (by weight) today. Now in the VSS, there is 27 percent plastic content. We will not go from 15 to 27 percent in one step, but proceed by gradual substitution, across the full range of Fiat cars."

One worry about extended use of plastics is that plastics, being derived from oil, might be subjected to sudden price increases or even interruption of supply. Fiat claims that the specific plastics used on the VSS are made from oil-processing subproducts that have few other applications; thus their availability should be assured.

Further, as these plastics are the result of a complex production cycle, the raw material cost is relatively unimportant in relation to the end product cost. The energy content of a plastic component (i.e., the energy contained in the material as well as the energy needed to process it) is approximately half of that of the equivalent steel component. In terms of energy conservation, plastic components are preferable, Fiat concludes.

In direct material costs, the VSS is calculated to be about $100 per car higher than for the current Ritmo. But there are savings in the amount of energy spent in building the car, saving on the tooling costs (compared with new tooling for a conventional shell), and savings in labor time. Together, these savings could amount to several times the rise in material cost.

Chapter 19
Peugeot Vera and Renault Eve

ABOUT MID-YEAR, 1981, TWO EXPERIMENTAL automobile prototypes were delivered to French government representatives. Two years earlier, the Department of Industry, under Andre Giraud, had decided to sponsor the construction of test cars that would do the same job as popular contemporary family sedans, but consume one-third less energy. He allocated $6 million for this purpose, and the Department of Industry signed two contracts, splitting the amount equally between the two contractors—privately owned Peugeot on one hand and state-owned Renault on the other.

The objective was to demonstrate means of taking a very large step toward better fuel mileage in the short almost immediate, term, almost within the normal new-model lead time of the French car companies. Spending government funds on such projects was justified by the fact that France imports all its motor fuel and pays for it in hard currency (dollars). The expectation was that French cars would quickly adopt the lessons of the project and consequently reduce domestic fuel consumption and come into great demand from other countries in similar situations so as to advance French exports and redress the balance of trade.

The timetable was short, and the contractors were therefore restricted to current technology. There was no time to undertake new research. They had to find better combinations using the state-of-the-art technology.

Peugeot calls its car Vera (Vehicule Econome de Recherche Appliquee; Applied Research Economy Vehicle). See Fig. 19-1.

THE VERACITY OF THE VERA

With its boxy shape, steep windshield, and four large doors, the Vera is not unusual in its appearance. Yet the Vera reveals certain things about engineering priorities in Peugeot organization.

Fig. 19-1. Peugeot Vera prototype.

The main attack came in the area of saving weight. Secondly, Peugeot engineers did a lot of work on reducing the rolling resistance. And thirdly, they tackled the matter of airflow. No rocket noise or no stretched tail spoil the sobriety of the Vera's looks and practical dimensions (170 inches long overall). Retaining the windshield and rear window angles from the 305 GL, Peugeot specialists have worked in the St. Cyr wind tunnel and come up with a combination of small devices that can help an aerodynamically unfavorable shape (such as a box).

The Vera body is not a new shape; it is a modified edition of the existing 305 body shell. Current production model 305 has a drag coefficient of 0.44. That is not a good starting point except in terms of making great percentual improvements.

The most eye-catching detail of the Vera's aerodynamic devices is the rear wing. It helps prevent airflow separation and reduces tail-end lift. The windshield and rear window are flush fitted, without exterior joints, for smooth airflow transitions. A front spoiler (Fig. 19-2) is part of the bumper. It alone is good for 0.015 points by deflecting the air and controlling air flow under the car (but Vera does not have a full, smooth undertray!).

In addition, the air vents were modified, the air-intake surface was reduced, and full-face wheel covers were extended right to the rim edges fitted. The front fender skirt was opened up for better air evacuation and the floor line was deflected downward at the rear to mask more of the rear tires. As a result, Vera's drag coefficient has been brought down to a creditable 0.318 (a 27-percent improvement).

MATERIAL SUBSTITUTION

The body is lightened by 221 pounds. Only the stressed portions of the body are made of steel. It was made with high-strength steel inserts in crucial areas, and unstressed parts were made of plastic or aluminum. The left doors are plastic and the right ones are aluminum giving Peugeot a chance to test both.

Inside it has front bucket-seat frames made of plastic (Fig. 19-3) and a plastic support bracket for the three pedals (Fig. 19-4). The master cylinder has an aluminum housing. Experimental Michelin plastic wheels and low-profile TR-E tires were adopted—saving 8½ pounds per wheel. The complete Vera—with fuel, oil, and water on board—weighs only 1632 pounds; that's 408 pounds less than the 305 GL.

"We realized we were going to save a lot of weight," said Pierre Saintigny, a test engineer at the Belchamp proving grounds, "so we determined that it would be possible to use a smaller and lighter engine." That meant using the 1124cc engine, from the little 104 model as a basis, instead of the 1472cc 305 unit. Both are four-cylinder designs with aluminum blocks and heads. Using the smaller block saved nearly 40 pounds.

Aluminum, magnesium, and plastic parts were substituted for heavier iron and steel components. Peugeot fitted a lighter flywheel, aluminum radiator, plastic camshaft cover, magnesium carburetor body, and thin-sheet stainless steel exhaust manifold. That saved another 35 pounds.

With the lightened power train, Vera could use the brakes, steering, and suspension parts also from the smaller 104 model. This resulted in a further weight saving.

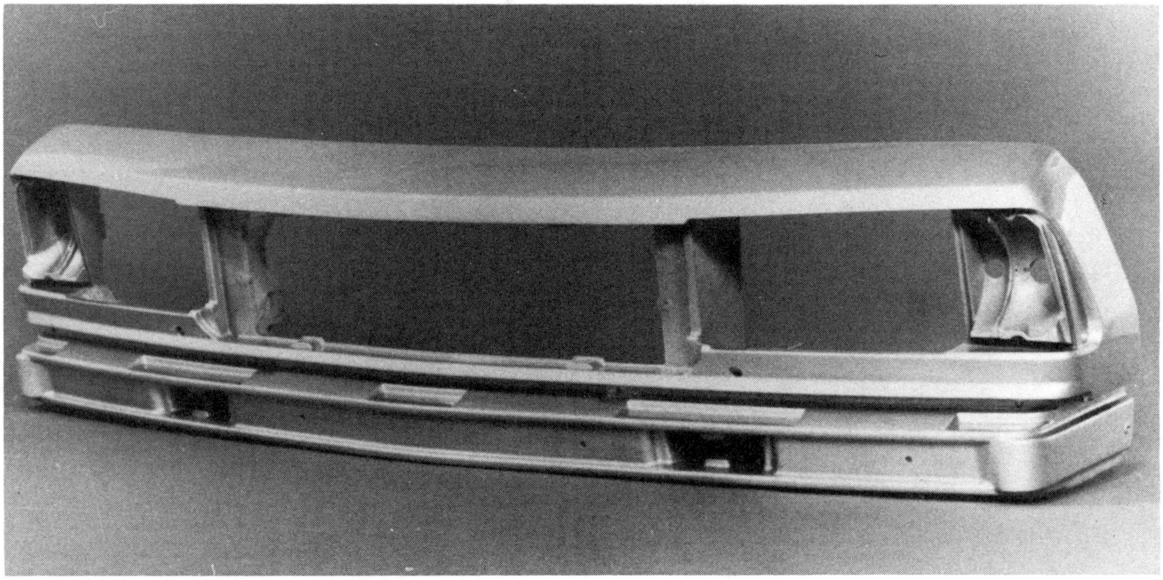

Fig. 19-2. Vera front bumper and nose section.

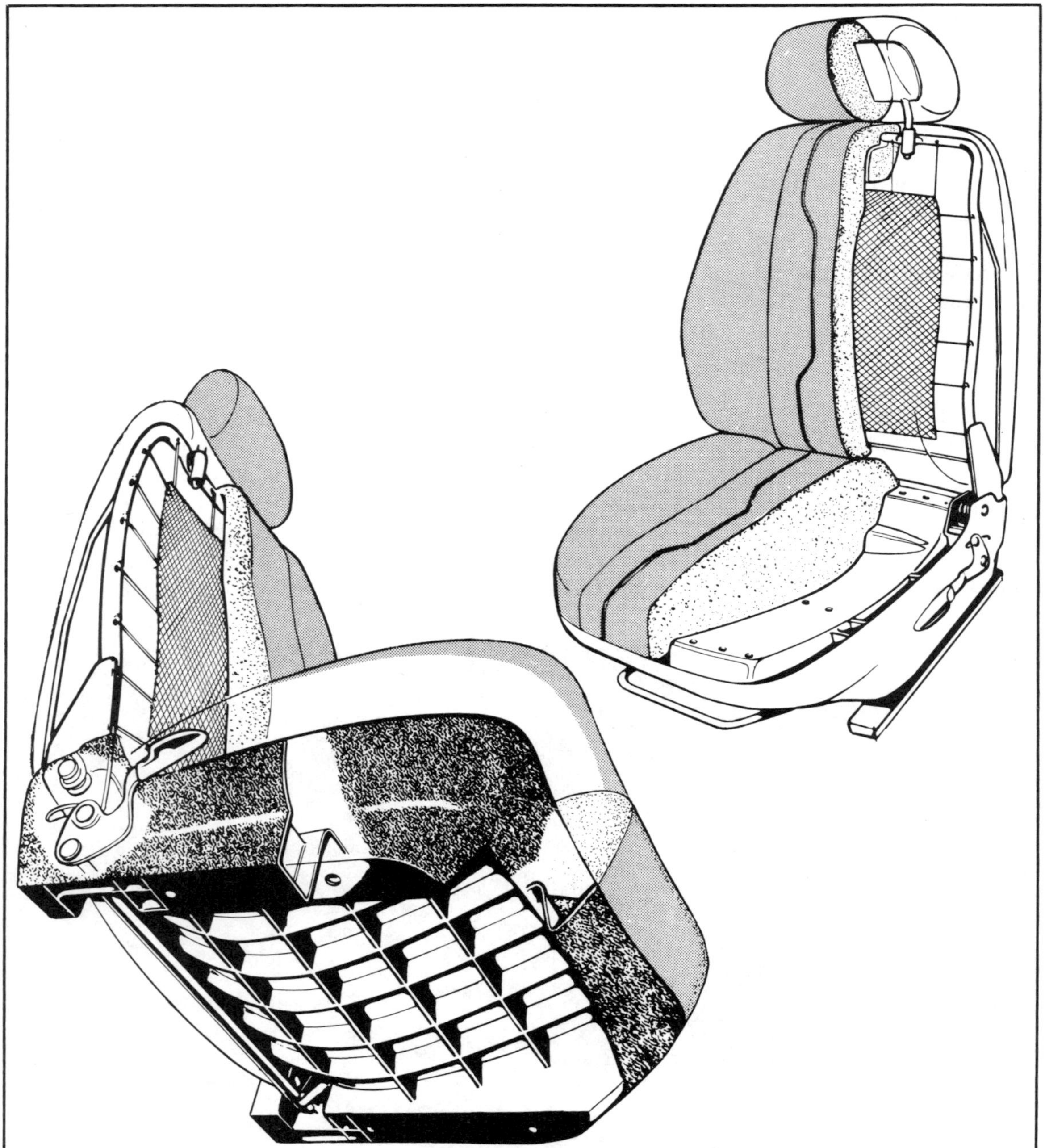

Fig. 19-3. Vera seat construction.

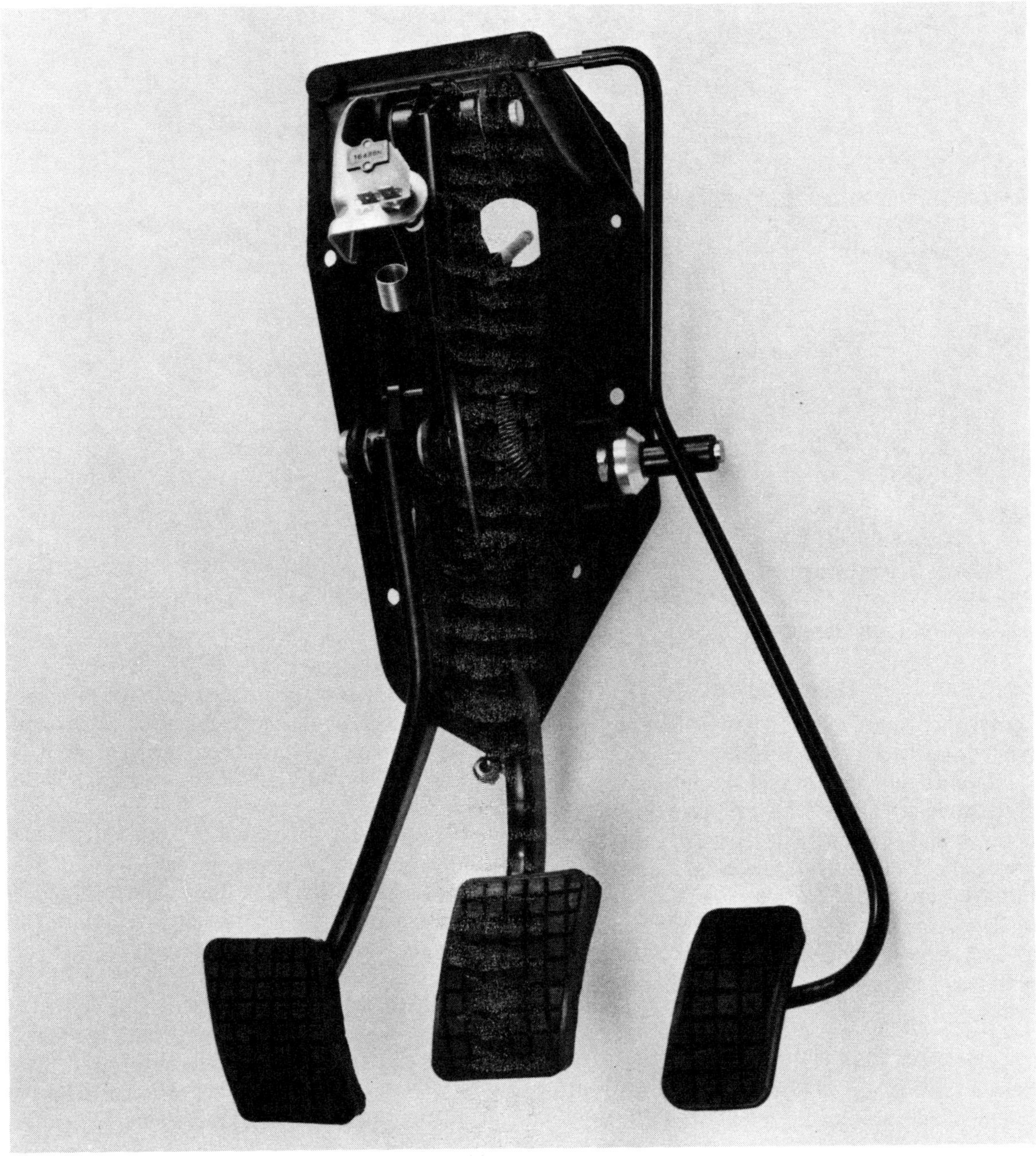

Fig. 19-4. The Vera pedal mounting bracket is plastic and pedal arms are aluminum.

Fig. 19-5. Testing the Renault Eve.

TEST RESULTS

I drove this car at Peugeot's vast Belchamp proving grounds to check out its performance for myself. It went from 0 to 60 mph in 14.4 seconds, and reached a top speed of 97 mph! Not having the time to make my own fuel-consumption tests, here are the figures from Peugeot's engineering report. In the European urban driving cycle, the Vera came out at 37.3 mpg; at constant 56 mph, 56 mpg; and cruising at 74.5 mph, 42 mpg. That's fully 33⅓ percent better, on average, than the 1981-model 305 GL sedan.

NEW LOOK FOR RENAULT'S EVE

Renault picked the R-18 as the basis for its Eve (Elements de Voiture Economique; Economy Car Elements). A technical consultant to Renault, Pierre Villaume, explains that the Eve project is conceived as a synthesis of all known and imaginable ways to save energy without giving up what has been achieved in the areas of safety or antipollution measures.

The Eve is built on the R-18 floor and a 96-inch wheel-base, with front-wheel drive (Fig. 19-5). Compared with the R-18 TL, a model that has proved more fuel efficient than the average car in its weight class, the Eve is 220.5 pounds lighter (with a curb weight of 2596 pounds).

The weight reduction has been achieved not by downsizing, but by material substitution. Steel is retained only for the underbody parts carrying the highest stress. The rest is aluminum. Pechiney's alloy 5182 is used for stressed elements, and the unstressed exterior panels are CP 609 alloy.

It is not easy to put truly aerodynamic shape on the R-18 wheelbase without doing damage to the car's practical aspects (roominess, entry/exit ease) or safety (visibility). The R-18 has a drag coefficient of 0.39.

Frontal area could be reduced by making the car lower, but that would mean a loss of seating room and comfort unless compensated for by adding length (which would add weight). Sagely, the Renault body designers agreed to keep the basic R-18 seating envelope while making the most of the potential for improving internal air flow (engine breathing, cooling, and interior ventilation).

Gaston Juchet, who did the R-15 and R-17 coupes in 1971/72, made up the initial sketches for Eve, and Pierre Mignon developed the design concept into a 1/5 scale model on the R-18 track-and-wheelbase proportions.

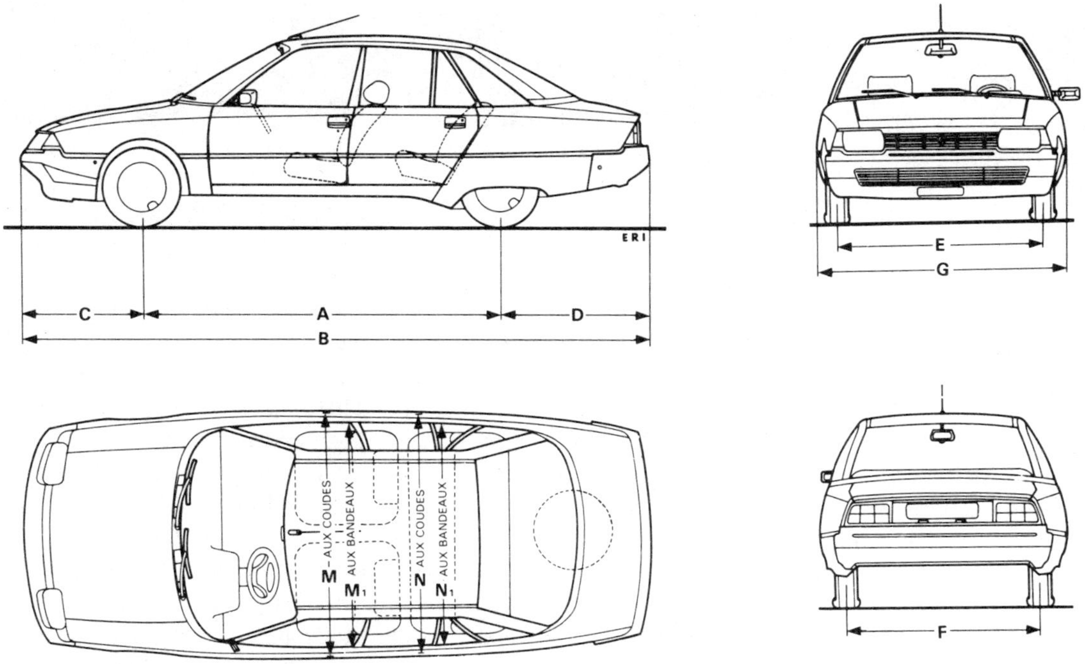

RENAULT EVE			
	Ag = 2509	Dg = 1044	M = 1440
	Ad = 2477	Dd = 1076	M_1 = 1430
		E = 1415	
	B = 4408	F = 1367	N = 1380
	C = 855	G = 1740	N_1 = 1380

Fig. 19-6. Dimensions of the Eve.

Fig. 19-7. A sharply raked windshield and a short deck make for smooth air flow over the Eve.

In plan view, the Eve body shows a tapering towards the rear, windows flush with the skin, and an absence of excrescences. Smooth transitions in airflow are assured by gentle angles from hood to windshield and windshield to roof. The front end presents a "soft" face, curved in three planes, compressed to avoid adding to overall length.

Tail end design is intended to avoid airflow separation and vortice formation. Rear window slope, deck length and angle, taillight panel, and bumper profile have been aerodynamically integrated to an optimal compromise within compact dimensions.

The first wind tunnel test showed a drag coefficient of 0.204 without internal air flow, realistic underside, or revolving wheels. A slightly longer nose brought the figure down to 0.19, and a slight widening of the lower body, permitting smoother curvature of the skin, gave a value of 0.168. This was rounded off to 0.17 when the windows were made more realistic. A full-scale model based on the final shape showed a drag coefficient of 0.21 (still with the air intakes blocked off and completely smooth undertray). The working EVE prototype came out at exactly 0.25. That means a 35-percent improvement in overall air drag (Fig. 19-7).

ONE-THIRD LESS FUEL PER MILE

Just how well does Eve work towards saving fuel? Take the R-18 TL for comparison. It gives 22.4 mpg on the DIN 70030 urban cycle (which includes several idling stages, in contrast with the LA4 test used by the DoT). At constant speeds, the car gives 36.7 mpg at 56 mph; it drops to 27 mpg at 74.5 mph.

In its first road tests, the Eve prototype gave 29.4 mpg in the urban cycle, and 52.2 and 39.2 in highway tests at constant speeds of 56 and 74.5 mph. Steady development work has resulted in substantial improvements, and Renault engineers more than reached their final targets: 33.6 mpg in the urban cycle, 61.8 mpg at 56 mph, and 44.8 mpg at 74.5 mph.

Those are impressive figures for a family sedan that can go from standstill to 60 mph in 15.5 seconds and reach a top speed of over 100 mph. The lessons from the Eve program will be incorporated, singly, if necessary, and wholesale, if possible, for future Renault production cars as soon as feasible.

Chapter 20

Quartz by Pininfarina

AUDI RELEASED ITS BOMB IN THE SPRING OF 1980. It had been over three years in the making, and it made big waves. The Audi Quattro, the 200-hp, four-wheel drive rally car (Fig. 20-1) is a limited-production model, built basically for competition, with an output of no more than 10 cars a day. Its impact on the world of motoring has been out of all proportion to its sales figures. The engine is a turbocharged five-cylinder unit mounted in front, slanting to the right.

The basic Audi 80 two-door body shell was modified for the Iltis-like chassis, receiving partial restyling by Hartmut Warkuss (who drew up a special paint job to emphasize its wedge shape). The standard notchback was turned into a fastback with a nonadjustable spoiler. Plastic-covered bumpers were adopted (with an integral air dam in front). Flexible plastic fender skirts allowed the Quattro to mount low-profile Pirelli P7 tires of 205/60 VR 15 size.

Careful development work by project manager Walter Treser kept the weight increase over the basic front-wheel drive Audi 80 down to 165 pounds. The car has a top speed of about 130 mph, and can go from standstill to 100 mph in under 18 seconds.

THE ITALIAN TOUCH

When Sergio Pininfarina and his chief designer, Leonardo Fioravanti, saw the Quattro, their immediate reaction was that the Quattro concept was particularly suitable as a basis for a new study in material substitution they were then planning. The idea grew into a project, which was named Quartz, to symbolize the goal of adapting aerospace materials and technology to the automobile.

With the Quartz, Pininfarina wants to demonstrate a practical example of how metal and nonmetallic materials can be combined in a body structure. Its principles are valid for bodies of many

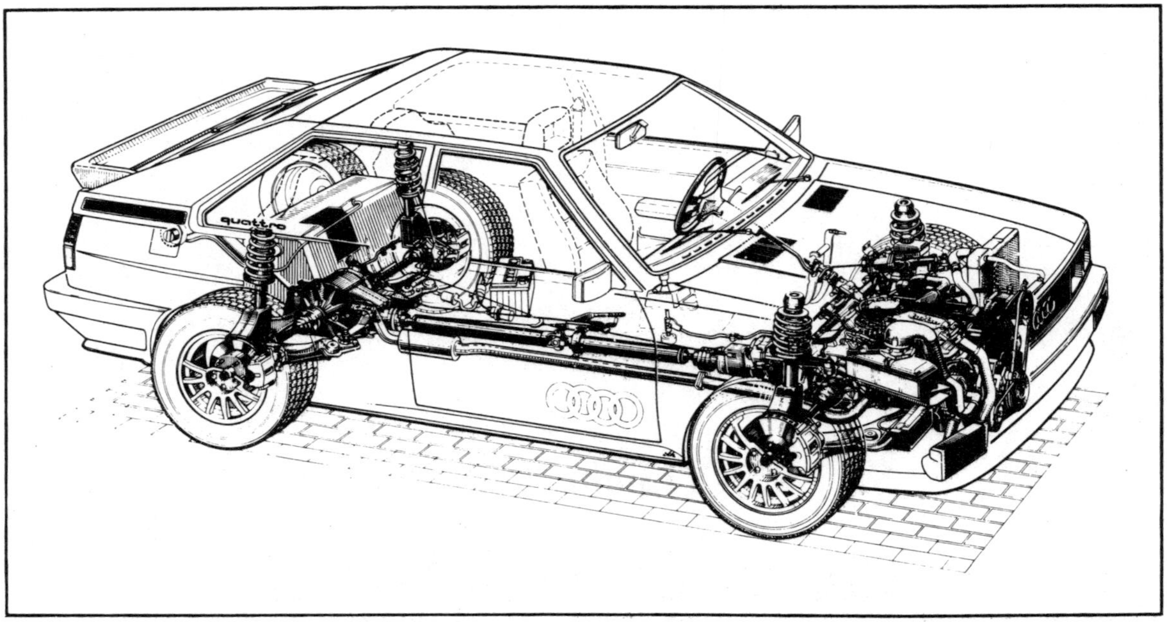

Fig. 20-1. The standard body on the Audi Quattro.

different shapes, but, as always, Pininfarina wanted a new creation. It had to be simple and it had to be highly individual (Fig. 20-2).

The result is a body with a girdling groove that divides the upper part from the lower (Fig. 20-3). This groove also performs the function of channel for the internal aerodynamics of the engine compartment: air enters at the front through the slit in the front part of the groove and is discharged in the area immediately behind the front wheel housing (Fig. 20-4).

Another feature of the car is the total absence of protuberances. The windows have no outside struts, the door handles (Fig. 20-5) are out of the way behind the door windows, and front and rear light clusters are integral with the body.

Out of respect for the basic theme, the dimensions of lighting clusters have been cut to the minimum (Fig. 20-6). The headlights are the result of a Carello study making use of elliptic optics and their frontal dimensions are very small while performance is undiminished. Their lens-type appearance and setting are the main features of the front of the car. For the sake of compactness, the rear part has minimal overhang, but this is still sufficient to give room to a trunk of normal size for this class of car.

AIR VENT BUILT INTO THE MIRROR

Other typical features are the outside rear view of mirrors that also serve for dynamic ventilation of the passenger compartment. Quarter lights were used to provide interior ventilation. Now the openable quarter-light has disappeared, mainly as a result of visibility regulations, and the new Saiag-patented rear view mirror provides for this function in a more modern key. Wind tunnel tests have confirmed its validity (Fig. 20-7). Used air exits from the compartment via a slit resulting from the special design of the rear side window (Fig. 20-8).

The Quartz prototype, first exhibited at the Geneva show in March, 1981, was notable for seven novel material applications:

☐ Front and rear bumpers in a sandwich structure of Kevlar/honeycomb/Kevlar.

☐ Hood, roof panel, and side panels in aluminum.

☐ Door panels in a sandwich laminate of steel/polypropylene/steel.

☐ Rear window in polycarbonate.

☐ Rear window frame in carbon fiber.

☐ Seat frames and steering wheel in carbon fiber.

☐ Linings and paddings in parachute fabric.

Kevlar is a special aramid fiber that can be used in an enormous range of industrial applications. It possesses an extremely low specific gravity and high tensile strength. The principal features of Kevlar can be summed up as follows. It has the highest strength/specific gravity ratio of all fibers; a rigidity/specific gravity ratio that is second only to graphite; excellent intrinsic sturdiness; good heat resistance and thermal stability; good resistance to chemical and atmospheric agents; and vibration-dampening power eight times higher than steel and four times that of other reinforced fibers.

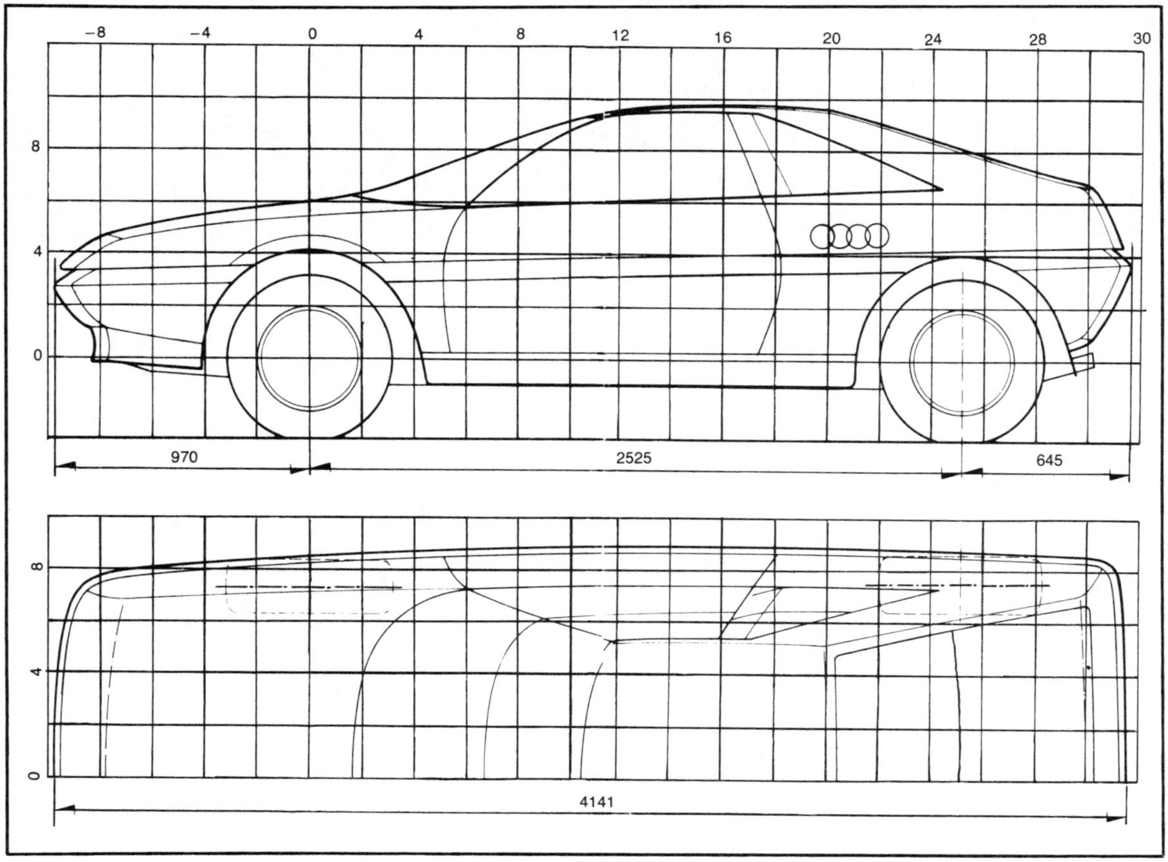

Fig. 20-2. A dimensional sketch for the Quartz.

Fig. 20-3. The Pf Quartz front and left side view.

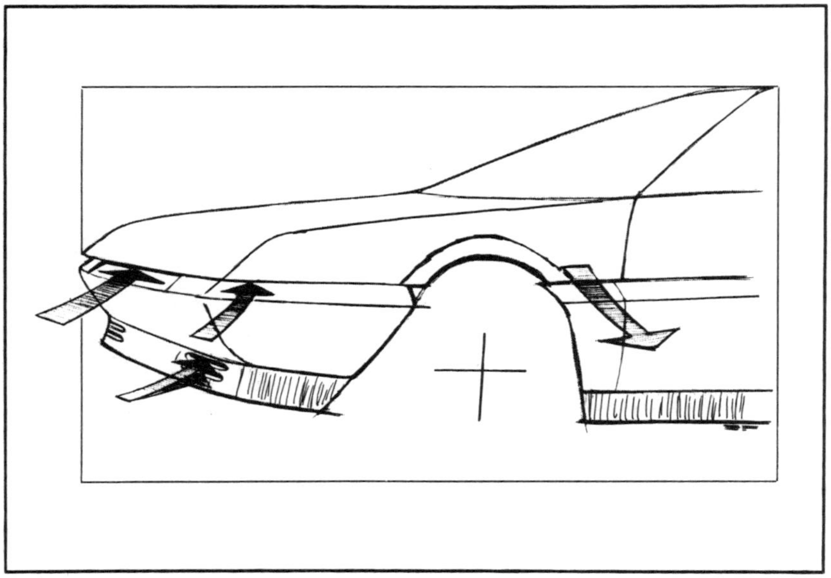

Fig. 20-4. The internal air flow path.

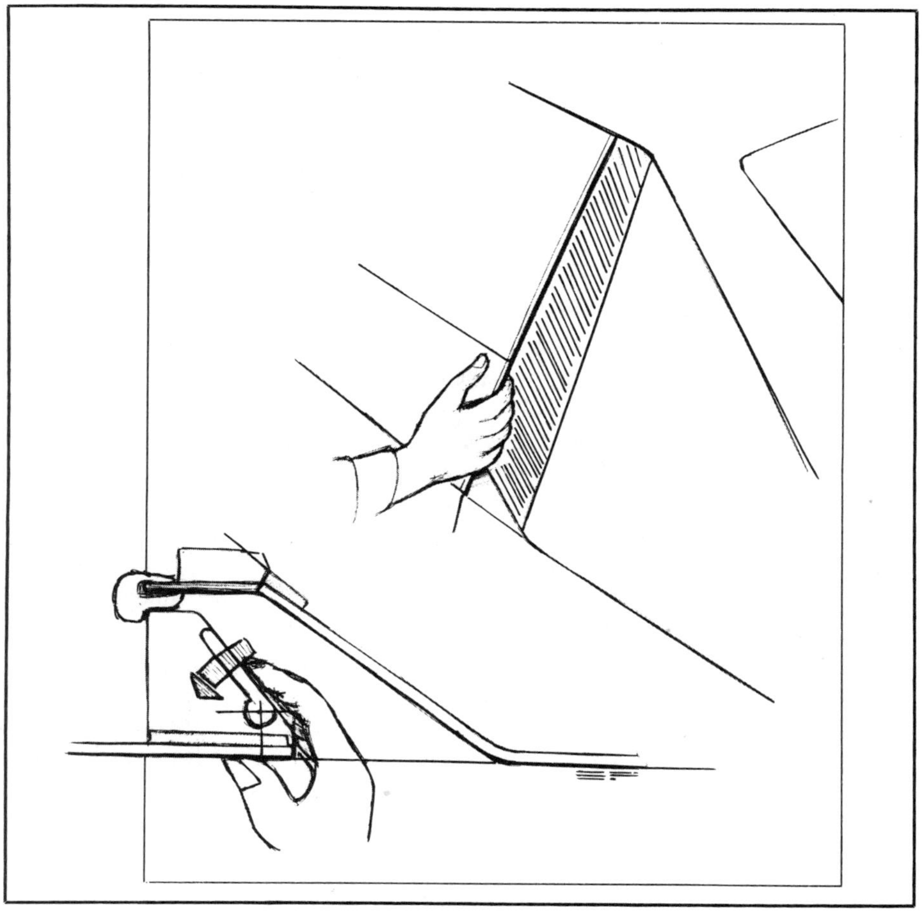

Fig. 20-5. A concealed door handle.

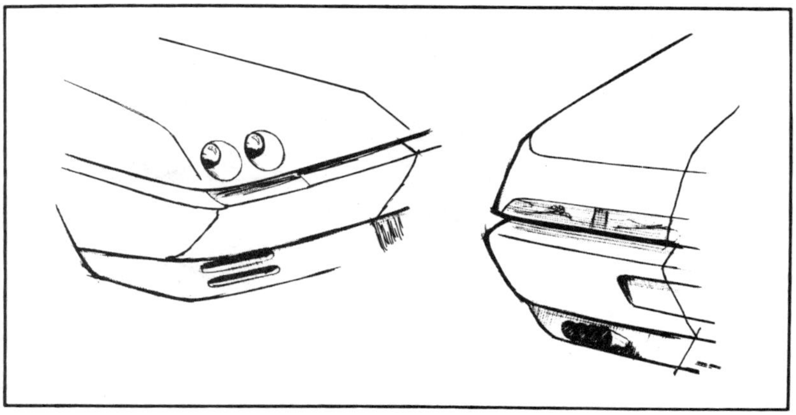

Fig. 20-6. The headlamp and taillight arrangement.

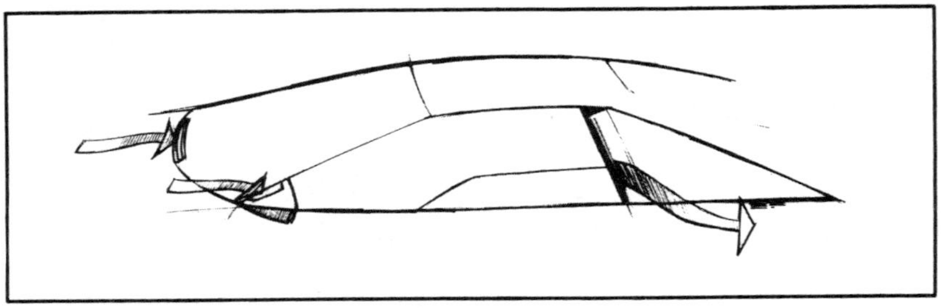

Fig. 20-7. Ventilation air intakes and evacuation.

USING AEROSPACE TECHNOLOGY

In the Quartz prototype, Kevlar has been used in front and rear bumpers, particularly as the ends of a sandwich with honeycomb panels in between. Overall characteristics are a very low weight and high compressive strength. In other words, a bumper capable of withstanding collisions of some magnitude without bending.

The honeycomb is a special cell configuration in aluminum or aromatic-polyamidic paper widely used in the aerospace industry. The main advantages of honeycomb construction are that it gives a very high strength/weight ratio; very high rigidity/weight ratio; high compressive strength; high fatigue strength; good resistance to wide heat fluctuations; good sound-absorbing properties; and a high exposed surface/occupied volume ratio.

Depending on the type of cell configuration employed, the material can be made to assume special curves and twists suitable for round covers, curved parts, and cylindrical surfaces. The center of the bumpers has been made by shaping honeycomb panels, whose weight is negligible, to the cross section of the bumpers.

Carbon fibers are relatively new in the automobile industry. Rolls-Royce has been developing them for use in turbojet aircraft engines—not for use in cars. On the newest Formula One Ferrari P-126, the frame makes use of carbon fibers. Their greatest advantage is low weight.

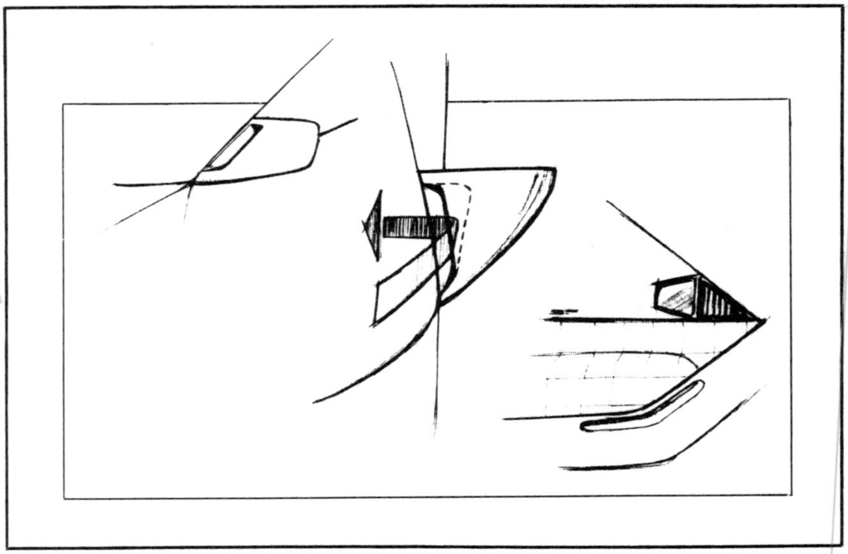

Fig. 20-8. Mirror arrangement on the Quartz.

Without any loss of mechanical strength, it is possible to achieve weight savings of the order of about 70 percent compared to steel and about 33 percent compared to aluminum.

These are their characteristics: low specific gravity; high rigidity; excellent corrosion strength; high fatigue strength; adequate heat stability; good electric conductivity; and low X-ray absorption. Taken together these qualities make for a vast range of possible car applications for mechanical parts such as drive shafts, engine bearings, springs, or parts of the bodywork such as internal reinforcements, bumpers, door panels, and hoods. As possible applications in carbon fiber fabric stratified with epoxy resin, the Quartz prototype suggests the rear window support frame, the seat forms, and the steering wheel.

LAMINATING FOR STRENGTH

Combined metal/plastic laminates are an innovation in the field of lightweight mechanical construction material. In a typical application, two fine layers of steel or aluminum are combined with a polymeric core that together form a structure whose properties are comparable with steel and aluminum as such. This product cuts weight by about 50 percent compared to an equivalent shell part of pressed steel. If we also remember their lowest purchasing and employment cost, metal-plastic laminates are thus also valid substitutes for sheet steel.

Such laminates have excellent thermal and acoustic insulation properties that can help make the car's interior quieter and more comfortable.

The forming of this type of material requires less powerful bending machines and presses for the same material thickness, but it must be remembered that the strength equivalent of standard plate can only be obtained with a total sandwich thickness greater than that of the reference plate. The quantity of steel by weight, however, is smaller.

A very important part is played by the plastic core and the laminate's performance depends on it. It has been observed that nylon cores appear to perform better in terms of flexing resistance, denting, recovery of flexibility, and resistance to collisions—particularly if combined with aluminum.

From the body assembly point of view, one disadvantage of metal-plastic laminates is that it is impossible to weld them together. For the moment, we have to use riveting, bonding or bolting. The latter solution has been in use for some time for front fenders. Bonded or riveted assemblies are usually adopted for doors and hoods.

Due to this disadvantage, Pininfarina chose to make the doors for the Quartz from a steel-polypropylene laminate. This highlights the material's drawability features with respect to the special side-panel section. Laminate thickness is only 1/25 of an inch; it consists of 60 percent plastic surrounded by two thin steel sheets.

PASSENGER COMPARTMENT DESIGN

The Quartz's interior has been designed in the spirit of the basic theme. Lightweight materials and linings are used in accordance with a design concept of maximum simplicity and practicality. A shelf with a section like that of a handrail runs along the sides of the doors and the rear sides to serve as arm-rest, grab handle, and controls housing.

The shelf continues along the front of the interior to become an instrument panel. This panel is unusual because there are no pro-tuberances. Instrumentation is flush with the surface and covered by a special screen, transparent beneath certain angles and dark beneath others, thus offering top instrument readability from the driving position and eliminating annoying instrument reflections on the windshield.

The flap of the draw houses the computer. With the flap open, the computer is displayed like a printer keyboard. Interior facilities are rounded off by secondary instrumentation that includes two visual indications of the differential lock located on the central console and four visual indications of low tire pressure.

The Quartz retains the Audi Quattro fuel tank, a T-shaped container located behind and between the rear seats. Its shape permits a trunk of more practical proportions and size (about 10 cubic feet). The rear seat backs can be folded together or separately. When both are folded, there is about 25 cubic of luggage space.

Chapter 21
Audi's Auto-2000 Project Car

CAN YOU IMAGINE THE SEATING BUCK FROM the styling studio—complete with carpeting, wiring connections, and all trim details—being lifted into a body shell from below as the normal way of assembling the car on the line? The engineers of Audi can; they propose to do almost exactly that.

The project car for the year 2000 features a plastic floor (Fig. 21-1) inserted into a sheet-steel body structure. That's a major innovation in itself. But to use the floor as a basis for a subassembly, including front and rear seats, is going a step beyond what Fiat planned for the VSS. This would do away with the present labor-intensive assembly operations used to fit the interior trim through the door openings. There would also be more access for other assembly work before the floor and roof panels are fitted to the vehicle.

The plastic structure was developed in cooperation with airframe manufacturers Messerschmitt-Bolkow-Blohm. It is a laminated sandwich with foam inserts. It is stiffer than a steel floor and 15.5 pounds lighter.

In keeping with M-B-B's aircraft-manufacturing techniques, it has outer layers of fiber-reinforced plastic sandwiching a polyurethane hard foam core for bracing. This construction combines weight with a high resistance to deformation. The plastic floor pan also incorporates the sill members. This finally puts an end to corrosion in this particularly exposed area of the auto body.

SPACIOUS FAMILY CAR

Audi displayed the Auto-2000 at the Frankfurt show in September, 1981. It is not purely Audi's own equipment. Its design and construction was partly paid for by West Germany's federal government. The West German Ministry for Research and Technology (BMFT) initiated a research program, in 1978, to promote the development of vehicles capable of being mass produced while at the same time possessing distinct advantages over present-

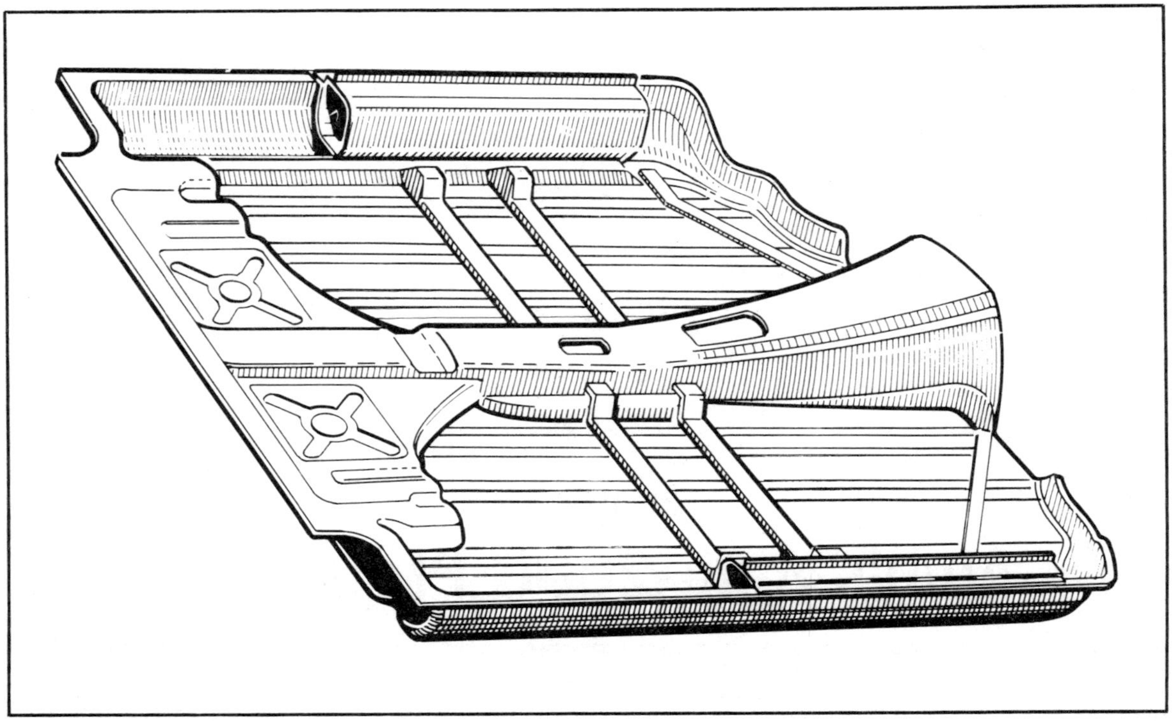

Fig. 21-1. Plastic floor insert.

day cars particularly in the areas of fuel economy, conservation of energy and raw materials, safety, and pollution control.

Along with the parent organization, Volkswagen, Audi participated in the BMFT program. Volkswagen chose the subcompact size, which corresponds to the bulk of its sales, while Audi opted for a much larger car. Why? Because the family sized car is Audi's main business, and the company expects that will continue to be the case.

The Auto-2000 is designed for the one-car family that remains the backbone of the West German car market. Such a car must be large enough to offer comfort for the whole family, even on a long trip, according to Heiko Barske, assistant chief engineer of Audi. Consequently, it has seating for five persons and a roomy trunk, and it is corresponding closely to the current Audi 100 in carrying capacity.

The engineers who developed the project car deliberately avoided the easier technical approach of sacrificing interior space, comfort, or performance to gain improvements in economy or pollution characteristics.

Built on a 107.3-inch wheelbase, the project car is 187.6 inches long, overall, with a width of 70.4 inches and a height of 55.4 inches, unladen. Its curb weight is 3682 pounds. That's heavy, but as Heiko Barske points out, "it gives 30-percent more miles to the gallon than any production car of that weight." Audi's real ambition for this project is to build a car for the top end of the midrange market that will meet the future demands of the environment by virtue of economy, low pollution levels, and sensible use of available materials.

Like the French engineers responsible for the Eve and Vera products, Audi's experts tackled the dual questions of material substitution and aerodynamics with equal emphasis.

SUBTLE BUT EFFECTIVE STREAMLINING

The body shape, due to Hartmut Warkuss, borrows something from the sleek NSU Ro-80 of 1967, but its drag coefficient is even lower—at 0.288. This low drag coefficient was achieved by attention to innumerable details. Smooth-surfaced plastic bumper elements are used front and rear. The car has a carefully calculated wedge shape with a low, gradually rising nose, a large, very fast windshield angle, and a relatively high tail.

The windshield and rear window are adhesive bonded and flush with the surface of the body. This design also contributes directly to the stiffness of the body structure as well as improves the surface air flow. The side windows are also flush so that there is nowhere for the air stream to separate and create turbulence. For the same reason, most of the underside of the car is a continuous smooth surface.

COST QUESTIONS UNANSWERED

The materials used for the body were chosen for optimum function with minimum weight. The body

Fig. 21-2. The Audi research car.

framework is made from sheet steel pressings because this material possesses the strength necessary for the rigidity of the passenger cell. Steel can also absorb large amounts of energy; that is one of the basic requirements for occupant protection in an accident.

Outside the passenger cell, the side members of the body structure are deformable according to calculated values. The front and rear fenders bolt on and are made of light alloy. Use of aluminum instead of steel for the door panels brought a disappointingly small weight saving. Doors still need steel guard beams for occupant protection against side collisions.

The roof is a light sandwich panel, but with a high resistance to longitudinal compression forces, and is mounted on the body structure from above complete with the pre-formed headliner molding. See Fig. 21-3.

The trunk lid is made of Kevlar fiber-reinforced epoxy as used in space craft and Formula One racing cars. It is far too expensive for mass-

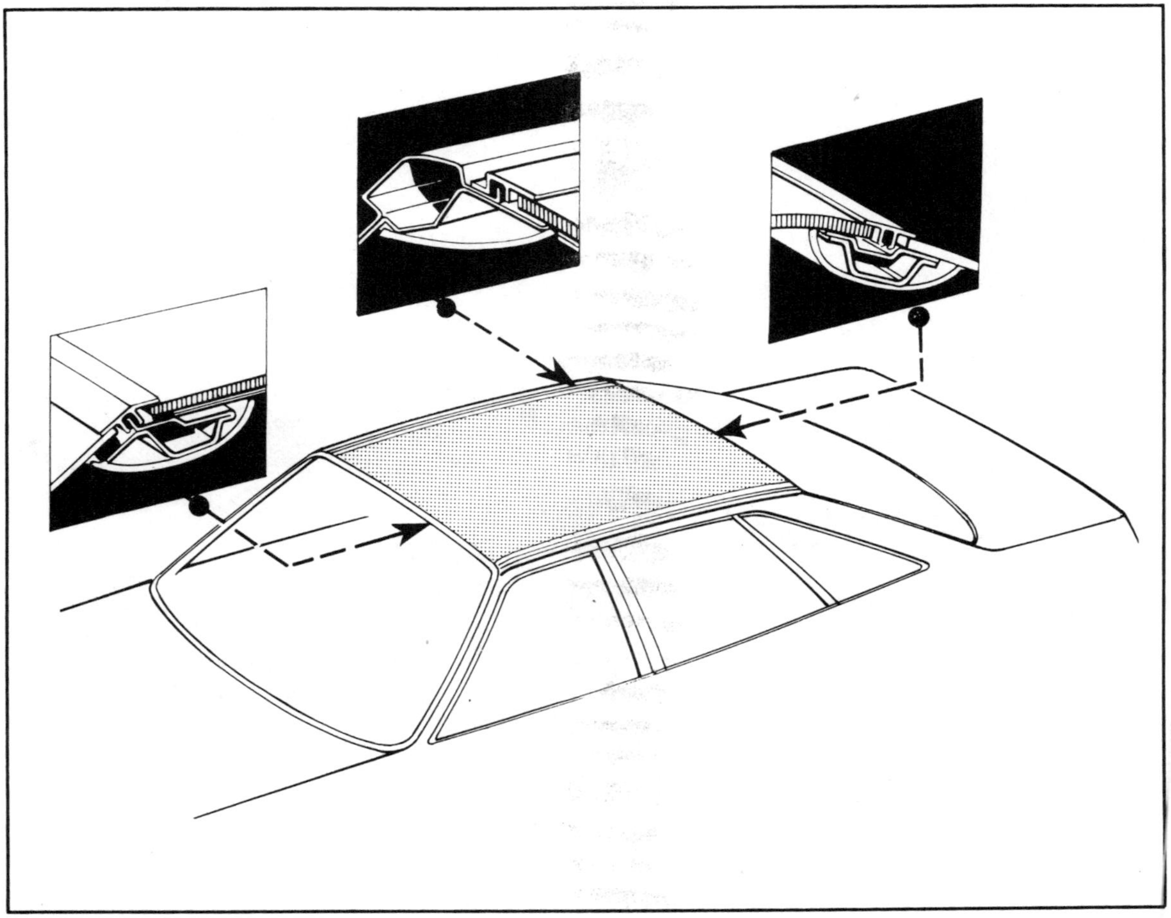

Fig. 21-3. Roof structure and mounting.

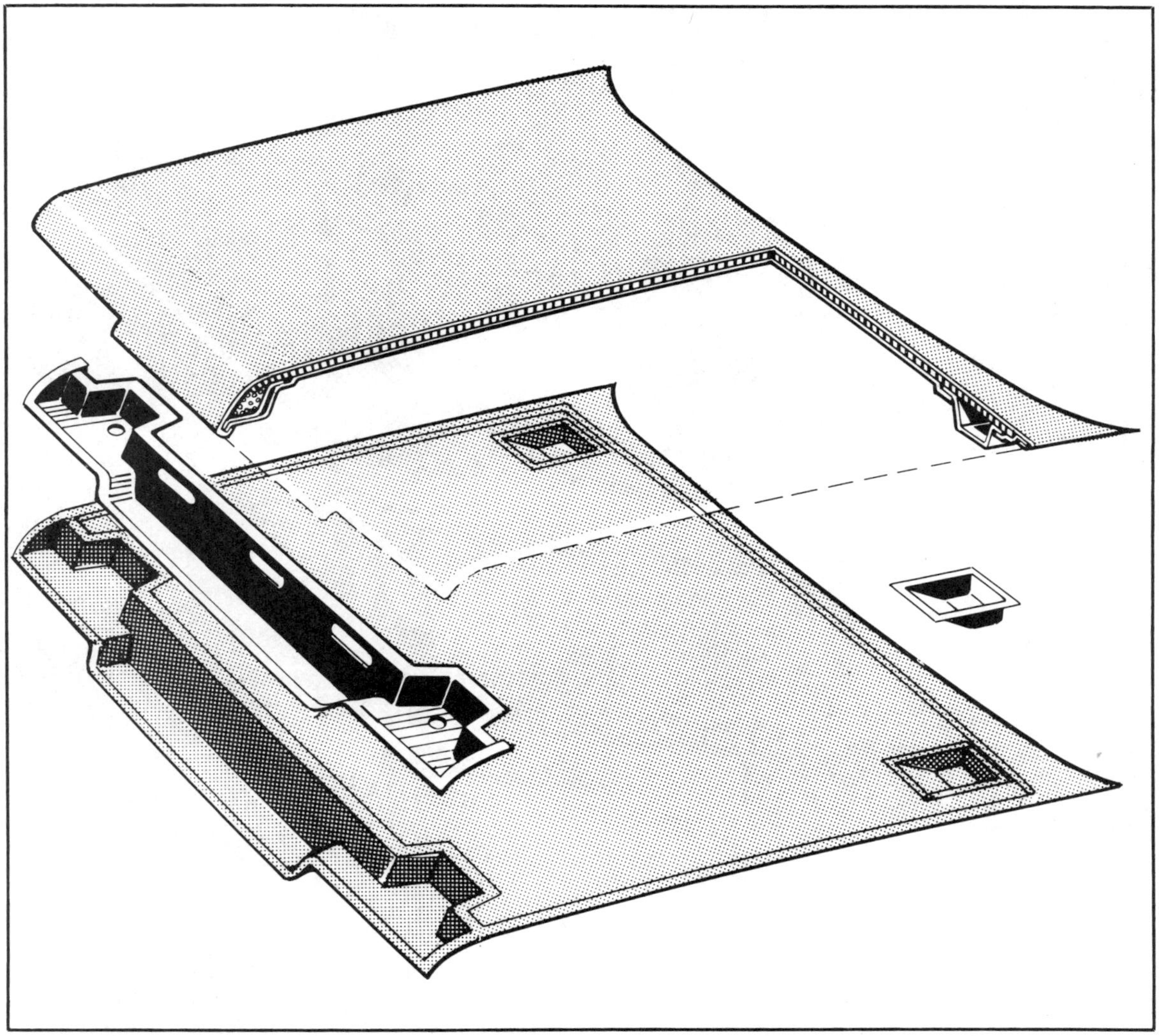

Fig. 21-4. Laminated hood construction.

produced cars. Audi engineers say they plan a less expensive solution with more common plastics.

The hood (Fig. 21-4) is an aluminum sandwich construction with the weight cut down to 15.5 pounds from the 46.5 pounds of a sheet-steel hood panel. The front hood is part of a noise-insulating capsule (Fig. 21-5) surrounding the engine, with a hexagonal paper insert between the two thin sheets of aluminum.

The sides and bottom of the engine capsule are made of foam rubber and have a thin plastic coating for easy cleaning. It is also designed to deform in those areas that involve the greatest risk of injury to a pedestrian in a collision.

Fig. 21-5. An encapsulated engine for noise control.

It has been possible to reduce the noise emitted by the Audi project car very nearly down to the level of the rolling noise from the tires.

For this large and heavy car, Audi uses a small 97.6-cubic-inch, four-cylinder engine that is turbocharged for a peak power output of 110 hp. Fuel economy at steady speeds is 47 mpg at 56 mph and 34 mpg at 74.5 mph. In the European urban-driving cycle, Audi's Auto-2000 has a fuel rating of 27.6 mpg.

Chapter 22

Research Car from the Academeme

FROM OCTOBER, 1979 TO SEPTEMBER, 1981, about 35 student engineers and instructors at four technical universities in West Germany worked full time to create a research car. They were sponsored in part by a grant from the BMFT. The project was coordinated by Professor Bert Breuer in Darmstadt.

The result is a roomy family car (Fig. 22-1) with sedan/wagon profile, diesel power, and front-wheel drive. The project came into being under the same BMFT program in which Audi, Mercedes-Benz, and Volkswagen participated. But in September, 1981 it was announced that only the Uni-Car (University-Car) qualified for continued BMFT funds. Support for the other projects was terminated. This left the car companies to pursue their research and development work independently.

Whether the Uni-Car is better than what the motor industry can come up with or not, it is worth a closer look. A lot of thought has gone into it, and some of the solutions are daring yet realistic. The universities placed greater emphasis on passive safety than the car companies. Throughout the car, its designers have aimed for new solutions in simplified maintenance, easier repairs, and longer life. Suitability for mass production also has been a fundamental requirement.

Foremost among the design objectives was a 25-mpg average fuel economy for a four seater with 900 pounds payload capacity and a range of 250 miles. Moreover, it should have a top speed of at least 90 mph and be able to accelerate from 0 to 60 mph in no more than 13 seconds. Other goals concerned noise, exhaust emissions, durability, recyclability, and energy use during manufacture. It is not a production prototype, but a carrier of ideas that can help put new solutions into the service of the industry.

The Uni-Car was designed for a 108.3 inch

Fig. 22-1. The Uni-Car in its garage at the Darmstadt Technical University.

wheelbase with a track of 57 inches in front and 54.3 inches in the rear. Its overall dimensions are 181.1 inches in length, 69 inches in width, and 54 inches in height—running on 175 HR-15 tires.

The weight target was 2756 pounds maximum (anything lighter would have been preferable). But during the construction period, it was found that the target could not be met, and the prototype ended up with a dry weight of 3032 pounds.

The team from the Stuttgart Technical University, led by Professor Ulf Essers, was assigned the body design, aerodynamics, ventilation, and noise control. The wind tunnel work and consequent modifications were carried out by Jürgen Potthoff.

Professor Hermann Appel and Professor Hans-Peter Willumeit of the Technical University of Berlin led the team that concentrated on body structure, ergometrics, occupant protection, interior layout, and nonaggressivity to pedestrians. Thus there was close cooperation between the two teams from beginning to end. Karosseriewerk Karmann was brought in as a subcontractor for the actual building of the body. The main structure is made of steel. The roof is a sheet-steel panel, but the doors and tailgate are made of aluminum to save weight. The hood is mainly plastic, and attached to an aluminum frame.

Once the basic interior space requirements had been formulated, body design was directed by aerodynamic principles. Four different shapes were tested on 1/5 scale models in the wind tunnel, and the Kamm-back winner (Fig. 22-2) was chosen for further development. The final model proved to have a Cx as low as 0.226. That led the team to the conclusion that a drag coefficient not exceeding 0.25 could be achieved in a full-size version of the same design. In fact, the basic shape was so favorable, and Potthoff's fine-tuning so successful, that the actual car has a drag coefficient as low as 0.24. "And further improvement is still possible," says Potthoff.

Fig. 22-2. A rear view of the Uni-car shows air extractors in the front fender and dual wipers on the rear window.

Fig. 22-3. Instrument panel and controls.

Among its notable design features is the smoothly rounded nose where bumper, lamps, and plastic hood blend into an unbroken surface. The body has a pronounced taper from the A-post to the rear end panel. The small, silenced air intake integrated with the bumper, air extractors in the forward edge of the front doors, faired-in outside mirrors, flush-fitting wheel covers and full-panel rear fenders, play small but vital roles.

Various refinements were added for their potential role in adding to active safety. For instance, the big housings for the outside mirrors also contain wiper mechanisms for the wiper arms that sweep the front door windows. The mirrors are not unusually big. The size of the housings is due to the requirement for allowing the mirrors to yield in impacts with pedestrians. At the same time, the springs and hinges must be sufficiently strong to prevent accidental damage, and the linkage could not be miniaturized.

The steering wheel is taken from a BMW. And VDO supplied the instrumentation using liquid-crystal displays (Fig. 22-3).

In addition to the energy-absorbing front end (Fig. 22-4) and the side-impact protection (Fig. 22-5) built into the doors, the Berlin team made some interesting decisions. They decided to fix the front seat in place and back it up with a steel bulkhead running the full width of the car (from B-post to B-post). That will guarantee adequate survival space even in a frontal collision at any speed where seat belts will save the occupants.

Fixing the seat made it necessary to provide a wide range of adjustment for the pedals as well as the steering wheel. The seat belts do not have their locks in the middle of the car—as is common practice because the B-post makes the most convenient anchorage point for the upper-torso end of the belt—but next to the doors so as to facilitate unlocking by outside persons coming to the rescue after an accident. The upper-torso anchorages take the form of a bracket on the roof-height crossbar forming part of the central bulkhead.

The team from the Technical University of Aachen, led by Professor Jürgen Helling, was responsible for most of the Uni-Car chassis (the brakes were done in Stuttgart). The chassis has all-independent suspension with automatic level control at the rear end, four-wheel disc brakes with ABS antilock, and recirculating-ball, low-energy-loss power steering. The level control is not just a comfort feature, but plays a part in the aero-

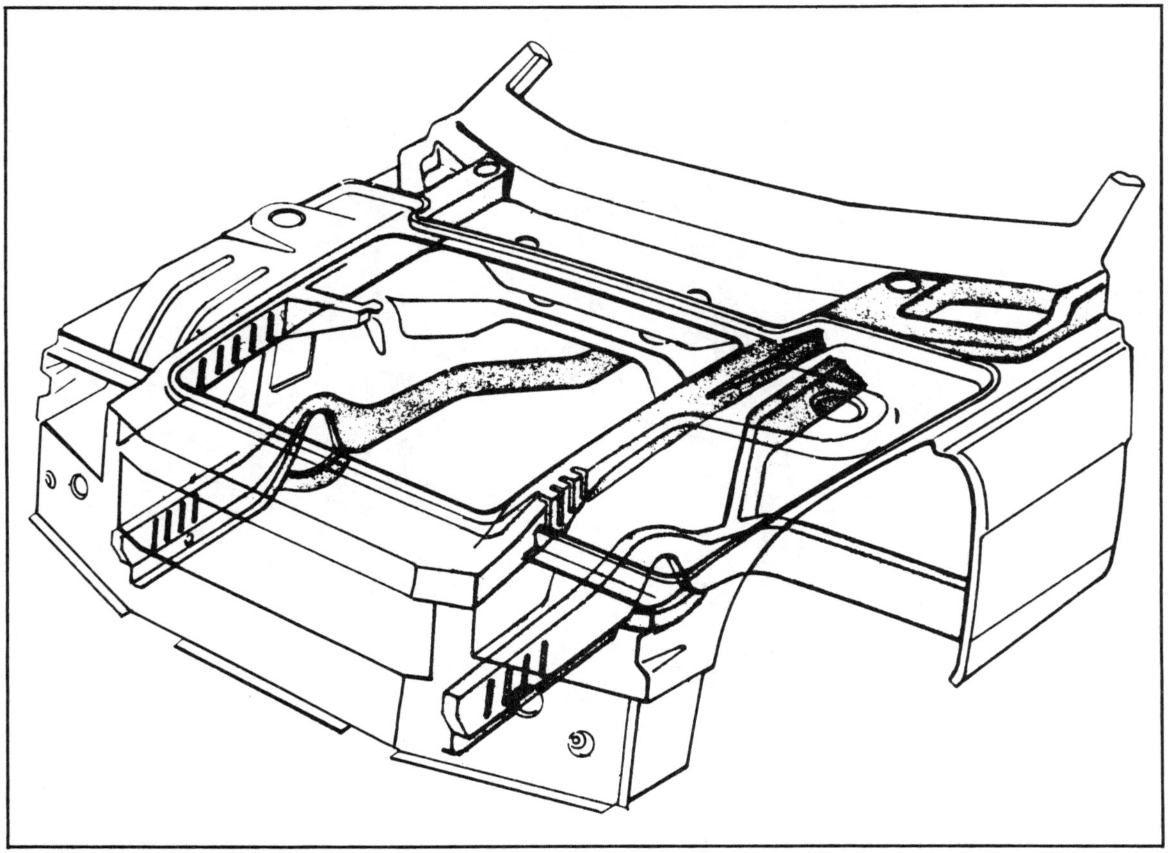

Fig. 22-4. Front end body structure.

dynamics because the body attitude can affect the air flow.

Responsibility for the engine rested with the Stuttgart team. They subcontracted with the big truck and engine makers of Munich, M.A.N. to build a 2.5-liter, four-cylinder diesel engine. It is turbocharged and delivers 98 hp. Fully encapsulated in its compartment, its noise pollution is insignificant. Because of the engine dimensions and the lack of width between the front wheel housings, it became impossible to mount it transversely as had been the original intention. Instead it was mounted longitudinally, and tilted at 45 degrees to the right to avoid putting bulges in the hood (Fig. 22-6).

The four universities agreed that with a front-mounted engine, it is logical to drive the front wheels even in a vehicle of this type and size. The

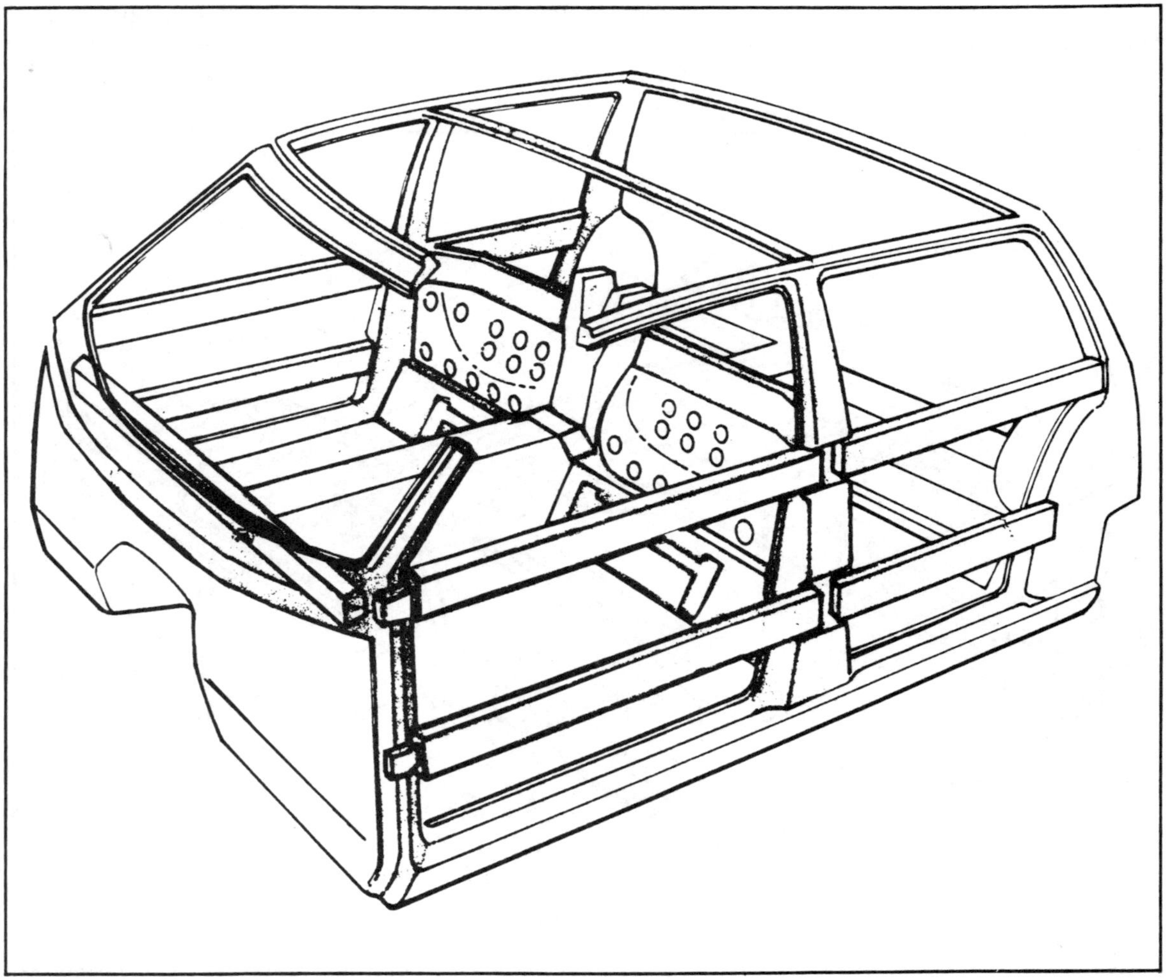

Fig. 22-5. Side and rollover protection for the occupants.

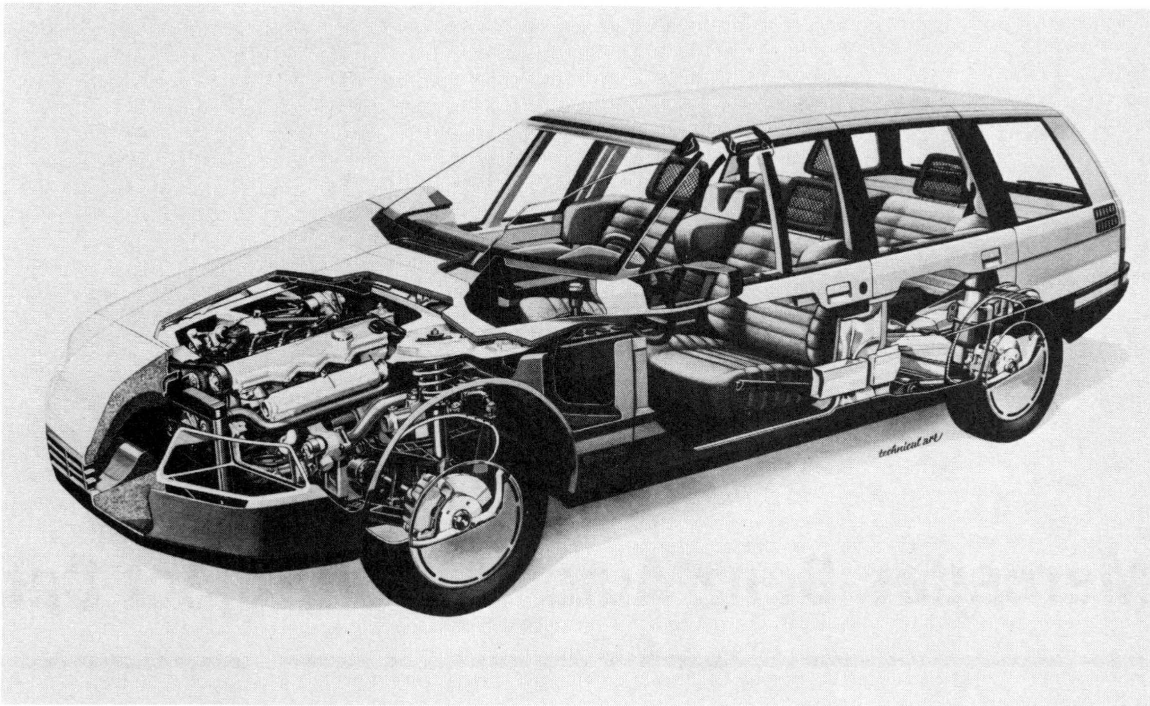

Fig. 22-6. A cutaway view of the Uni-Car.

transmission was Aachen's task, and the choice fell on an experimental stepless drive under development by Van Doorne (Transmatic).

Tests with the four prototypes assembled at the Darmstadt Technical University began early in 1982. The Uni-Car vastly surpassed its top-speed target—reaching 120-125 mph without difficulty. The acceleration target also was comfortably met. But the fuel economy? Considering the weight and payload capacity of the Uni-Car, it is outstanding. At a constant 56 mph it gives 51 mpg. Cruising at 74.5 mph it gives 39 mpg. And in the urban driving cycle it is a real winner at 30 mpg.

Professor Bert Breuer at Darmstadt is pleased with the results, so far, and foresees two important consequences of this work. First, he looks forward to a greater readiness in the automobile industry to accept new technology from academic sources. Second, he expects the Uni-Car to generate more interest in auto engineering at the high-school level and bring in higher numbers of engineering students to the universities.

Chapter 23

Mercedes-Benz Auto-2000 Project Car

As was the case with the experimental safety cars built by Mercedes-Benz, the company decided to rely on concepts and components from its current production models to a maximum for the research vehicle the firm proposed to present for the BMFT program.

The result is a car that shares the present S-class chassis in respect of wheelbase and chassis components, but incorporates a number of experimental features in a body that represents an extrapolation of the design trends at Mercedes-Benz.

In keeping with the company's product philosophy, it is large. But it is not a gas-guzzler (as further examination will demonstrate). If it looks extravagant at first glance, it is nevertheless an example of rational and purposeful design for its functions. According to Werner Breitschwerdt, technical director of Mercedes-Benz; "The projected touring sedan is the logical answer to prognoses which foresee greater mileages being covered in business travel and by industrial transport, and heavier holiday traffic which will result from people having more leisure time.

"The Mercedes-Benz research car, designed to accommodate a number of occupants which is expected to be twice as large as the present-day average, must be able to offer plenty of room and a very comfortable ride, adequate luggage space, and sufficient engine power despite lower energy consumption, and at the same time greater economy, greater environmental protection and safety: an assignment which presented a real challenge to our engineers.

"The best answer to the future problems of energy and the environment is not to be found merely in the reduction of the amount of room, the weight of the car or the size of the engine. In order to obtain worthwhile results, energy-saving measure, for instance, must be directed towards the typical uses to which the consumer puts the car."

COMPROMISE ON DESIGN OBJECTIVES

Participation in the BMFT program taught the company something new in the administration of research work. Before, Mercedes-Benz conducted research and development programs only within narrowly defined areas. Consideration of the impact of a single new development on the car as a whole followed at a later date, and only in the context of total-vehicle reviews and planning. Because of the terms of this contract, Mercedes-Benz was led to abandon its former practice and take a new approach. There was full coordination right from the start among a variety of basically different aspects of advanced engineering and design, and combining them into a single overall concept car.

The main design objectives were to create a roomy four-door sedan with improved fuel economy. That's a difficult compromise in itself. It should also contain advances in the areas of both active and passive safety and be suited for compliance with even stricter exhaust emission standards. Noise levels should be brought down below existing standards. In addition, it should incorporate new ideas in instrumentation and informative displays having a potential for reducing the stress on the driver.

Built on a 115.6-inch wheelbase, the Auto-2000 weighs 3530 pounds. It is 201 inches long overall, and has a width of 71.7 inches and a height of 52.2 inches (Fig. 23-1).

THREE ALTERNATIVE POWER SYSTEMS

The engine is mounted in front, under a majestic hood, with an automatic transmission and drive to the rear wheels. Three different types of power systems are being developed. Why? It is not a new concept; it is just an extension of common practice. Just as some cars are offered with a small six, a big six, or a V-8 engine in the same basic vehicle, according to the owner's choice of performance

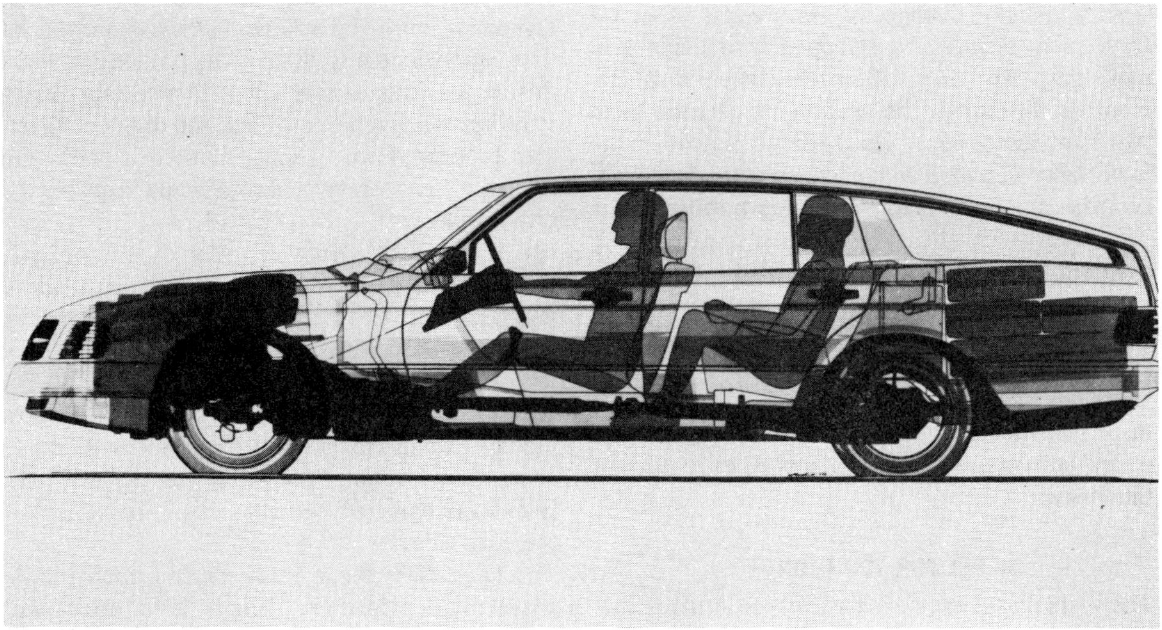

Fig. 23-1. The Mercedes-Benz Auto 2000 layout.

level, Mercedes-Benz merely goes further in its specialization of power systems according to the intended use of the car.

The first power system is a gas turbine; the second a V-six diesel engine; and the third is a light-alloy gasoline-fueled V-8.

For the gas-turbine system, Mercedes-Benz chose an open-cycle, twin-shaft, regenerative type with a centrifugal compressor and two axial-flow turbines. It runs on diesel fuel, uses a can-type combustor, and a single, horizontal ceramic disc type of revolving regenerator.

The gas turbine power package occupies more space than a V-8, mainly due to the voluminous ducting, but weighs less. Can it be economical in operation? The big weakness of gas turbines is that they are most efficient when working under full power. Unfortunately, the average car runs on full power only for brief periods. The picture changes somewhat when you remember that Mercedes-Benz does not make average cars.

Statistics of distances covered and the speed of travel shows that cars equipped with engines of more than two liters' (121 cubic inches) displacement do three-quarters of their mileage on main roads and motorways. This tends to play down the importance of part-load fuel consumption and must count as a point in favor of the gas turbine.

Progress made in ceramics research enable the hitherto maximum working temperature to be raised so that, compared with present-day vehicle turbines, it has been possible to reduce fuel consumption by at least 20 percent. Over the next 10 years the Mercedes-Benz researchers foresee the materialization of a gas turbine with an output of 150 hp and an average fuel economy of 23 to 28 mpg for this heavy car.

DIESEL FOR TOP ECONOMY

The V-six diesel engine is basically a cutoff production V-8 block. It is very short and could permit some reduction in hood length for the car. Otherwise the power package is similar to the compact 380 SE V-8 engine.

The spark-ignition engine planned for the Auto-2000 is a production-model V-8 taken from the 380 SE, with economy-measure modifications. It gives an average fuel consumption of 24 to 28 mpg. The turbocharged diesel, with similar power and performance, falls in the 28 to 33 mpg range.

This project also includes a four-speed, automatic transmission with electronic control. For each of the three power systems, the driver can select three different programs (shift patterns) suited for three sets of traffic conditions. Transmissions and engine functions are linked electronically via the same control unit.

EMPHASIS ON PASSIVE SAFETY

The passenger compartment is basically the same as in the S-class sedan. The front is taken from the SEC coupe. The Kamm-back rear section was designed for improved airflow, but it has drawbacks. The rear window is difficult to produce, and the design does impair rear vision to a certain extent. Priority was given to air flow; the drag coefficient was brought down to 0.26. But for a production model, vision (safety) and cost would probably dictate a redesign.

In the body structure, a lot of material substitution was done to save weight. Front fenders, nose section, and bumper are made of energy-absorbing polyurethane plastic that is reinforced with fiberglass with foam backing (Fig. 23-2).

The hood works as a stressed structural member when closed, and participates in impact absorption. It is hinged on the side, a la Buick 1941. It does not open to alternate sides; it opens only to the curb side.

For safety, there is more cross ribbing in the passenger compartment. The doors interlock with the side members and the B-post base. In addition

Fig. 23-2. A Mercedes-Benz research car at the Frankfurt auto show in 1981.

to the impact bars in the doors, there are large impact zones for the hips of all occupants. The side windows are made of impact-resistant acrylic glass.

Material substitution in the chassis has two innovative forms of expression. The propeller shaft is a one-piece tube made of carbon-fiber-reinforced plastic that replaces the two-piece steel unit (with a center bearing). Secondly, the front stabilizer bar is not solid steel; being tubular, it saves one-third of the weight.

The suspension uses the optional hydropneumatic spring units from the S-class cars. The difference is that light-alloy parts have been substituted for ferrous-metal ones wherever possible.

Heat-pipe techniques are utilized for heating and air conditioning. Borrowing techniques used for cooling electronic equipment, the Mercedes-Benz specialists found methods and devised hardware to transfer heat from the engine coolant to air wanted for interior heating.

The same heat exchanger is capable of transferring unwanted heat from the passenger compartment to the refrigerant in the air-conditioning unit. Heat pipes like this are flat heat exchangers that could fit in many places (even inside the doors).

On the subject of noise abatement, Dr.

Breitschwerdt explained that, "To cater for the possibility of further noise reduction in vehicles with piston engines, we have encapsulated parts of the research car. We are investigating the problem of heat dissipation from the encapsulated space and the engine cooling circuit, bearing in mind that the maintenance needs to be kept as easy and convenient as possible and we are trying to find the most suitable sound-absorbing materials."

The gas-turbine has different noise problems, mostly of an aerodynamic nature, due to the vast amounts of air flowing through the system.

Extensive use is made of computers, linked up to form an information system. Control monitor, distance/consumption reckoner, route planner, radar vehicle spacing system, and communication system give the driver all the information he requires for operation, maintenance and traveling. A map electronically recorded in the route planner gives the driver the shortest possible route between starting point and destination.

A new item is the central information computer that controls the function and the data flow of the individual computers. To distract the driver's attention as little as possible from what is happening on the road, the computers and the communication system are controlled and adjusted by no more than four push buttons on the steering wheel. Only the information necessary for operation and safety and the information specifically requested is displayed to the driver.

Outsiders can only guess with regard to the importance of the Auto-2000 project car for future production designs from Mercedes-Benz. What seems certain is that it will begin to have some effect even in the short term. The reason is that much of the technology, if deemed cost-effective, can be applied immediately. For example, you don't have to wait for ceramic gas turbine parts before you put a hole in the front stabilizer bar. Material substitutions should occur, like pearls on a string, as the soundness of each application is proved out.

Chapter 24

The Volkswagen Auto-2000 Project Car

RESEARCH PROTOTYPES, IN MY OPINION, ARE more interesting the less futuristic and more realistic and closer to production they are. "A basic change in the car requires accurate aim and precise timing," says Ulrich Seiffert, Volkswagen's director of research, "it takes maybe 15 years to go from idea to production. The definition phase for a project can take up to five years. New jobs are written into a 10-year time frame, with specific milestones that must be reached by certain dates in the interim. Programs are eliminated if they fail to reach the milestones."

Dr. Seiffert's philosophy of research is result-oriented. "All our projects have goals. We never start anything with the idea of 'let's just do research.' We must see an end in our research, whether it's 20 years away or 50. We tend to set high goals in the definition phase, and then set the time table. Every new idea, every new part, is aimed for use in a production car."

In associating itself with the BMFT Auto-2000 program, Volkswagen decided on a concept that offers the optimum compromise between factors that are frequently seen in conflict with each other.

☐ Development of a scenario reflecting the probable economic development, probable government regulations, and the probable energy situation in the 1990s. Conclusions are drawn from this scenario in regard to its effects on the automobile.

☐ Engineering concepts are investigated during project implementation with a view to proving or disproving their practicality.

☐ Solutions are evaluated, adapted, and translated into a composite, well-balanced motor vehicle design concept.

The prognoses and projections outlined here led to the following conclusions:

☐ The disproportional rise in fuel prices should be countered with vehicle engineering measures aimed at lower fuel consumption (weight reduction, reduction of aerodynamic drag, increased efficiency of powerplant and transmission).

☐ The high share of labor in maintenance and repair services should be lowered by engineering measures, sectional repairs, and longer service intervals.

☐ In addition to engineering measures resulting in lower fuel consumption, alternative fuels, such as methanol and ethanol, should be taken into account in order to provide for some relief of the crucial petroleum supply situation.

☐ Compliance with all standards worldwide should be ensured with a base model that would require the least possible expense for modifications.

☐ The design of the base model should permit shortrange response to any fluctuations in the specification profile (economy and comfort design).

Many difficult choices present themselves whenever multiple objectives have to be combined into one product. First, it becomes a question of priorities. Are some objectives more important than others or do they all carry equal weight? What happens when ideas for energy-saving measures conflict with comfort and safety requirements?

Fig. 24-1. The Volkswagen Auto 2000.

Fig. 24-2. The Volkswagen Auto 2000.

Comfort and safety are elements of first priority; but only to a degree. How much comfort? How much safety? A certain minimum, clearly, is indispensable, and cannot be compromised.

Out of such reasoning, the Volkswagen engineers developed a scale for value factors they could attach to the various design objectives. The value factors were set in accordance with processes known from cost/effectiveness analyses.

With regard to comfort, the acceptable minimum is determined by market research. It depends on what level of comfort the car buyer can get at what price in the markets aimed at. The same goes for ease of driving. Environmental compatibility, on the other hand, is determined by the trend of legislation.

In respect of safety, the aim is not to reach the ultimate in occupant protection (or pedestrian protection), but at a more close-to-home concept that Dr. Seiffert prosaically calls "no loss of safety." In practice, that means no backward steps in safety can be tolerated, and progress in safety can be made when possible conflicts with fuel-consumption targets or cost considerations are resolved.

The VW 2000 (Figs. 24-1 and 24-2) is a two-door, four-seater sedan built on the VW Golf wheelbase (with closely similar overall dimensions). It has a low-drag, lightweight body, and a number of different power trains are being tried out.

"You see how it all goes together," said Dr. Peter Walzer, one member of VW's research department. "Your weight savings are additive to the energy savings of the low-drag form. Because the car is lighter and disturbs the air less, its power requirement is lower, and performance levels can be maintained with a smaller, lighter, and more economical engine."

During the vehicle conception phase of this project, particular thought was given to new production technologies and assembly processes. Special consideration was given to the following parameters which significantly affect the assembly concept:

—Job environment.
—Easy and simple repairs.
—Bonding technologies.
—Recyclability.
—Production figures.
—Structural concept.
—Corrosion protection.
—Compatibility of materials.

These parameters are not separate; they are significantly interdependent. A careful analysis was to be performed, therefore, so that a ranking order could be established as a basis for the production concept.

Vehicle body investigations were conducted for the replacement of sheet steel components by other materials. A departure from the customary (completely welded) integral sheet-steel body design and a kind of hybrid structure was considered where sheet steel is used for the supporting elements only and alternative materials replace the paneling and some other essential components such as the roof.

The advantages of this concept would be not only reduced vehicle weight by the use of substitute materials, but also comparatively simple and less complex body-in-white. This again would simplify production and permit more effective corrosion protection.

Investigations were performed on the following versions; vehicles made mainly from steel (including new, high-strength); chemical materials such as fiberglass-reinforced plastics; carbon filament reinforced plastics; thermosetting plastics; and aluminum.

The same basic considerations apply to the production of the chassis. Weight reduction by material substitution appears useful primarily in the area of heavy castings and major chassis component elements. Some of the chassis components made of alternative materials were projected and designed. These components are intended to be built so that the potential offered by use of lightweight materials can be determined if the components are designed with particular consideration of the material properties.

The chassis also could be assembled by teams that would make the larger subassemblies (which then could be easily mounted to the body). This approach could substantially reduce the amount of overhead assembly work.

Component assembly could be performed in parallel to the line assembly. This would apply to doors, roof, front and rear hood, front portion, and rear portion. Consideration was also given to the problems of corrosion protection and ease of repair in addition to the humanization of jobs and the creation of assembly teams. Particular consideration was given to easy assembly and disassembly of vehicle front, rear, and sides in view of the list of standard collision cases (Fig. 24-3).

The assembly of the interior equipment could be performed similar to that of the body-in-white. The individual components such as seats, dashboard, and electrical system would be largely preassembled. This would leave only a few steps for final assembly.

Even though the advantages of this concept appear obvious, a number of problems remain to be

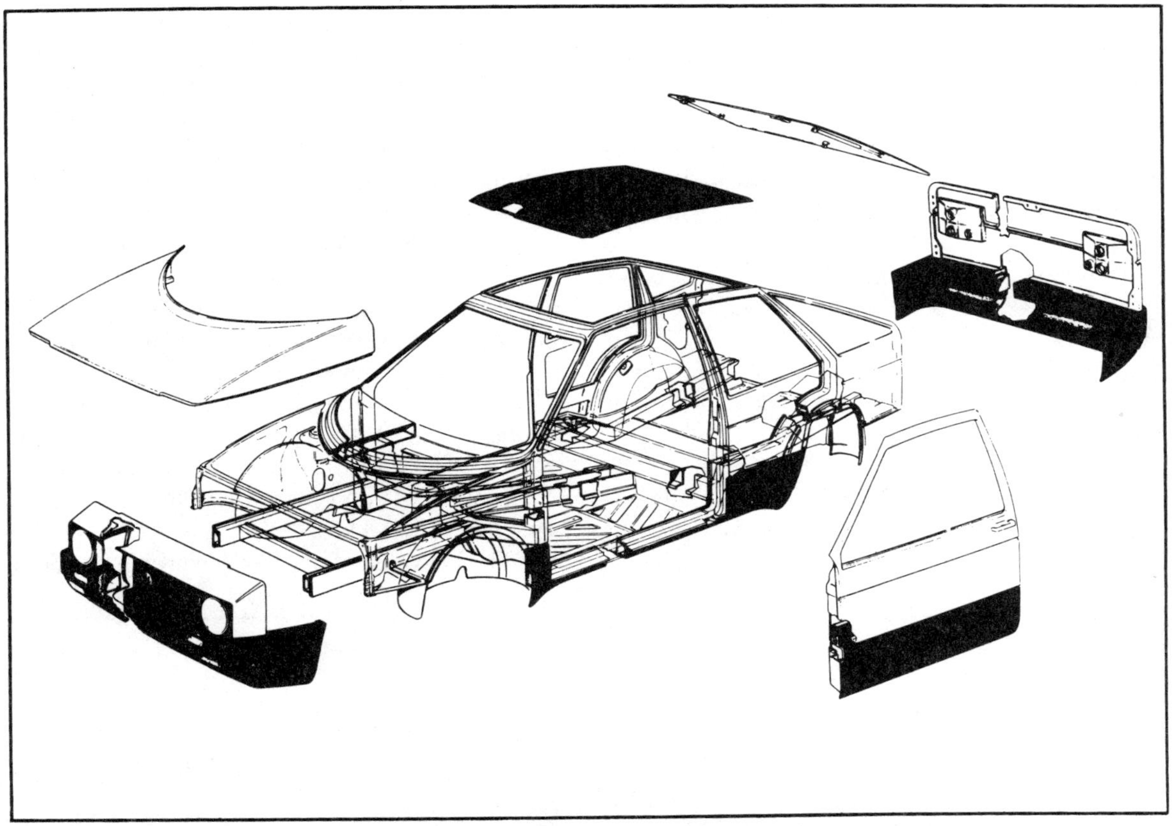

Fig. 24-3. The body structure and subassemblies.

solved. A way must be found to properly bond different materials in the production process where consideration is given to the specific properties of each material (modulus of elasticity, coefficient of expansion). This vehicle concept will lead to a loss in body stiffness because of the smaller number of welded sheet metal components. Nevertheless, there must be sufficient flexural and torsional stiffness. This means that the loss must be made up for by the supporting structure. Extensive calculations that would employ the Finite Element Method are called for so that the necessary vehicle body strength is reached while the consumption of materials is kept low for reasons of weight and cost.

This strong interest in plastics and lightweight metal alloys does not mean that all steel components will be replaced; new steel grades are being developed that offer a potential for weight reduction. The use of steel offers the advantage in that present production facilities would not require any retooling because they are laid out for steel. A change to other materials would require major investments.

Calculations must be performed with the use of material parameters such as the modulus of elasticity, strength, specific weight, etc., in order to determine the effects on weight reduction produced by alternative materials. It was found, for instance, that the replacement of a steel component by an aluminum component of the same flexural strength resulted in a 50-percent weight reduction. Replacement by chemical materials leads to weight savings of between 25 percent and 45 percent, depending on the material. The use of high-strength steel has resulted in weight reductions obtained by replacing as many base model components as possible with alternative materials.

"You get to a point where it costs too much money to save any more weight," says Harald Schimkat, body structure and safety engineer. "How much lightness is the car buyer willing to pay for?"

Two principal approaches have been used to save weight in the Auto-2000. With the surface-volume ratio, a great effort was made to reduce the total body surface to a minimum without giving up any useful space. Less surface means less material, less weight, and lower cost. The average VW today has a surface/volume ratio of about 0.45. Schimkat thinks it's possible to get to 0.57. "And the VW 2000 will actually be a littler roomier than the Golf, with more luggage space," says Styling Director Herbert Schaefer (Fig. 24-4).

The second approach is material substitution. That means, in practice, taking out steel and putting in aluminum and/or plastics. "A 17-percent weight reduction in the Golf body is a realistic target to be reached by material substitution alone," says Schimkat. That would mean removing enough steel to bring the car's total steel content down from 69 to an even 50 percent, while raising the plastics content from 10 to 18 percent and the aluminum content from 3 to 11 percent.

The high cost of plastics is a problem. "But if

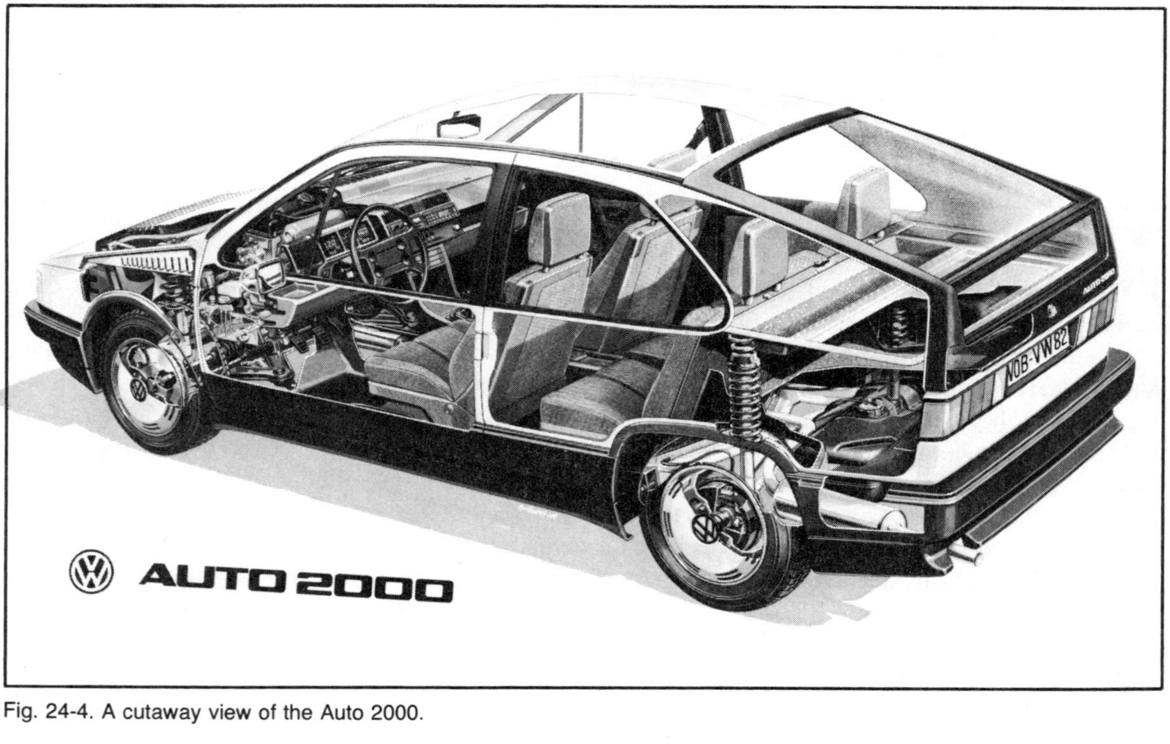

Fig. 24-4. A cutaway view of the Auto 2000.

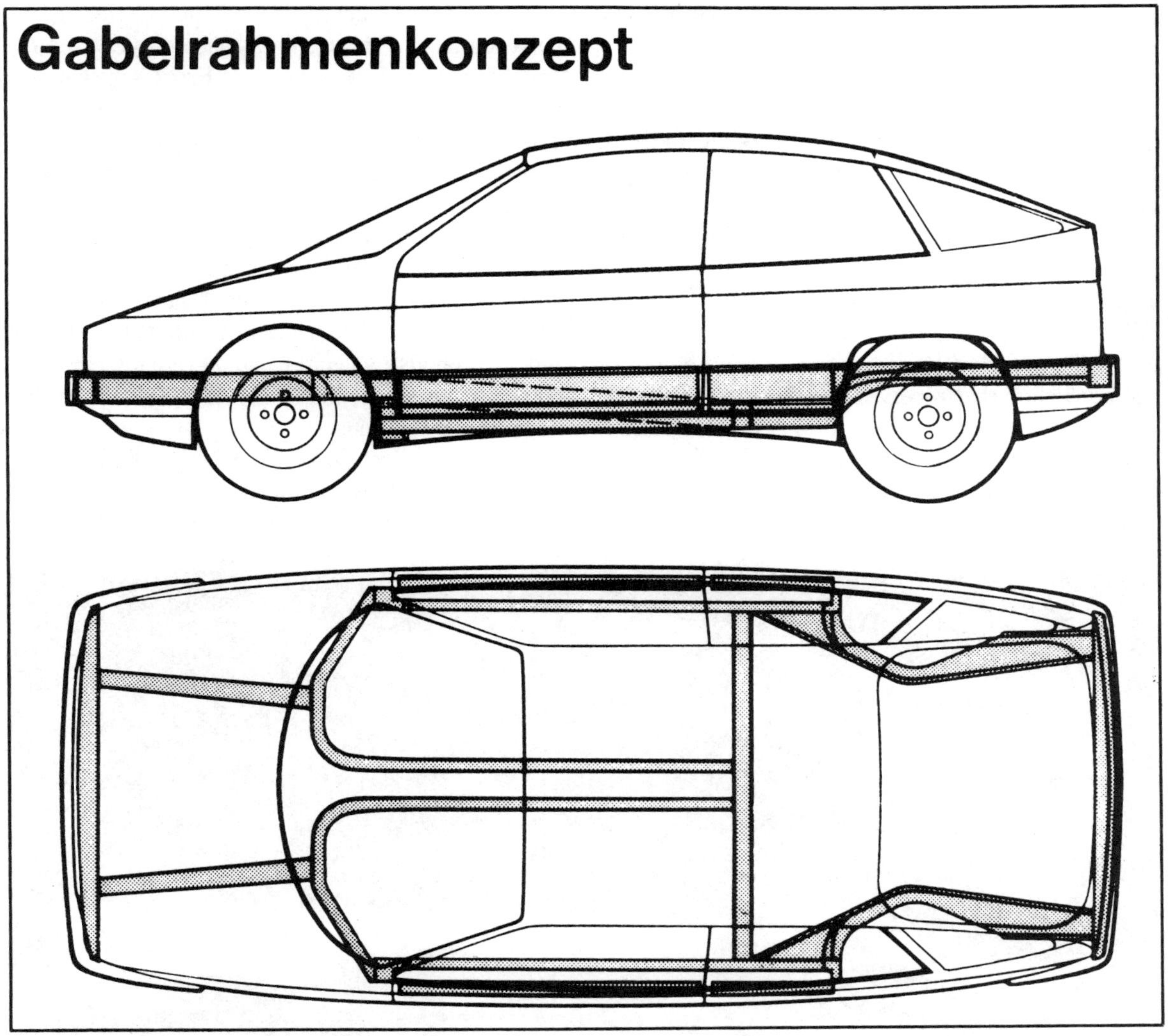

Fig. 24-5. The fork-frame concept.

you can combine several parts into one, the cost of going to plastics will come down. Or if you can use one sheet of plastic to replace three layers of steel," Schimkat points out.

"Material substitution is possible for all parts," says Chassis Engineer Andreas Bauer. "We are working on a rear axle in plastics, using finite element analysis. It means a total redesign. It will do the same job, but it will look quite different. When we've done the rear axle, we can carry over the same techniques to other parts."

The rear axle is made up of two identical halves bonded together (with integrated aluminum end plates). It is 50 percent lighter than the all-steel production version. The cost comes about halfway between steel and aluminum.

With independent front suspension, the opportunities for material substitution are restricted. Still, the VW engineers found it possible to make the lower part of front suspension struts from cast aluminum. The higher cost is partly offset by the simplified design and reducing machining requirements.

Many of the safety ideas from the ESVW II and IRVW prototypes are part and parcel of the VW 2000. They include a fork-frame reinforcement of the body shell that receives and distributes all side and frontal impact forces at one level, without bending moments. The beams of the fork frame are at bumper height and thus provide effective all-around protection. See Fig. 24-5.

The body structure is made up of a floor group subassembly, a front-end subassembly, two side-panel assemblies, and the roof. Front and rear sections are mainly plastic (with integrated bumpers). The trunk lid is an injection-molding piece.

Corrosion protection can be further facilitated by making the body-in-white and the cavities more accessible. This accessibility could be achieved by installing most of the body components during final assembly, i.e., after painting. Many of the most corrosion-prone, sheet-metal components such as the wheel housings could also be replaced by more corrosion-resistant materials. The Auto-2000 has plastic wheel arches.

The complete power train is installed in a subframe prior to final assembly. This facilitates automated assembly and decentralized production.

Fig. 24-6. The low-drag body was pulled out of the wind tunnel.

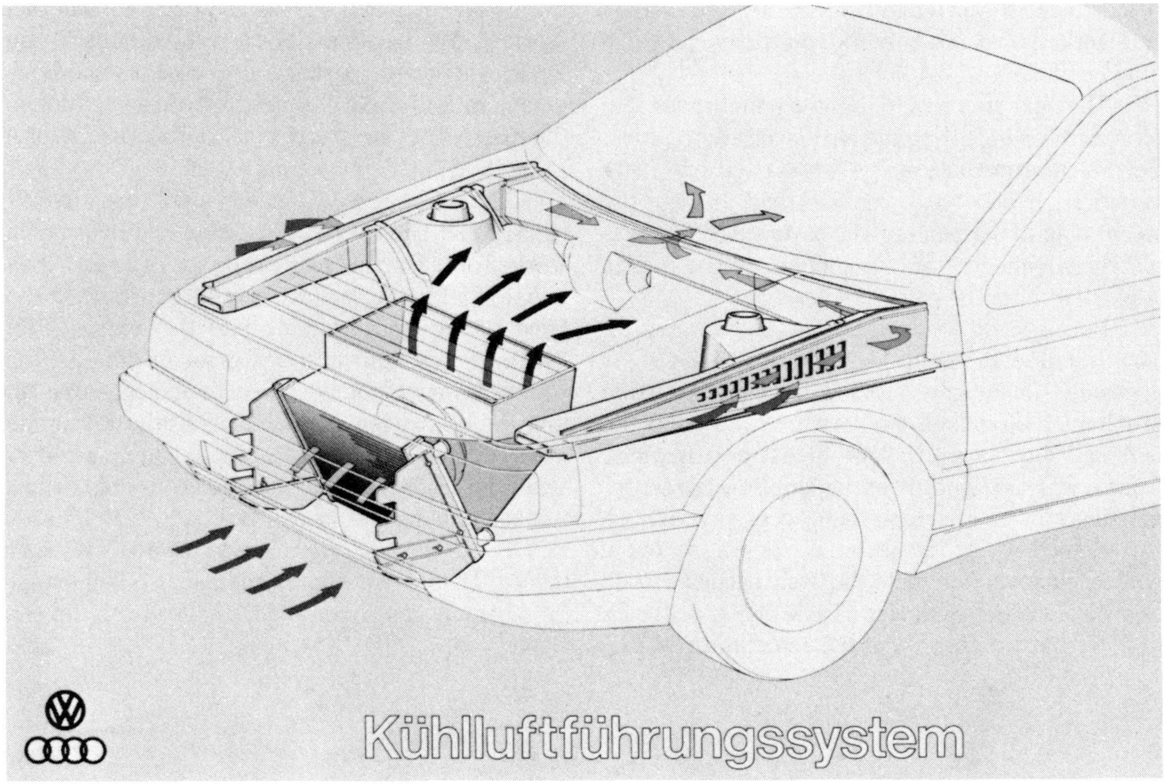

Fig. 24-7. Cooling air flows through a forward-slanted radiator and is ducted upward for low-resistance evacuation.

The engines for the VW Auto 2000 are based on current production units. One is a four-cylinder gasoline engine and another is a three-cylinder diesel. They are combined with various experimental transmission systems—automatic and manual.

The Auto-2000 has all-plastic seats with integral headrests (saving 8.8 pounds per seat when compared with the current production model). Even the wheels are plastic on this prototype.

Overall dimensions of the Auto-2000 are similar to the 1980 VW Golf: 158 inches long, 65.7 inches wide, and an unladen height of 53.2 inches. The prototype weighs 1720 pounds; that is 70.6 pounds lighter than today's production-model Golf.

Fuel economy depends on the specific power train. The gasoline-powered Auto-2000 has a top speed of 115 mph and offers 0 to 60 mph acceleration in 10-11 seconds. Its average fuel mileage is 30 to 35 mpg. The diesel is designed with higher priority on economy and lower priority on performance. It gives an average mileage of 45 mpg. Still it has a top speed of 90 mph and can go from standstill to 60 mph in 18 to 20 seconds.

The shape of the VW Auto-2000 body was created in a wind tunnel (Fig. 24-6). The starting point was an "ideal form" suggested by W. Klemperer of the Zeppelin Werft in 1922. It was a flat-bottomed bullet with wheels that showed a drag coefficient of 0.15 in the wind tunnel. The VW research team shortened it to fit the Golf chassis and worked over the corners until tests showed a drag coefficient of 0.16.

The next step was to turn the bullet shape into a car without upsetting the airflow characteristics. Several intermediate steps were tried out (with variations in the hood, roofline, and tail end designs). Out of it all emerged a basic shape that was clearly intended as an automobile. It had a drag coefficient of 0.18.

But it was a long way from being a practical car. It failed to meet legal requirements. The windshield angle was so rakish that light-refraction problems jeopardized the forward view, and the high tail did not permit an adequate rear view. There was not sufficient installation space for existing power trains and their accessories.

Remedies for these drawbacks were found while remaining faithful to the basic shape as much as possible. The result was a basic model with a Cx of 0.24. But a model is not the same as a real car, even with revolving wheels and ducts for internal air flow. The basic model evolved without regard for appearance or producibility. Stylists and body engineers will make changes that add something to the drag. Still, the styled body in its present form has a drag coefficient of only 0.26.

The internal airflow needs were carefully calculated and natural paths for flow and evacuation worked out. That led to the adoption of a very small radiator with the airflow ducted upwards to the hood rather than ramming into the engine (Fig. 24-7). This design cut the internal air drag in half.

This must not be taken to indicate that all future Volkswagen cars will slavishly follow the shape of the Auto-2000. "A 30-percent drag reduction from current production cars is possible, with a variety of different shapes and body types," says Rolf Buchheim, who is manager of the VW wind tunnel. That leaves a lot of room for styling exploration and poses a continuing challenge to the designers.

Index

Index

S

T

U

V

W

Z

Edited by Steven Bol: